Studies in Computational Intelligence

Volume 907

Series Editor

Janusz Kacprzyk, Polish Academy of Sciences, Warsaw, Poland

The series “Studies in Computational Intelligence” (SCI) publishes new developments and advances in the various areas of computational intelligence—quickly and with a high quality. The intent is to cover the theory, applications, and design methods of computational intelligence, as embedded in the fields of engineering, computer science, physics and life sciences, as well as the methodologies behind them. The series contains monographs, lecture notes and edited volumes in computational intelligence spanning the areas of neural networks, connectionist systems, genetic algorithms, evolutionary computation, artificial intelligence, cellular automata, self-organizing systems, soft computing, fuzzy systems, and hybrid intelligent systems. Of particular value to both the contributors and the readership are the short publication timeframe and the world-wide distribution, which enable both wide and rapid dissemination of research output.

The books of this series are submitted to indexing to Web of Science, EI-Compendex, DBLP, SCOPUS, Google Scholar and Springerlink.

More information about this series at http://www.springer.com/series/7092

Santosh Kumar Das · Shom Prasad Das ·
Nilanjan Dey · Aboul-Ella Hassanien
Editors

Machine Learning Algorithms for Industrial Applications

Springer

Editors
Santosh Kumar Das
School of Computer Science
and Engineering
National Institute of Science
and Technology (Autonomous)
Berhampur, Odisha, India

Shom Prasad Das
School of Computer Science
and Engineering
National Institute of Science
and Technology (Autonomous)
Berhampur, Odisha, India

Nilanjan Dey
Department of Information Technology
Techno India College of Technology
Kolkata, West Bengal, India

Aboul-Ella Hassanien
Information Technology Department
Cairo University, Faculty of Computer
and Artificial Intelligence
Giza, Egypt

ISSN 1860-949X ISSN 1860-9503 (electronic)
Studies in Computational Intelligence
ISBN 978-3-030-50640-7 ISBN 978-3-030-50641-4 (eBook)
https://doi.org/10.1007/978-3-030-50641-4

This Springer imprint is published by the registered company Springer Nature Switzerland AG
The registered company address is: Gewerbestrasse 11, 6330 Cham, Switzerland

Organization

List of Reviewers

Abhilekh Nath Das	NIST, Berhampur, Odisha, India
Abhishek Kumar	Swami Vivekanand Subharti University, Meerut, Uttar Pradesh, India
Ajay Kumar Mallick	IIT(ISM), Dhanbad, India
Ambedkar Kanapala	IIT(ISM), Dhanbad, India
Arijit Karati	National Sun Yat-sen University Kaohsiung, Taiwan
Arun Prasad Burnwal	GGSESTC, Bokaro, Jharkhand, India
Asish Kumar Roy	NIST, Berhampur, Odisha, India
Bhabani Sankar Gouda	NIST, Berhampur, Odisha, India
Debashis Das	Techno India University, West Bengal, India
Harsh Nath Jha	Asansol Engineering College, Asansol, West Bengal, India
Jaydeb Mondal	NIST, Berhampur, Odisha, India
Jayraj Singh	IIT(ISM), Dhanbad, India
Jeevan Kumar	R.V.S College of Engineering and Technology, Jamshedpur, India
Jitesh Pradhan	IIT(ISM), Dhanbad, India
Madhuri Malakar	NIT Rourkela, India
Mahendra Prasad	IIT(ISM), Dhanbad, India
Manoj Kumar Mandal	Jharkhand Rai University, Ranchi, India
Mukul Majhi	IIT(ISM), Dhanbad, India
Nabajyoti Mazumdar	Central Institute of Technology, Kokrajhar, Assam, India
Nabanita Dutta	VIT University, Vellore, Tamilnadu, India
Neha Verma	Swami Vivekanand Subharti University, Meerut, Uttar Pradesh, India
Nishant Jain	IIT(ISM), Dhanbad, India

Praphula Kumar Jain	IIT(ISM), Dhanbad, India
Preeti Gupta	Inderprastha Engineering Technology, Ghaziabad, Uttar Pradesh, India
Priye Ranjan	NIST, Berhampur, Odisha, India
Ranjit Kumar	Galgotias College of Engineering and Technology, Greater Noida, India
Ranjit Kumar Behera	NIST, Berhampur, Odisha, India
Raunak Chandak	NIST, Berhampur, Odisha, India
Ruchika Padhi	NIST, Berhampur, Odisha, India
Sagar Samal	NIST, Berhampur, Odisha, India
Sai Charan Bharadwaj	VIT University, Vellore, India
Smita Silla	BPUT Rourkela, India
Soumen Nayak	SOA University, Bhubaneswar, India
Sourav Samanta	University Institute of Technology, The University of Burdwan, West Bengal, India
Subhra Priyadarshini Biswal	NIST, Berhampur, Odisha, India
Sunil Kumar Gautam	Institute of Advanced Research, Gandhinagar, Gujarat, India
Susmita Mahato	NIST, Berhampur, Odisha, India
Vishwas Mishra	Swami Vivekanand Subharti University, Meerut, Uttar Pradesh, India

Preface

In the last few decades, the applications of machine learning increased rapidly. Its main reason is that the world is rapidly moving toward the big data and data analytics. It brings the huge opportunities to the students, professionals, researchers, academicians, and industry people to face the new challenges in the design and innovation of machine learning algorithms. Machine learning algorithm is used to design an intelligent system that learns from dataset to achieve the purpose of the users such as prediction, classification, and extraction of useful data from a large dataset. So, data-driven decision-making system increases rapidly, and it incrementally uses the traditional domains as well as some new domains. Here, some of the real-life industrial applications of machine learning are given as follows: automated transportation, e-mail classification and spam mail filtering, environment protection, financial services, image recognition, information retrieval, innovations in banking, language identification, medical diagnosis, etc. In all of the above-mentioned applications, data is the crucial parameter for industrial application, and it is the main key for unlocking the value of industry. Hence, to enhance the area of industrial application, there is a need of some novel or innovative ideas in terms of machine learning algorithms.

Objective of the Book

This book contains some machine learning algorithms for industrial applications. It basically deals with the design, implementation, and enhancement of different system of industries. The main aim of this book is to enhance the applications of industries in terms of modernization and efficiency. This book is edited for industry people along with academicians and researchers.

Organization of the Book

The book contains 17 chapters that are organized into four sections as follows. **Section 1** contains four chapters that outline several techniques of natural language processing based on industry applications. **Section 2** contains six chapters that highlight some works related to computer vision including image processing. **Section 3** contains four chapters that illustrate some algorithms related to data analysis and prediction. **Section 4** contains three chapters that demonstrate the algorithms related to decision-making systems.

Section 1: Natural Language Processing (Chapters 1–4)

This section outlines some methods related to natural language processing as classification, automated categorization, and data prediction and recommendation. Short descriptions of these chapters are as follows.

Chapter 1

This chapter illustrates a hybrid feature selection method for Urdu news article classification. The proposed method is the fusion of two machine learning techniques such as latent semantic indexing and support vector machine. First method is used to extract essential features of the Urdu text. Second method is used to classify the text from dataset.

Chapter 2

In this chapter, an automated document categorization model is illustrated with the help of K-means and Native Bayes classifier. First technique is used to prepare a training dataset. Second technique is used as a statistical text classifier. In this method, the selected representative documents are considered as the initial centroids, then create a supervised classifier on the initially categorized set.

Chapter 3

In this chapter, an automated categorization and mining tweets are designed for disaster management. In this work, machine learning is applied on tweets that are generated during the course of a disaster, to categorize them into different stages of that disaster and extract the metadata.

Chapter 4

In this chapter, the collected ratings on air transport management are given by customers from different sites. There are ratings on seat comfort, cabin staff, food beverage, inflight entertainment, and many more, which is further combined to give the overall rating through which recommendation is done.

Section 2: Computer Vision (Chapters 5–10)

This section highlights various computer vision algorithms and techniques based on industrial applications. These methods are illustrated in terms of classification, learning, conversion, and detection. Short descriptions of these chapters are given as follows.

Chapter 5

The objective of this chapter is to reconstruct large continuous regions of missing or deteriorated parts of an image using extreme machine learning. In this method, the dataset trained based on region surrounding the corrupted region. Each image is divided into two sections: the missing part which is to reconstruct and the context.

Chapter 6

In this chapter, an efficient and highly scalable parallel architecture to segment input images containing tabular data with and without borders into cells and reconstruct the tabular data while preserving the tabular format. The performance improvement thus made can be used to ease the tedious task of digitizing tabular data in bulk. The same architecture can be used for regular OCR applications to improve performance if the data is in huge quantities.

Chapter 7

In this chapter, the background and working of few shot learning algorithms are explained. The problem statement for few shot classification and segmentation is described. It includes the recent advances in the application of few shot learning to medical imaging tasks such as classification and segmentation are explored.

Chapter 8

This chapter proposes a designing of classifier that uses fewer labeled samples as possible for classification is highly desirable. It uses active learning which is a branch of machine learning that finds most uncertain samples in an iterative way from unlabeled dataset resulting in relatively smaller training set to achieve adequate classification accuracy.

Chapter 9

This chapter proposes a novel approach is presented to convert handwritten images to computerized text document. In this approach, first the handwritten characters are extracted from the input image using digital image processing techniques and then these characters are recognized by using machine learning techniques.

Chapter 10

In this paper, a general regression neural network-based model along with some image in painting techniques is being used. Each image is divided into two sections: the missing part that is to be reconstructed and the context. The network would work identically for arbitrary removals not just for regions having particular shapes such as square or rectangles.

Section 3: Data Analysis and Prediction (Chapters 11–14)

This section highlights various data analysis and prediction-based algorithms based on industrial applications. These methods give several new directions to the readers as well as researchers. Short descriptions of these chapters are given as follows.

Chapter 11

This chapter aims at distinguishing reviews as positive or negative from the content of the online customer reviews submitted by the previous customer and providing a recommendation. In this chapter, the authors compared three machine learning algorithms, namely logistic regression, stochastic gradient descent, and random forest classifier. It also predicts the accuracy of recommendation done by machine learning techniques.

Chapter 12

This chapter proposes a meta-heuristic-based approach to locate critical failure surfaces under prevailing conditions and constraints. The reliability and efficiency of the approach are examined by investigating the benchmark case studies. The outcome results indicate that the proposed approach could acquire acceptable performance over existing methods and attain a better solution quality in terms of accuracy and efficiency.

Chapter 13

In this chapter, a real-world dataset is considered for the study, where the sales revenue of restaurant is predicted. A second stage regression model is built upon base regression models which are linear regression, ridge regression, and decision tree regressor. Based on the results obtained, some analysis and prediction are performed.

Chapter 14

In this chapter, there are more other machine learning algorithms which can analyze the anomalies in the system with the help of predictive control model. The predictive control hybrid model is the new socket of study where the researchers can forecast to shrink the vigor loss of resource and time and can make the system flawless. Thus, there is a big challenge before the researchers regarding the application of machine learning for detecting the anomalies in the pumping system.

Section 4: Decision-Making System (Chapters 15–17)

This section illustrates some decision-making systems based on machine learning algorithm with the context of industrial applications. Short descriptions of these chapters are given as follows.

Chapter 15

In this chapter, a fast accessing non-volatile, high performance with high density and optimized array is designed for machine learning processor. It overcomes the limitations of traditional memory technologies which are not able to cope with machine learning processor because of high density and low power requirement.

Chapter 16

In this chapter, the proposed system uses various sensors to detect distinctive parameters which are regularly checked via the IoT network. And in case any parameter is triggered, the system will automatically stop. These systems will likewise lessen the overload on the government on maintenance and safety-related issues of the existing framework with lower expenses.

Chapter 17

In this chapter, a novel nature-inspired algorithm based on the pollination process of plants is used for locating the critical surface. The quantitative evaluation of stability analysis in terms of factor of safety demonstrated the performance of the approach. The findings indicate the appropriate performance over current methods and declare the optimum solution.

Berhampur, India — Santosh Kumar Das
Berhampur, India — Shom Prasad Das
Kolkata, India — Nilanjan Dey
Giza, Egypt — Aboul-Ella Hassanien

Contents

Natural Language Processing

A Hybrid Feature Selection Approach Based on LSI for Classification of Urdu Text 3
Imran Rasheed, Haider Banka, and Hamaid Mahmood Khan

Automated Document Categorization Model 19
Rakhi Patra

Automated Categorization and Mining Tweets for Disaster Management 37
Rakhi Patra

Sentiment Analysis in Airline Data: Customer Rating Based Recommendation Prediction Using WEKA 53
Praphula Kumar Jain and Rajendra Pamula

Computer Vision

Image Inpainting for Irregular Holes Using Extreme Learning Machine 69
Debanand Kanhar and Raunak Chandak

OCR Using Computer Vision and Machine Learning 83
Ashish Ranjan, Varun Nagesh Jolly Behera, and Motahar Reza

Few Shot Learning for Medical Imaging 107
Jai Kotia, Adit Kotwal, Rishika Bharti, and Ramchandra Mangrulkar

Hyperspectral Remote Sensing Image Classification Using Active Learning 133
Vimal K. Shrivastava and Monoj K. Pradhan

A Smart Document Converter: Conversion of Handwritten Text Document to Computerized Text Document 153
Ranjit Kumar Behera and Biswajeet Padhi

GRNN Based an Intelligent Technique for Image Inpainting 167
Debanand Kanhar and Raunak Chandak

Data Analysis and Prediction

Content-Based Airline Recommendation Prediction Using Machine Learning Techniques 185
Praphula Kumar Jain and Rajendra Pamula

A Meta-heuristic Based Approach for Slope Stability Analysis to Design an Optimal Soil Slope 195
Jayraj Singh and Haider Banka

An Application of Operational Analytics: For Predicting Sales Revenue of Restaurant 209
Samiran Bera

Application of Machine Learning Algorithm for Anomaly Detection for Industrial Pumps 237
Nabanita Dutta, Palanisamy Kaliannan, and Umashankar Subramaniam

Decision Making System

Fast Accessing Non-volatile, High Performance-High Density, Optimized Array for Machine Learning Processor 267
Divya Mishra, Abhishek Kumar, Vishwas Mishra, Shobhit Tyagi, and Shyam Akashe

Long Term Evolution for Secured Smart Railway Communications Using Internet of Things 285
Shweta Babu Prasad and P. Madhumathy

Application of Flower Pollination Algorithm to Locate Critical Failure Surface for Slope Stability Analysis 301
Jayraj Singh, Ravishankar Kumar, and Haider Banka

About the Editors

Santosh Kumar Das received his Ph.D. degree in Computer Science and Engineering from Indian Institute of Technology (ISM), Dhanbad, India, in 2018 and completed his M. Tech. degree in Computer Science and Engineering from Maulana Abul Kalam Azad University of Technology (erstwhile WBUT), West Bengal, India, in 2013. He is currently working as an assistant professor at School of Computer Science and Engineering, National Institute of Science and Technology (Autonomous), Institute Park, Pallur Hills, Berhampur, Odisha, India. He has more than eight years teaching experience. He has authored/edited of two books with Springer, Lecture Notes in Networks and Systems, and Tracts in Nature-Inspired Computing. He has contributed more than 27 research papers. His research interests mainly focus on ad hoc and sensor network, artificial intelligence, soft computing, and mathematical modeling. His h-index is 13 with more than 480 citations.

Shom Prasad Das received his Ph.D. degree in Computer Science and Engineering from Biju Patnaik University of Technology, Rourkela, Odisha, India, in 2018 and completed his M. Tech. degree in Computer Science and Engineering from Biju Patnaik University of Technology, Rourkela, Odisha, India, in 2010. He is currently working as a professor at School of Computer Science and Engineering, National Institute of Science and Technology (Autonomous), Institute Park, Pallur Hills, Berhampur, Odisha, India. He has more than eighteen years experience in teaching as well as industry. He has contributed several research papers. His research interests mainly focus on data science, big data analytics, machine learning, software engineering and data warehousing, and data mining.

Nilanjan Dey is an assistant professor in the Department of Information Technology at Techno International New Town (formerly known as Techno India College of Technology), Kolkata, India. He is a visiting fellow of the University of Reading, UK. He is a visiting professor at Duy Tan University, Vietnam. He was an honorary visiting scientist at Global Biomedical Technologies Inc., CA, USA (2012–2015). He was awarded his Ph.D. from Jadavpur University in 2015. He is the editor in chief of the International Journal of Ambient Computing and Intelligence, IGI Global. He is the series co-editor of Springer Tracts in nature-inspired computing, Springer Nature, series co-editor of advances in ubiquitous sensing applications for health care, Elsevier, and series editor of computational intelligence in engineering problem solving and intelligent signal processing and data analysis, CRC. He has authored/edited more than 50 books with Springer, Elsevier, Wiley, and CRC Press and published more than 300 peer-reviewed research papers. His main research interests include medical imaging, machine learning, computer-aided diagnosis, data mining, etc. He is the Indian ambassador of the International Federation for Information Processing (IFIP)-Young ICT Group.

Aboul-Ella Hassanien is the founder and head of the Egyptian Scientific Research Group (SRGE) and a professor of Information Technology at the Faculty of Computer and Artificial Intelligence, Cairo University. He has more than 1000 scientific research papers published in prestigious international journals and over 50 books covering such diverse topics as data mining, medical images, intelligent systems, social networks, and smart environment. He won several awards including the Best Researcher of the Youth Award of

Astronomy and Geophysics of the National Research Institute, Academy of Scientific Research (Egypt, 1990). He was also granted a scientific excellence award in humanities from the University of Kuwait for the 2004 Award and received the superiority of scientific in technology, University Award (Cairo University, 2013). Also he was honored in Egypt as the best researcher in Cairo University in 2013. He also received the Islamic Educational, Scientific and Cultural Organization (ISESCO) prize on Technology (2014) and received the State Award of excellence in engineering sciences 2015. He holds the Medal of Sciences and Arts from the first class from President of Egypt in 2017.

Natural Language Processing

A Hybrid Feature Selection Approach Based on LSI for Classification of Urdu Text

Imran Rasheed, Haider Banka, and Hamaid Mahmood Khan

Abstract The feature selection method plays a crucial role in text classification to minimizing the dimensionality of the features and accelerating the learning process of the classifier. Text classification is the process of dividing a text into different categories based on their content and subject. Text classification techniques have been applied to various domains such as medical, political, news, and legal domains, which show that the adaptation of domain-relevant features could improve the classification performance. Despite the existence of plenty of research work in the area of classification in several languages across the world, there is a lack of such work in Urdu due to the shortage of existing resources. In this paper, First, we present a proposed hybrid feature selection approach (HFSA) for text classification of Urdu news articles. Second, we incorporate widely used filter selection approaches along with Latent Semantic Indexing (LSI) to extract essential features of Urdu documents. The hybrid approach tested on the Support Vector Machine (SVM) classifier on Urdu "ROSHNI" dataset. The evaluated results were used to compare with the results obtained by individual filter feature selection methods. Also, the approach is compared to the baseline feature selection method. The proposed approach results show a better classification with promising accuracy and better efficiency.

Keywords Urdu Text Classification · Latent Semantic Indexing · Machine Learning · Singular Value Decomposition · Support Vector Machine

---

I. Rasheed (✉) · H. Banka
Department of Computer Science and Engineering, Indian Institute of Technology (ISM), Dhanbad, Dhanbad, India
e-mail: imranrasheed@cse.ism.ac.in

H. Banka
e-mail: haider.banka@gmail.com

H. M. Khan
ALUTEAM, Fatih Sultan Mehmet Vakif University, Beyoglu, Istanbul, Turkey
e-mail: hmkhan@fsm.edu.tr

S. K. Das et al. (eds.), *Machine Learning Algorithms for Industrial Applications*, Studies in Computational Intelligence 907,
https://doi.org/10.1007/978-3-030-50641-4_1

1 Introduction

Text classification may be assumed as one of the most important research areas due to the availability of a high volume of electronic texts on the world wide web. Generally, high-quality information (termed as features) is extracted from text documents, which aims to specify text documents with unknown classes by certain classifiers. The role of text classification is to convert open-ended text into categories. Text classification is included in various applications, such as (a) video and audio music genres classification [1], (b) classification of reading materials for students of different levels [2], (c) email filtering for spam [3], (d) classification of news articles, (e) sending user inquiries to the appropriate department (technical support, customer service, sales, etc.) and (f) Sentiment analysis.

Text classification and feature selection have given more attention to current research. Researchers emphasize choosing the feature to classify text for different languages (such as Arabic, Urdu, Telugu, Hindi, and Punjabi). Suitable feature subsets enable machine learning algorithms to train faster and give much better accuracy than a full set of features for the same algorithm. Additionally, it can reduce the overfitting and complexity of the model and make it easy to interpret. Several techniques and methodologies can be used to make the subset feature space and help models to perform better and more efficiently. Methodologies are broadly classified in three categories: Filter methods [4], Wrapper methods [5], and Ensemble methods [6]. Filter approach [4] is used to define a subset of features from high dimensional datasets without using a learning algorithm. Whereas, the wrapper approach [5] uses a learning algorithm to assess the accuracy of a particular subset of features while categorization. Several studies have explored the role of soft computing techniques in selecting features in recent years. Among them, the most effective and widely used techniques are meta-heuristic or nature-inspired approaches [7–9]. That's why it quickly affects several other applications, particularly in an ad-hoc wireless network which is one of the variations of a wireless network [10, 11].

The main key point in the proposed method is the classification of texts. When document classification is done manually, all the articles must be read by people, then labeled and retained. To do the classification job, it requires lots of specialists with rich experience and specialized knowledge. This method has some drawbacks, including a high cost, long cycle, and low efficiency. It's hard to meet real needs. However, by categorizing text, people can not only locate the information needed more accurately, but can also search the information quickly. On the other hand, the automatic classification of texts is a method of supervised learning. A relative model is developed mathematically between documents labels and attributes of documents during the training of documents. After that, relevant models are used to assign respective labels to given unknown documents. So, here we show how a document can be automatically categorized.

To our knowledge, the current work is the first comprehensive study to use a hybrid approach to classify Urdu news. In this article, First, we propose a hybrid features selection approach for text classification of the Urdu dataset of 29,931 news articles that followed the TREC standard, with 16 different categories. Secondly, to achieve maximum accuracy of the machine learning algorithms, we integrate commonly used filter feature selection methods with the Latent Semantic Indexing (LSI) method. The filter methods are used, such as Chi-Square (CHI), Information Gain (IG), and Gain Ratio (GR). Lastly, we chose a machine learning technique to classify news content into sixteen predefined classes, namely, international, sports, the MuslimWorld, economic, political, science and technology, International, National, etc.

This paper is arranged as follows: Sect. 2 outlines some related work on the classification of text, while Sect. 3 offers a brief introduction to the language of Urdu. Section 4 defines the methodology, acquisition, and pre-processing of the text collection. A brief overview of the dimensionality reduction and feature extraction method described in Sect. 5, whereas Sect. 6 describes the brief summary of classification algorithms. Result and discussion are reported in Sect. 7. Section 8 concludes this paper and suggests some ideas for future work.

2 Related Works

The proposed method attempts to merge the LSI method and the filter approaches into one system, where features are selected in two stages. The first phase uses singular value decomposition to transfer data from original wide-dimensional space to a new low-dimensional space, and the second phase uses methods IG, GR, and CHI as a filter approach.

Choosing the feature is a critical pre-setup activity for any text mining application in advance to the classification tasks. Though there are plenty of well-developed statistical and mathematical feature selection and classification techniques today. Several feature selection methods have recently been suggested for text categorization, but primarily applied to European languages. Young and Jeong [12] used the Naive Bayes classifier to implement a new scaling approach for feature selection. This system was tested on a large collection of news articles. It was readily so successful that it surpassed all the other existing ranking methods, such as Information Gain similarly Bidi and Elberrichi [13] proposed a feature selection technique for text classification based on Genetic Algorithm (GA). Initially, a detailed study of different GA based feature selection technique is provided where each technique consists of some text representation method. Later, it provides details performance evaluation using some existing techniques such as Support Vector Machine (SVM), K-Nearest Neighbour (KNN), and Naive Bayes (NB) to outperform its results.

The author collected Urdu news headlines data for classification. The data is trained on the SVM classifier, followed by a pre-processing stage such as normalization, stemming, and removal of stopwords [14] while Wahab et al. [15] dealt with an extensive review of various state-of-the-art machine learning methods for classifying text. Ali et al. [16] present a comparative study of Urdu text classification using widely used classifiers, i.e., SVM and NB on 26,067 documents. In their experimental study, they observed that SVM produced higher accuracy than the Naive Bayes classification. There are six different categories in the list, including News Sports, Economy, Culture, Consumer Information, and Personal Communication. Zia et al. [17] examined and evaluated the five well-known techniques of feature selection such as IG, CHI, GR, and oneR using NB, KNN, and DT on two Urdu test collections, i.e., EMILLE and Naive. The result reveals that the SVM and KNN classifier shows better performance with IG method whereas for Roman-Urdu, Bilal et al. [18] analyzed three classification models to detect the opinion of the people about certain things. The experimental results showed that Naive Bayes performed better than the Decision tree and KNN in terms of accuracy, recall, and F-measure.

Recently soft computing techniques i,e. neurocomputing, fuzzy logic, genetic algorithm, etc. are gaining growing popularity for their remarkable ability to handle real-life data like a human being in an environment of uncertainty, imprecision, and implicit knowledge. Gunal and Serkan suggested a mixture of filter and wrapper approaches [19]. In his study, features are first chosen by four filtering methods (DF, MI, IG, CHI) and then merged as an input of a Genetic algorithm (GA) in the second phase. The SVM and DT classifiers are being used for feature subset evaluation, where the fitness of the subset depends on the macro-and micro-average F-measure. In other studies, Fang et al. [20] explored the efficiency of a combination of IG, MI, DF, CHI filter methods with a GA, and Lei and Shang used information gain with GA as a text categorization approach for feature selection [21]. These studies have suggested that the hybrid feature selection approach can efficiently reduce text dimensionality and significantly improve the efficiency of categorization.

Mukhtar et al. [22, 23] proposed a technique for Urdu sentiment analysis using supervised machine learning techniques. Primarily, data is acquired from various blogs of 14 different genres and then apply three well-known classifiers (such as KNN, decision tree, and SVM) individually and in combination. Finally, he concluded that the KNN classifier outperforms than other classifiers on this dataset in terms of accuracy, F-measure, precision, and recall. Later, he proposed a lexically based technique on multiple domains for Urdu sentiment analysis and found that the lexical based method outperforms than supervised machine learning technique in terms of accuracy, F-measure, precision, and recall besides time and efforts used. On the other hand, Sohail et al. [24] proposed a text classification for poor-resource language like Urdu by performing lexical normalization of terms using phonetic and string similarity. The author also explores the supervised feature extraction technique to obtain category wise, highly discriminating features for improving classification performance. While Khan et al. [25] presents a machine learning approach for Urdu word segmentation with the help of conditional random fields (CRF). Besides, this method also helps in reducing compound and redundancy words. Word segmentation is useful in many applications such as information retrieval, part-of-speech (POS), named entity recognition (NER), sentiment analysis, etc. In contrast, Puri and Singh [26] utilized the combined strength of SVM and fuzzy logic to classify the Hindi text documents. In comparison, SVM is used to classify data, and fuzzy logic is used to decrease the uncertainty related information. Based on the latent Dirichlet allocation approach Anwar et al. [27] proposed a speculative study for forensic examination in Urdu text.

3 Urdu Language Structure

Urdu is a widely spoken language in the Indian subcontinent, with over 300a million speakers spread all over the world. Urdu belongs to a Perso-Arabic cluster of languages [28] and is mainly composed of words from Arabic, Persian, and Sanskrit. Being a national language of Pakistan, has over 300 million speakers spread across the globe with a large chunk of the population reside in the Indian subcontinent [29]. Urdu is originally derived from the Perso-Arabic script of Iran and is written from right to left like Arabic or Persian and is characterized in the Nasta'liq format [30, 31]. The family tree of Urdu trace back to the mix of Indo-European, Indo-Iranian, and Indo-Aryan lingo evolution [32]. Urdu is known to have rich and complex morphology [32, 33] with its syntax structure is composed of a combination of Persian, Sanskrit, English, Turkish and Arabic. Despite having a large number of Urdu speakers, it is deprived of language resources. Urdu still does not have a massive and reliable text collection of domains of general interest. Moreover, an extensively large collection is necessary to conduct Information Retrieval research. Although Urdu is served with lots of dictionaries, it still a deficiency of WordNet-like semantic terms.

4 Proposed Method

This system serves as a filter to mute out unimportant or redundant features. In this paper, we proposed a novel hybrid technique for features selection of Urdu text. The set of features extracted from well-known feature selection methods discussed by Rasheed et al. [34]. Selected features from these methods were then combined to form a subset of essential features to improve the performance of classifiers on Urdu text. Improving the feature selection set will surely enhance the overall performance. The proposed model consists of different modules, namely, dataset, pre-processing, feature extraction & selection, classification, and performance evaluation. Figure 1 shows various modules used in our system, and the details of each module are discussed in the sections below.

4.1 Urdu Dataset

A standard collection is necessary to obtain good accuracy in the classification of a statistical system. A standard collection is necessary to perform any Natural Language Processing (NLP) tasks. For this reason, a reasonably large collection of 29,931 news articles is compiled from 'Daily Roshni,' a Srinagar, India, Urdu newspaper. Original news data were gathered from December 2010 through August 2012. The encoding schemes in original articles were non-standard and font-based. Therefore,

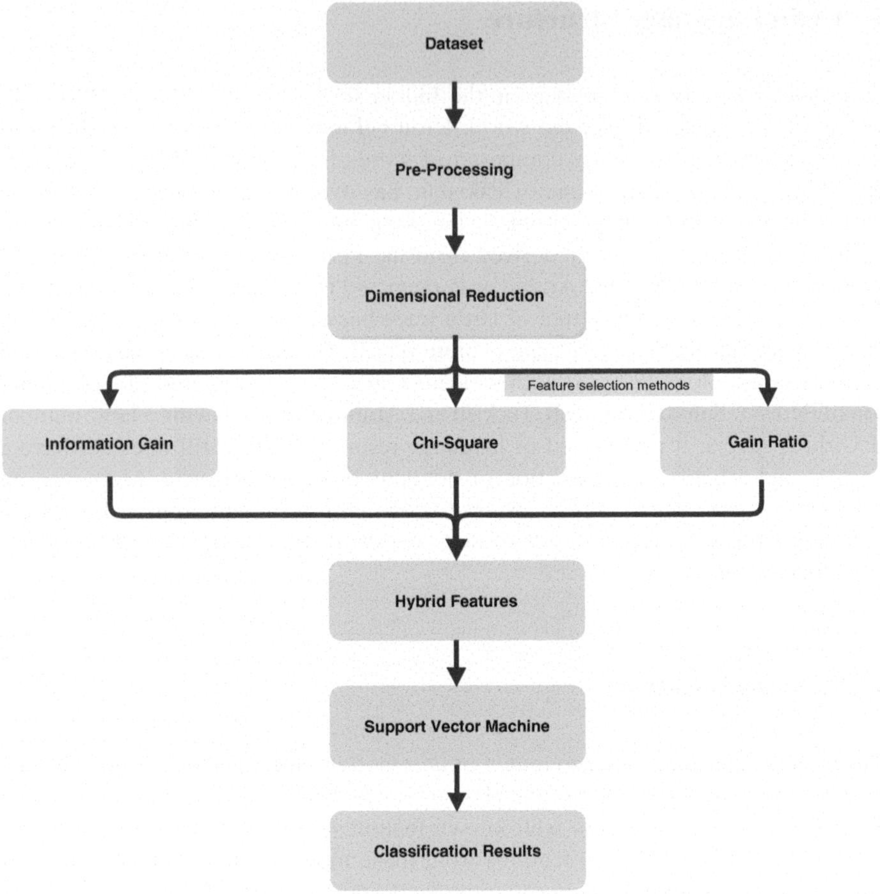

Fig. 1 Proposed system of Urdu text classification

the documents were changed into UTF-8 encoding to homogenize the corpus. The ultimate collection of 29,931 articles was, therefore, distributed into sixteen different categories, including sports, politics, social, economic, science and technology, etc. with varied sizes. The collection is then distributed into sixteen different categories, such as international, sports, the Muslim world, political, and entertainment, etc. The distribution of documents for each class is outlined in Table 1. The entire collection consists of 6,95,606 unique terms and 1,11,12,892 tokens.

Table 1 Analysis of documents at various levels of Pre-processing stages

Categories	Documents	Tokens	Distinct word
Articles	871	1005399	78307
Sports	2513	575682	42519
MuslimWorld	4381	1659057	71585
Health	1142	575804	39763
UPNews	720	160689	23284
International	2354	626969	49529
LocalNews	6919	1830731	69771
Culture	671	339762	28386
Social	1067	589646	57912
Political	2973	1590481	65190
Entertainment	1034	205169	26700
National	3974	1560220	68531
Miscellaneous	900	157374	26551
Opinion	66	101809	20458
Economical	157	62670	12420
Science&Technology	189	71430	14700

4.2 *Pre-processing*

In the current module, different methods were adopted to do away with extraneous and noisy data from the collected set before applying any operation as discussed in [35]. This section is composed of four subtasks: tokenization, stopwords removal, normalization, and diacritics removal.

(a) Tokenization
It is the process of analyzing or segregating a character's sequence in a given document into words known as tokens. Tokens can be words, numbers, identifiers or punctuation, etc. [36]. The tokenizer creates tokens out of a string by reading delimiters like /-"[]():?<>!.

(b) Normalization
Text normalization is the process of rearranging multiple, comparable representations of characters into their standard form. An alphabet has more than one Unicode as it belongs to duo languages, i.e., Urdu and Arabic language. Such alphabets must be replaced by alternate Urdu alphabets to stop creating multiple copies of a word.

(c) Diacritics Removal
Diacritics are the non-functional terms that are used to ease text reading. In principle, they are chosen to add significance to the text to make them meaningful to the readers. The diacritics attach great importance in the text mainly when they appear in isolation since many beginners end up making mistakes while reading Urdu text. During the pre-processing stage, all the diacritics are removed to homogenize the text.

(d) Stopwords Removal
Stop words are the words that occur most in any language. These words are used to complete the sentence structure, although they mean nothing individually. Before implementing any algorithm, stopwords are taken off from the text to reduce the size of the vocabulary. Stop words are listed as unfavorable, and they do not participate in the indexing process [37, 38]. In our case, the list of stopwords is collected in the entire corpus based on the word frequency. If the word frequency reaches the threshold value, then it will be viewed as stop-words. Eventually, we reviewed the entire list diligently and removed every keyword from the list. Some of Urdu, stopwords are shown in Fig. 2 and most frequent words are shown in Table 2.

Fig. 2 List of some stopwords

بعض بلکہ بج بچے بخیر میرا
با بشرطیکہ بعد بلکل بلوجود باہر مگر
بغیر جاتا جاتی جاتے جانی جانے

Table 2 Most frequent words in the collection

Categories	Documents	Tokens	Distinct word
Terms	Frequency	Terms	Frequency
(ke) كي	489816	(kya) کیا	127063
(ki) کی	381123	(es) اس	123749
(mein) میں	342848	(keh) کہ	106472
(hai) ہي	219942	(hain) ہیں	89816
(nay) ني	209272	(bhi) بھی	74120
(aur) اور	207902	(ek) ایک	70113
(se) سي	199404	(kar) کر	57351
(ko) كو	190695	(yeh) یہ	51310
(ka) کا	151971	(in) ان	48128
(par) پر	131688	(nahi) نہیں	43474

Note: Urdu fonts are not supported in LaTeXso we used Arabic fonts here.

4.3 *Text Representation Model*

Computers are perfect for numbers but not for textual data. Term Frequency-Inverse Data Frequency (TF-IDF) is one of the techniques most commonly used for processing and representation of textual data. It is a method for creating a vector representation of a document which is proposed by Sparck Jones [39, 40]. Getting a vector representation of a document to enable you to compare documents to their similarity by calculating the distance between the vectors. So that we do many tasks such as document categorization, text summarization, etc. Equation 1 is the general formula of TF-IDF used for term scoring.

$$tf - idf(t_{u,v}) = tf(t_{u,v}) \times log(N/df(t_u)) \quad (1)$$

where TF is the frequency of the term t_i in document d_j and DF(t_i) is the number of documents that contain term t_i and N is the total number of documents in collection.

5 Dimensionality Reduction and Feature Selection

In today's world of Data mining and IoT, we are quickly filled with extremely high dimensional rich datasets. The dimensionality reduction of the term-document matrix is a crucial job. These techniques are typically used while solving machine learning problems to obtain better features for classification or regression tasks. The high-dimensional feature space is transformed to substantially cut the size of the feature space in a way that maximizes the variance of the data in the lower dimensional space. Hence, in our proposed method, one of the most popular probabilistic methods, Latent Semantic Indexing (LSI), is used to project the data from a high-dimensional space into a new low-dimensional space [41, 42]. Based on term co-occurrence, the LSI method developed a relation between terms in a term-document matrix using the principle of Singular Value Decomposition (SVD) [43] method to transform a term frequency matrix into a singular matrix. It selects only λ maximum values of the singular value vector, mapping the text vector from the term space to a λ-dimensioned semantic space. The impact of LSI in dimensionality reduction in text classification is noteworthy.

The choice of features becomes very crucial for performing any machine learning task and data sciences, especially when dealing with high-dimensional datasets. Since certain features could be unrelated to the dependent variable or less critical, their excessive inclusion in the model leads to

1. Increasing the sophistication of a model and make it more difficult to understand.
2. Increase the amount of time for a model to get trained.
3. Result in a weak model that is incorrect or less confident.

It, therefore, gives an urgent need to perform feature selection.

In the present research, the most widely used probabilistic distribution methods such as Chi-Square (CHI) [44], Information Gain (IG) [45] and Gain Ratio (GR) [46] used for the selection of features. These methods also called filtering methods. They take the inherent characteristics of the features (i.e., the significance of the feature) calculated through statistical analysis rather than cross-validation results. In terms of classification, these methods not only reduce the total dictionary size but also removes the unimportant features that improve the performance of the classifier. The outcome of each machine learning classifier is highly dependent on the set of features selected.

6 Classification Algorithms

Dataset typically contains important details which are used to make quick decisions. It is challenging for any program to make smart decisions without classifying such datasets. Thus, classification algorithms simplify the task by extracting useful models and highlighting essential categories of data. All relevant documents may be grouped into classes that are supervised, unsupervised, and semi-supervised. There are a variety of methods for classifying texts such as K-Nearest Neighbour (KNN), Support Vector Machine (SVM), Naive Bayes (NB), Artificial Neural Networks (ANN) and Decision Trees (DT). However, in the current study, only SVM and NB were chosen for document classification.

6.1 Support Vector Machine Classifier

Support Vector Machine (SVM) classifier is one of the supervised machine learning technique that is proposed for text classification. In n-dimensional space, the SVM treats input data as two sets of vectors. In that space, SVM generates a separate hyperplane, maximizing the margin between two datasets [47]. The main advantages of SVM include high accuracy and less prone to overfitting. Also, when compared to other machine learning classifiers, the SVM technique gives good results for text classification problems due to its quick ability to improvise a solution. Two types of configurations undertaken in SVM:

(a) Training Mode: In this mode, scores of each feature are determined, and these scores are used to train the SVM to classify the documents into different categories.
(b) Testing Mode: In this mode, For each feature of new documents, scores are evaluated using test datasets, and each document is categorized into different classes.

7 Result and Discussion

7.1 Evaluation Measures

To study the performance of a classifier, we have used three common measures, i.e., Precision, Recall, and F-measure. The three measures concerning the positive class can be defined as

Precision (P): It is the fraction of retrieved documents that are relevant.

$$P = \frac{Number\ of\ correct\ positive\ predictions}{Number\ of\ positive\ predictions} \quad (2)$$

Recall (R): It is the fraction of relevant documents that are retrieved.

$$R = \frac{Number\ of\ correct\ positive\ predictions}{Number\ of\ positive\ examples} \quad (3)$$

F-measure (F): It is the harmonic mean of Precision and Recall.

$$F = \frac{2 \times P \times R}{P + R} \quad (4)$$

7.2 Result Analysis

In this study, the proposed system, on the “Roshni” dataset. It is a standard Urdu Unicode-based dataset containing 29,931 documents. A document distributed in different categories is unequal. For example, categories such as opinion, science & technology, economical, culture, articles, UPNews, and Miscellaneous contain considerably fewer documents (i.e.,$<$1000) compared to other significant classes, which lead us to discard them entirely for smooth operation. As categories label with large frequencies of allotted documents is bound to dominate the statistical test activity.

For each document, a term (or feature) vector in the vector space model is built and assigns a score to each term using the TF-IDF scheme. The term weight will be set as 0 if the term is absent in the document. The term vectors created above are split into a training set and test set using k-fold cross-validation. It is used to tune hyper-parameters in machine-learning algorithms. It means we split data into ten equal parts (i.e., k = 10). 90% training set is used to build, and the rest of the 10% data is used for testing. The test set is used for assessment purposes; on the other hand, the training set term vectors are composed to create a term-by-article matrix M.

The pre-processing and LSI implementation is done by using NLTK and scikit learn library in Python. The optimum value of λ is determined, taking into account that there is a balance between the number of features (λ) and the overall discriminatory power of the features compared. Thus, λ is set to 1000 by LSI. Chi-Square (CHI), IG, and GR are added for a further reduction. On the baseline, results are not as satisfactory as expected. So, we combined the features extracted from the above three methods to select the best features to improve the efficiency of the classifier's result. The selected attributes are then trained on SVM, a commonly known machine learning classifier.

We have shown the effectiveness of classification techniques trained on top 500 ranked features sets, whereas the baseline results were calculated on top ($\lambda = 1000$) features as the initial set. Experimentation of SVM with the proposed method shows quiet better results than the other feature selection methods and gets 62.57% accuracy with ten-fold cross-validation while the proposed method achieved almost 10% higher efficiency when compared with baseline method as shown in Tables 3 and 4 respectively. The accuracy of the classifier with each feature selection method is shown in Table 5.

Table 3 Performance of SVM without using filter method

Feature selection methods	Accuracy (%)
HFSA	62.57
Information Gain	59.95
Chi-Square (CHI)	54.68
Gain ratio	52.50
Baseline	51.54

Table 4 Category wise performance of SVM without using filter method

Category	Recall	Precision	F-Measure
International	0.304	0.415	0.351
National	0.404	0.411	0.408
Social	0.346	0.339	0.342
Entertainment	0.562	0.623	0.591
Political	0.599	0.403	0.482
Health	0.444	0.371	0.404
MuslimWorld	0.555	0.552	0.553
LocalNews	0.584	0.694	0.634
Sports	0.841	0.744	0.789
Average	0.515	0.526	0.516

Table 5 Performance of SVM with different filter methods on Roshni dataset

Category name	Gain Ratio (GR)		Chi-Square (CHI)		Information Gain (IG)		HFSA	
	Precision	Recall	Precision	Recall	Precision	Recall	Precision	Recall
International	0.253	0.524	0.275	0.498	0.471	0.212	0.589	0.284
National	0.363	0.490	0.423	0.499	0.470	0.429	0.438	0.464
Social	0.417	0.377	0.419	0.326	0.742	0.293	0.744	0.288
Entertainment	0.665	0.613	0.667	0.676	0.774	0.695	0.777	0.730
Political	0.624	0.394	0.674	0.432	0.623	0.596	0.595	0.667
Health	0.496	0.268	0.470	0.317	0.567	0.288	0.636	0.285
MuslimWorld	0.637	0.545	0.622	0.569	0.543	0.801	0.619	0.773
LocalNews	0.663	0.694	0.619	0.738	0.536	0.859	0.551	0.870
Sports	0.922	0.726	0.901	0.758	0.897	0.882	0.920	0.914
Average	0.571	0.525	0.574	0.547	0.610	0.600	0.640	0.626

8 Conclusion

In this experimental study, the effectiveness of the hybrid features selection approach was explored that incorporated three filtering methods with the LSI method. The results of the proposed method were more effective in reducing dimensionality. Also, they could produce a higher categorization precision as compared to each filter approach. Urdu text classification was carried out using well-known SVM classifiers, and the results were found to be comparatively satisfactory with accuracy as high as 62.57%. Since the dimension reduction effect of LSI in text classification is remarkable. Still, it can filter out essential characteristics for a rare category in the entire document collection, which leads to an impact on the performance of the classifier. Therefore the task of further enhancing its classification with various classifiers is indeed necessary to make it available to the readers. Besides, the collection of features is a significant step in this study, which must be explored and improved in the future. Another recommendation for the future is to include more documents in the collection.

Acknowledgements We truly thank the chief editor of 'Daily Roshni' Mr. Zahoor Ahmad Shora for his outstanding contribution involuntary sharing of raw data for the collection.

Appendix

The following list of common acronyms is used in this article.

[Abbreviations]	Word/Phrase
[TF-IDF]	Term Frequency-Inverse Document Frequency
[HFSA]	Hybrid Feature Selection Approach
[SVM]	Support Vector Machine
[ANN]	Artificial Neural Networks
[KNN]	K-Nearest Neighbour
[NLP]	Natural Language Processing
[NER]	Named Entity Recognition
[POS]	Part of Speech
[LSI]	Latent Semantic Indexing
[CHI]	Chi-Square
[IoT]	Internet of Things
[NB]	Naive Bayes
[DT]	Decision Tree
[IG]	Information Gain
[GR]	Gain Ratio
[GA]	Genetic Algortihm
[MI]	Mutual Information

References

1. Chen, K., Gao, S., Zhu, Y., & Sun, Q. (2006). Music genres classification using text categorization method. In *2006 IEEE Workshop on Multimedia Signal Processing* (pp. 221–224). IEEE.
2. Miltsakaki, E., & Troutt, A. (2008). Real time web text classification and analysis of reading difficulty. In *Proceedings of the Third Workshop on Innovative Use of NLP for Building Educational Applications* (pp. 89–97).
3. Pandey, U., & Chakravarty, S. (2010). A survey on text classification techniques for e-mail filtering. In *2010 Second International Conference on Machine Learning and Computing* (pp. 32–36). IEEE.
4. Bommert, A., Sun, X., Bischl, B., Rahnenführer, J., & Lang, M. (2020). Benchmark for filter methods for feature selection in high-dimensional classification data. *Computational Statistics & Data Analysis*, *143*, 106839.
5. Aladeemy, M., Adwan, L., Booth, A., Khasawneh, M. T., & Poranki, S. (2020). New feature selection methods based on opposition-based learning and self-adaptive cohort intelligence for predicting patient no-shows. *Applied Soft Computing*, *86*, 105866.
6. Hoque, N., Singh, M., & Bhattacharyya, D. K. (2018). EFS-MI: an ensemble feature selection method for classification. *Complex & Intelligent Systems*, *4*(2), 105–118.
7. Dey, N., & Ashour, A.S. (2016). Antenna design and direction of arrival estimation in metaheuristic paradigm: a review. *International Journal of Service Science, Management, Engineering, and Technology (IJSSMET)*,*7*(3):1–18
8. Dey, N., Ashour, A., & Bhattacharyya, S. (2020). *Applied Nature-Inspired Computing: Algorithms and Case Studies*. Springer
9. De, D., Mukherjee, A., Das, S. K., & Dey, N. Nature inspired computing for wireless sensor networks

10. Das , S.K., & Tripathi, S. (2018). Adaptive and intelligent energy efficient routing for transparent heterogeneous ad-hoc network by fusion of game theory and linear programming. *Applied Intelligence, 48*(7), 1825–1845.
11. Das, S. K., & Tripathi, S. (2019). Energy efficient routing formation algorithm for hybrid ad-hoc network: a geometric programming approach. *Peer-to-Peer Networking and Applications, 12*(1), 102–128.
12. Youn, E., & Jeong, M. K. (2009). Class dependent feature scaling method using Naive Bayes classifier. *Pattern Recognition Letters, 30*(5), 477–485.
13. Bidi, N., & Elberrichi, Z. (2017). Feature selection for text classification using genetic algorithms. In *Proceedings of 2016 8th International Conference on Modelling, Identification and Control (ICMIC 2016)*
14. Ahmed, K., Ali, M., Khalid, S., & Kamran, M. (2016). Framework for Urdu news headlines classification. *Journal of Applied Computer Science & Mathematics, 10*(1), 17–21.
15. Khan, W., Daud, A., Nasir, J. A., & Amjad, T. (2016). A survey on the state-of-the-art machine learning models in the context of NLP. *Kuwait Journal Science, 43*(4), 95–113.
16. Ali, A.R., & Ijaz, M. (2009). Urdu text classification. In *Proceedings of the 7th International Conference on Frontiers of Information Technology* (pp. 21:1–21:7). ACM
17. Zia, T., Akhter, M. P., & Abbas, Q. (2015). Comparative study of feature selection approaches for Urdu text categorization. *Malaysian Journal of Computer Science, 28*(2), 93–109.
18. Bilal, M., Israr, H., Shahid, M., & Khan, A. (2016). Sentiment classification of Roman-Urdu opinions using Naïve Bayesian, Decision Tree and KNN classification techniques. *Journal of King Saud University - Computer and Information Sciences, 28*(3), 330–344.
19. Günal, S. (2012). Hybrid feature selection for text classification. *Turkish Journal of Electrical Engineering and Computer Science, 20*(Sup. 2), 1296–1311.
20. Fang, Y., Chen, K., & Luo, C. (2012). The algorithm research of genetic algorithm combining with text feature selection method. *Journal of Computational Science and Engineering, 1*(1), 9–13.
21. Lei, S. (2012). A feature selection method based on information gain and genetic algorithm. In *2012 International Conference on Computer Science and Electronics Engineering* (Vol. 2, pp. 355–358). IEEE
22. Mukhtar, N., Khan, M. A., & Chiragh, N. (2018). Lexicon-based approach outperforms Supervised Machine Learning approach for Urdu Sentiment Analysis in multiple domains. *Telematics and Informatics, 35*(8), 2173–2183.
23. Mukhtar, N., & Khan, M. A. (2017). Urdu sentiment analysis using supervised machine learning approach. *International Journal of Pattern Recognition and Artificial, 32*(02), 1851001.
24. Sohail, O., & Karim, A. (2018). Text Classification in an Under-Resourced Language via Lexical Normalization and Feature Pooling Text Classification in an Under-Resourced Language via Lexical Normalization and Feature Pooling.
25. Khan, W., Subhan, F., Khan, A., Khan, S. N., Ullah, A., & Ullah, B., et al. (2018). Urdu word segmentation using machine learning approaches. *International Journal of Advanced Computer Science and Applications, 9*(6), 193–200.
26. Puri, S., & Singh, S. P. (2018). Hindi text document classification system using SVM and Fuzzy. *International Journal of Rough Sets and Data Analysis, 5*(4), 1–31.
27. Anwar, W., Bajwa, I. S., Abbas Choudhary, M., & Ramzan, S. (2019). An empirical study on forensic analysis of Urdu text using LDA-based authorship attribution. *IEEE Access, 7*, 3224–3234.
28. Hardie, A. (2003). Developing a tagset for automated part-of-speech tagging in Urdu. In *Corpus Linguistics 2003*.
29. Daud, A., Khan, W., & Che, D. (2016). Urdu language processing: a survey. *Artificial Intelligence Review,* pp. 1–33.
30. Humayoun, M., Hammarström, H., & Ranta, A. (2006). *Urdu morphology, orthography and lexicon extraction*. Chalmers tekniska högskola.
31. Sharjeel, M., Nawab, R.M.A., & Rayson, P. (2016). Counter: corpus of Urdu news text reuse. *Language Resources and Evaluation,* pp. 1–27.

32. Abbas, Q. (2014). Exploiting language variants via grammar parsing having morphologically rich information. In *Proceedings of the EMNLP'2014 Workshop on Language Technology for Closely Related Languages and Language Variants* (pp. 36–46).
33. Gupta, V., Joshi, N., & Mathur, I. (2015). Design & development of rule based inflectional and derivational urdu stemmer 'usal'. In *2015 International Conference on Futuristic Trends on Computational Analysis and Knowledge Management (ABLAZE)* (pp. 7–12). IEEE.
34. Rasheed, I., Gupta, V., Banka, H., & Kumar, C. (2018). Urdu text classification: a comparative study using machine learning techniques. In *2018 Thirteenth International Conference on Digital Information Management (ICDIM)* (pp. 274–278). IEEE.
35. Rasheed, I., & Banka, H. (2018) Query expansion in information retrieval for Urdu language. In *2018 Fourth International Conference on Information Retrieval and Knowledge Management (CAMP)* (pp. 1–6). IEEE.
36. Karthikeyan, M., & Aruna, P. (2013). Probability based document clustering and image clustering using content-based image retrieval. *Applied Soft Computing, 13*(2), 959–966.
37. Ahmad, Z., Orakzai, J.K., Shamsher, I., & Adnan, A. (2007). Urdu Nastaleeq optical character recognition. In *Proceedings of World Academy of Science, Engineering and Technology* (Vol. 26, pp. 249–252). Citeseer.
38. Riaz, K. (2008). Concept search in Urdu. In *Proceedings of the 2nd PhD workshop on Information and Knowledge Management* (pp. 33–40).
39. Jones, K.S. (1972). A statistical interpretation of term specificity and its application in retrieval
40. Jones, K. S. (2004). A statistical interpretation of term specificity and its application in retrieval. *Journal of Documentation, 60*, 11–21.
41. Deerwester, S., Dumais, S. T., Furnas, G. W., Landauer, T. K., & Harshman, R. (1990). Indexing by latent semantic analysis. *Journal of the American Society for Information Science, 41*(6), 391–407.
42. Thomo, A. (2009). *Latent semantic analysis (tutorial)* (pp. 1–7). Canda: Victoria.
43. Moohebat, M., Raj, R. G., Thorleuchter, D., & Kareem, S. B. A. (2017). Linguistic feature classifying and tracing. *Malaysian Journal of Computer Science, 30*(2), 77–90.
44. Forman, G. (2003). An extensive empirical study of feature selection metrics for text classification. *Journal of Machine Learning Research, 3*, 1289–1305.
45. Yang, J.O., & Pedersen, Y. (1997). A comparative study on feature selection in text categorization. In *ICML* (p. 35).
46. Mitchell, T. (1997). *Machine learning* (Vol. 45(37), p. 2). Burr Ridge: McGraw Hill.
47. Gu, Q., & Song, Z. (2009). Image classification using SVM, KNN and performance comparison with logistic regression. *CS44 Final Project Report*.

Automated Document Categorization Model

Rakhi Patra

Abstract The aim of this work is to build a generic model of Document Clustering that automatically groups together the related documents. Model is built with unsupervised and supervised learning with the assumption of no prior knowledge of the given domain. No manual effort is required for creating the training document set, instead the proposed model automatically generates training document. After that, it uses those for categorizing text documents. In the proposed model, the entire process is broadly divided into two steps. First, the initial classification is done in an unsupervised way. Apply K-means algorithm on the unlabeled documents in order to prepare the training dataset. Text documents are represented here as feature vector format where keywords extracted are considered as a feature. Here the selected representative documents are considered as the initial centroids. In step 2, create a supervised classifier on the initially categorized set. The categorized documents resulted from the previous step are used to train the supervised classifier. Naive Bayes classifier will be used as a statistical text classifier which uses word frequencies as features.

Keywords Unsupervised Grouping · Text Classification · Document Organization · Text Mining · Text Categorization · K-means · Naïve Bayes

1 Introduction

Applications of text mining are increasing rapidly with growing number of textual information in different fields, such as Email messages, news articles, customer, employee feedback, insurance claims, patient reports in hospitals, clinics, as well as a huge number of resources for scientific abstracts. In text classification, a set of related documents are grouped together which is known as document clustering. The objective of this paper is to leverage machine learning algorithms with open source

R. Patra (✉)
Department of Software Engineering, Birla Institute of Technology and Science, Pilani, Pilani 333031, Rajasthan, India
e-mail: rakhi.patra@gmail.com

S. K. Das et al. (eds.), *Machine Learning Algorithms for Industrial Applications*, Studies in Computational Intelligence 907,
https://doi.org/10.1007/978-3-030-50641-4_2

technology to create an automated document grouping model in unsupervised way that classify texts without the need for manual labeling of the category of documents. The main preliminaries of the proposed method are unsupervised learning, K-means, training set, information retrieval, WordNet and corpus. Automatic document clustering has played an important role in many fields like sorting of news articles, emails etc. that saves manual effort to process documents. Unsupervised discovery with K-means algorithm is used here to identify the main topic or theme of the documents within the corpus. The domain is considered here as unknown and no training data is available. The keyword list is constructed directly from the documents collection itself. It is assumed that the source documents collection contains documents on different topics. By different topics we mean topic whose keyword vocabularies may or may not be related. Text documents are transformed into vector form based on a list of index keywords. Vectors derived from text documents are processed by well-known numerical procedures as TF-IDF of cluster analysis.

The roadmap of the manuscript is given as: literature review discussed in Sect. 2. Section 3 describes the domain and background study. Section 4 illustrates the proposed work. Section 5 describes the experimental framework. Section 6 describes experimental result with discussion. Section 7 concludes the paper.

2 Related Work

The main aim of the document clustering is to put the related documents into same cluster. There is huge number of text documents available every day in digital form. So, the demand of automatic organizing documents is increasing. In recent years, an extensive number of studies have covered different methods of text feature extractions, dimensionality reduction, algorithms and techniques of text classification, and evaluations methods. Many machine learning approaches have achieved good results in text processing and its classification. The success of these learning algorithms relies on their capacity to understand unstructured text data. In paper [1] M. Ikonomakis, S. Kotsiantis & V. Tampakas, present a survey of a wide variety of algorithms of general process of text classification that describes some techniques for classification and finally evaluation of effectiveness of algorithms. Regarding vector representation the paper discussed about stop word filtration stemming as preprocessing step and then possible word vector representation with boolean word indicators and word count. In the section "feature selection", various methods to choose a feature subset such as Best Individual Features, Sequential Forward Selection, Information Gain and their performance are discussed. The same section pointed out how feature selection is inefficient in a very large training corpus and Instance Selection is used to speed up the process in this situation. After representing documents in vector format, the paper describes the literature survey on how different text classifiers are proposed using machine learning techniques. In this context, the paper presents a literature review that discussed some methods like precision, recall & accuracy to check effectiveness of classifier. Anuradha Purohit, has described [2] a text classification system implemented with some intelligent techniques. They used Porter stemmer algorithm to filter unnecessary words from text documents. After filtration, the association rule is used to derive feature sets from each text documents with rest of the words. The derived features are then processed with the concept of Naive Bayes Classifier to calculate the probability of derived word sets. In this paper, abstracts from different research papers have been used as corpus data sets for training and testing of the proposed classifier. Total 40 abstracts of four classes of papers from DBMS, Operating System, Java and Data Structure were considered for their experiment. Their classification has achieved 75% accuracy from the experiment. In [3] paper, Daniel has used the Sequential Minimal Optimization (SMO) implementation of Support Vector Machine (SVM) method as classifier algorithm. For their experiment they use collection of news stories published by Reuters Press for one year covering the period from 20.07.1996 through 19.07.1997. After applying a standard stop-word filter and extracting the word stem, each document was represented as a vector of words. In the vector the value of the word's weight was computed with Term Frequency TF(d,t) where term t occurs in document d. The weight was of binary representation that it can take values between 0 and 1. Feature selection methods were applied to reduce the dimensionality of the vector which was important to work on large collections of documents. In this paper Daniel has present a comparison study of three feature selection methods based on the OneRAttribute, Information Gain and Gain Ratio. According to Daniel best results were obtained with the Information Gain method. In paper [4], author Niladri Biswas described components of text mining framework such as preprocessing, Tokenization, POS tagging, Syntactical Parsing, Information Extraction. In information extraction, Niladri described the concepts of topic tracking, summarization, categorization, feature/term selection, entity extraction, concept extraction, theme extraction, Clustering. The paper had shown importance of text mining and its business application in various area such as Human Resource Management, Customer Relationship Management (CRM), Market Analysis, sentiment analysis, warranty or insurance claims, diagnostic medical interviews, etc. In this paper various machine learning algorithm and their suitability in proper for business applications are discussed with examples. Finally the paper has listed down available commercial & open source text mining tools.

Compared to similar studies in text classification, this work is different in a way that text classification is described in unsupervised way to get rid of manual labeling of text.

3 Domains and Background Study

3.1 Text Mining

In this subsection, some common numerical weighting factor in text mining are described as follows.

a) **TF(d, t)-Term Frequency:** This function indicates how many times a term appears in a document. It is the ratio of number of times the word appears in the document and the total number of words the document has. Consider, d = documents and t = terms. Using given values TF(d, t) can be calculated as given in Eq. 1:

$$\begin{aligned} \mathrm{TF(d,t)} &= 0 \quad \mathrm{freq(d,t)} = 0 \\ &= 1 + \log(1 + \log(\mathrm{freq(d,t)})) \quad \mathrm{freq(d,t)} > 0 \end{aligned} \tag{1}$$

b) **IDF(d, t) – Inverse Document Frequency:** This function helps to determine, whether the term is commonly used or rare across all documents. It is calculated by find out the division of total number of documents in the corpus by the number of documents which contain the term and then calculate logarithm on the division result. IDF of any term t can be calculated as given in Eq. 2.

$$\mathrm{IDF(t)} = \log\left((1 + |\mathrm{d}|) / |\mathrm{dt}|\right) \tag{2}$$

where |d| is total no. of document and |dt| is no. of documents in which term, "t" is present.

c) TF_IDF(d, t): This function returns a numerical measure to calculate how important a word is in whole corpus of documents. It is calculated as product of TF and IDF defined as Eq. 3.

$$\mathrm{TF\_IDF} = \mathrm{TF(d,t)} * \mathrm{IDF(t)} \tag{3}$$

3.2 Preprocessing

First and important step of text mining is preprocessing of text. As we are going to find relation between the texts we should have only text data in hand before proceeding further. The preprocessing level consists of below steps (Fig. 6):

(a) **Remove Special Character**: First step is to get rid of all the special symbols such as $, @, _, -, +, – etc.
(b) **Remove Stop Words**: It includes words like "since", "of", "a", "the" and many more which are frequently used in the text but provide little bit information about the context.
(c) **Tokenize Data**: This is a process to split text into some smaller pieces or elements such as symbols, phrases, keywords and words.

3.3 Vector Representation of Documents and Feature Selection

This section describes the vector representation of a document. A document can be considered as series of words. Each document can be then represented by an array of words. The list words, the training set have is called vocabulary, or feature. Feature selection is a significant step in text classification which comes before creating the training set. It is the process of selecting a specific subset of the words from the entire feature set or vocabulary and using only those words in the classification algorithm. A document can be represented as feature vector. Say there is a set of t terms t for d documents. Each document is presented as a vector v in the t dimensional space. The ith coordinate of v indicates the frequency of the ith term in the document. Vector can have binary representation by assigning the value 1 in i^{th} position if the feature-word appears in the document without considering the number of occurrences and 0 if the document does not contain the word. Vector can have nominal representation as well if the term weight is calculated with TF-IDF. In this case, weight values can lie between 0 and 1. Feature-scoring methods conclude feature subset by ranking the features by their independently determined scores, and then select the top scoring features. In this paper, keywords that reflect the topic of the documents are selected based on feature scoring methods.

The general method of text classification stages which are discussed above are shown in Fig. 1.

4 Proposed Work

In this work a generic model of document clustering is described for a set of documents in a particular domain. For example domain could be scientific documents or technical documents (Fig. 1) present in a local file system. This model can also be applied in other domains e.g. classifying patient's clinical reports (Fig. 2).

Document Classifier can be used in many different applications and industries to parse and categorize text documents. Here are some examples:

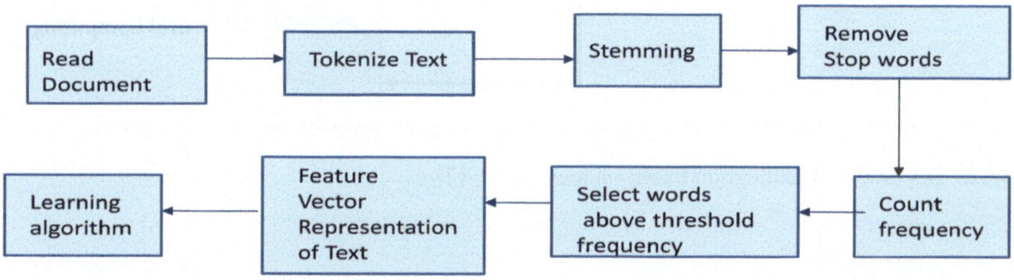

Fig. 1 Text classification components

(a) Banking and finance: Classify and cluster investments, loan applications and mortgage documents.
(b) Government: Sort documents into categories. Useful for Law industry, where there is a dealing with large number documents every day.
(c) Customer Service: Analysis of customer feedback or complaints.
(d) Insurance: To analyze claims proof documents in any kind of insurance company such as automobile or health insurance etc.
(e) Human Resource/Talent Solution: Analyze resumes/CVs and applications to derive deeper meaning.
(f) Higher Education: Useful for Academia to sort various documents.

In this paper text classification is done in unsupervised way. Unsupervised learning technique is proposed to overcome the difficulty of creating a high-quality labeled training dataset which is an important input to high-accuracy supervised learning techniques. The method used here to classify text is a combined implementation of K-means and NaiveBayes. Figures 3 and 4 describe how a training model is created and how the model is used for classifying an unknown document. The steps of the proposed method shown in Algorithm 1. The framework of the text document clustering shown in [6].

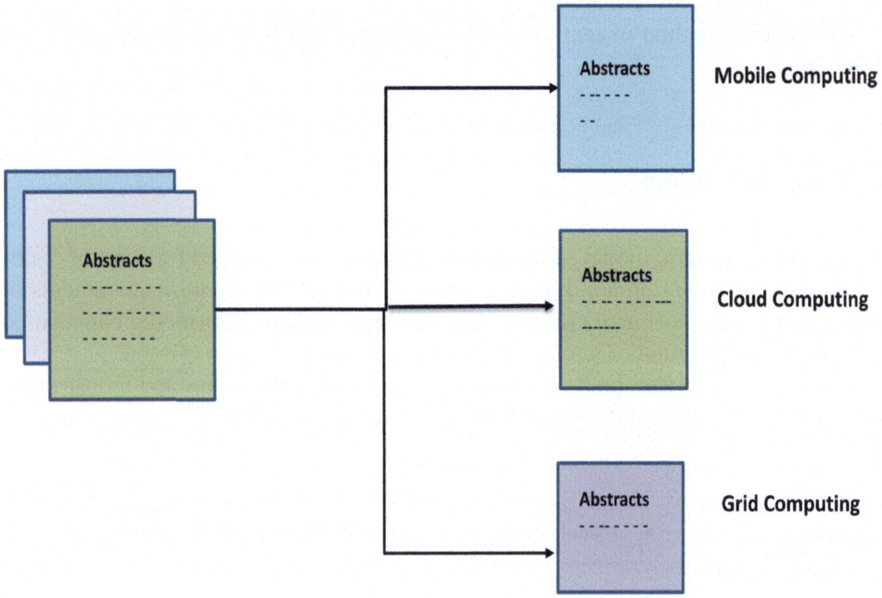

Fig. 2 Document classification based on technology [7]

Fig. 3 High level steps of proposed method

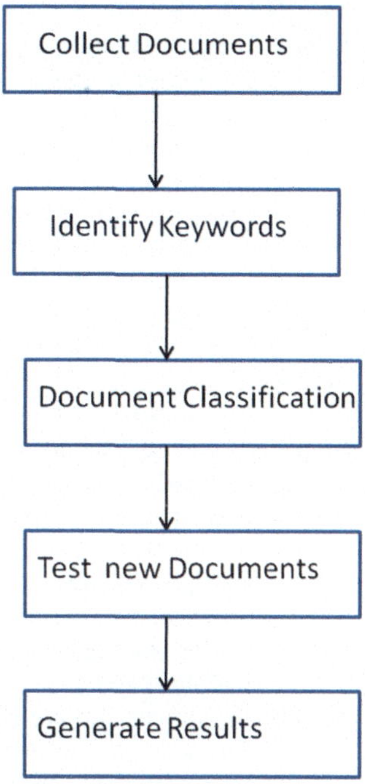

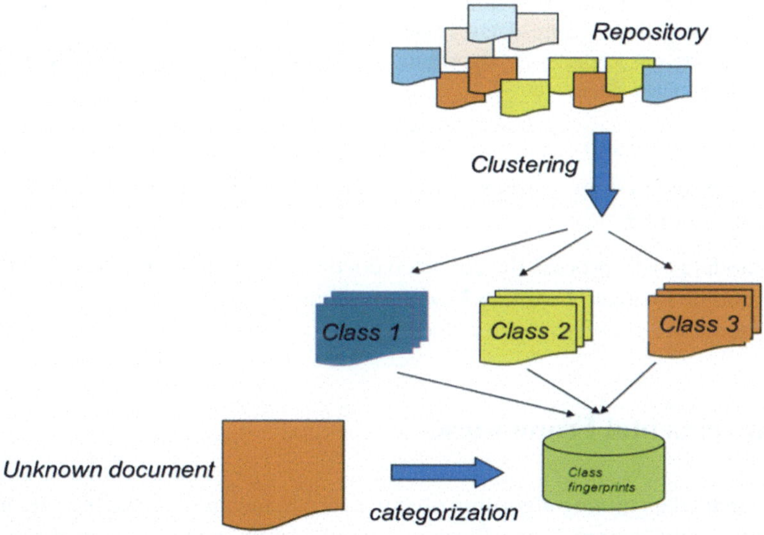

Fig. 4 Document categorization [9]

Algorithm1: The proposed method.

Step 1: Pre-processing:
1.1. Remove Special Characters, digits, numbers.
1.2. Remove stop words.
1.3. Tokenize data.
1.4. Stemming

Step 2: Identify Distinct Key Words of each category.

Step 3: Calculate Term Frequency TF (d, t). [where d= document, t = term]

Step 4: Calculate Inverse Document Frequency IDF (t).

Step 5: Calculate Term Frequency–Inverse Document Frequency TF-IDF (d, t). Prepare keyword list based on highest TF-IDF score.

Step 6: Once the keyword list is prepared, select the representative documents of each category based on highest TF-IDF weight.

Step 7: Apply K Means Clustering with TF-IDF Weights, where the document selected in above step is considered as initial centroids.

Step 8: One important step for text clustering is to consider how the text content can be represented in the form of mathematical expression for further analysis and processing. For transformation a feature space is constructed. Each dimension in the feature means one term, which comes from the key words of the document. Then each document is represented as one vector in this feature space.

Step 9: Documents are compared with each other to find out the similarity measure. The goal of the document clustering scheme is to maximize intra cluster similarity between documents, while minimizing inter cluster similarity. Fig 5 depicts the proposed framework of clustering.

Step 10: Once the training dataset with labelled documents are created, apply a traditional supervised algorithm (e.g. "Naïve Bayes") to classify a new unlabelled document.

The proposed system consists of three modules as shown in Fig. 5 and system flow shown in Fig. 6.

(a) A module to preprocess the collected documents and to extract key features
(b) A module to create Learning Model of training datasets
(c) A module to classify new text documents

5 Experimental Framework

The two main data mining algorithm used here are K-means to create training model of document cluster and Naïve Bayes to classify an unknown document based on class of document cluster.

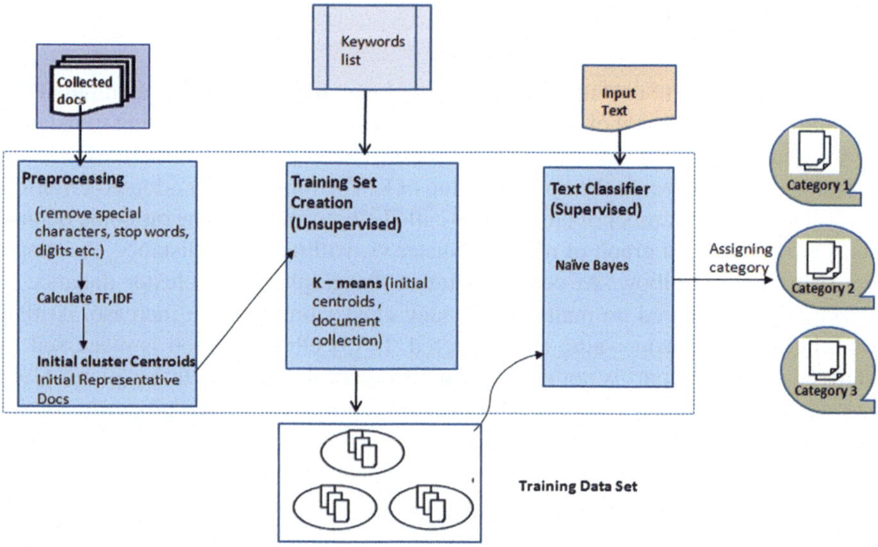

Fig. 5 Architectural components of proposed method [8]

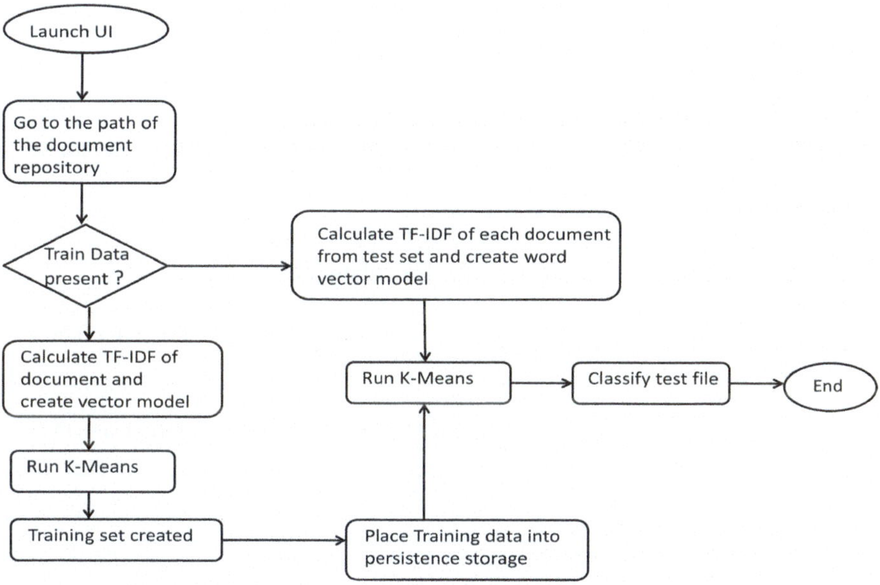

Fig. 6 System flow chart

5.1 K-Means Algorithm

K means requires us to provide the number of cluster to be detected as an input at the beginning. Correct choice of k, the number of cluster is often challenging to balance between the accuracy of data point's distribution and the number of cluster. There are several methods to decide the optimal value of k. The most widely adopted criterion to select the proper number of cluster is so called elbow method. The rationale behind this is as if you plot a graph of number cluster vs. within cluster distance, the graph gets shape like an elbow. At certain point in the graph within cluster distance is no longer gets changed no matter how many cluster number you increase. At that point, number of cluster value is considered. In [5] elbow method is discussed in detail. However this part is not explored in this work and kept as future plans. Here K is taken as predefined constant value which is apparent from the properties of the document set.

The main idea is to define k centroids, one for each cluster. The centroid of a cluster is formed in such a way that it is closely related (in terms of similarity function; similarity can be measured by using different methods such as cosine similarity, Euclidean distance, Extended Jacquard) to all objects in that cluster. How K-Means is applied in this solution of text document clustering, is described in Algorithm 2.

Algorithm 2: Working principle of K-means algorithm.

Step 1: Choose k number of clusters to be determined
Step 2: Choose k documents randomly based on TF-IDF score as the initial cluster center
Step 3: Repeat
    3.1: Assign each document to their closest cluster
    3.2: Compute new clusters, i.e. Calculate mean points.
Step 4: Until
    4.1: No changes on cluster centers (i.e. Centroids do not change location any more)
        OR
    4.2: No object changes its cluster (We may define stopping criteria as well)

5.2 Document Representation

Documents can be represented using the TF-IDF score.

(a) The class "DocumentParser" is defined to parse the text documents and split them into tokens. This class will interact with "TfIdf calculator" class to calculate the TFIDF.

(b) Define "DocumentVector" class whose instance holds the document and its corresponding representation on vector space.
(c) Class "DocumentCollection" represents all the documents to be clustered.
(d) Define a class "Centroid" in which documents are assigned during the clustering process.

5.3 *Finding Similarity Score*

Now the documents are ready with a numerical form that can be mutually comparable. Once the data model is ready, we need to compute the distances between the documents. To measure the similarity between two documents, cosine similarity is used. Cosine similarity between two vectors is computed by dividing the dot product of two vectors with the product of their magnitudes. This method is named as "FindCosineSimilarity". Here is how the method works. Let's say there are 2 documents, A and B. The mentioned method takes two arguments of vector representation from document A and B, and returns the similarity score as either 1 or 0, indicates document A and B are completely similar or dissimilar respectively. Cosine similarity of two vectors is computed by dividing the dot product of the two vectors by the product of their magnitudes.

- Define a class "CosineSimilarity" with method "FindCosineSimilarity" to perform the task.

5.4 *Extracting Keywords*

Create the keyword list for the given document set as follows.

(a) Most frequently occurred terms are selected from a given document based on their highest TF-IDF value.
(b) Execute step (a) for each document.
(c) Create a whole list of most frequently used terms across all documents.
(d) Get the count of each term of how many times it appears in the list. It implies that a most frequently used term of one document, is common with other documents as well in the set, which best represents the context of the given domain of which document set belongs to. Thus the term becomes the keyword of the whole document sets.
(e) Say, number of cluster defined by the user as n. Select n number of keywords from step (d).
(f) Once the keywords are determined for the initial set of documents of a given domain, they are stored in serialized object format for future references.

5.5 Creating Initial Representatives Documents Sets

In this step some documents are selected as initial centroids of K-means algorithm. Initial position of the centroids can have some influence on the result of the K-means algorithm. While creating this model with K-means, initial centroids are chosen as the relevant documents which have highest TF_IDF value based on the Keywords identified in previous step. These techniques ensure the right choices of initial centroids. Sometimes the documents containing keywords of a certain category but having the special features of the other category is considered as error documents. To remove such error documents, the representative documents can be ranked by computing the weight of each sentence. However this part is not considered in this work and kept open for future enhancement.

5.6 Preparing Document Cluster

This process follows the below steps.

1. Initializing cluster center
   Creating Initial Representative Documents sets
2. Identifying closest cluster center
   We can retrieve the index of the closest cluster center for each document with this function. Cosine similarity has been used here to identify the closeness of document. The array 'similarityMeasure' contains the similarity score for the document object for each cluster center. The array index where the value is maximum is considered as the closest cluster center of the given document.
3. Selection the new position of the cluster center
   Once each document is assigned to its closest cluster center, the mean of each cluster center is being recalculated, which indicates the new position of cluster center (centroid).
4. Stop Condition
   Repeat step 2 & 3 until the one of the following termination condition is met.
   a. A fixed number of iterations have been completed.
   b. Assignment of the documents to its clusters won't be changed between iterations.
   c. Centroids do not change between iterations. This criterion ensures that the clustering is of a desired quality after termination. In practice, we need to combine it with a bound on the number of iterations to guarantee termination.

   In this paper termination condition 'a' is considered.

5.7 Apply Object Serialization on Train Set

Initial set of labeled documents after running K-Means (Unsupervised method), are kept in persistent storage with Java's serialized object format. In the subsequent run these set will be used as training data set.

5.8 Naive Bayes Text Classification

Multinomial Naive Bayes (NB) model is a probabilistic supervised learning method. Once the training set of documents are ready, naïve bayes algorithm is applied to determine the class on new document comes in. To apply multinomial NB and train multinomial NB, is shown in Algorithms 3 and 4.

Algorithm 3: Apply multinomial NB
ApplyMultinomialNB (C , V , prior , condProb, d) 1. W ← Extract Tokens From Doc (V , d) 2. **for** each class c in classes C 3. do score[c] ← prior [c] 4. **for** each term t in W 5. do score[c] += condProb [t] [c] 6. **return** $argMax_c$

Algorithm 4 : Train multinomial NB
TrainMultinomialNB (Classes C, Document Collection D) 1. V←Extract Vocabulary (D) 2. N←CountDocs (D) 3. **for** each class c in classes C 4. do N_c←CountDocsInClass (D,c) 5. prior [c] ←N_c /N 6. $text_c$← Concatenate Text of All Docs In Class (D ,c) 7. **for** each term t in vocabulary V 8. doT_{ct}← Count Tokens of Term ($text_c$,t) 9. **for** each term t in vocabulary V 10. do conditional probability condProb [t] [c] ←T_{ct} + 1/ 11. **return** prior, conditional probability

6 Experimental Work and Discussion

Java1.6 is used here as programming language to implement these algorithms. To store the training model build with Kmeans, Java's object serialization feature is used. User interface is build up with Java Swing. Figure 7 describes the class structure of the proposed solution.

6.1 Data Set

The first use case of document clustering is selected here as to cluster a set of technical abstract papers. Abstracts from different research papers have been collected to prepare for training data sets and testing of the proposed method. Three classes of papers on mobile computing, cloud computing and grid computing were considered here for the experiment. Total 25 abstracts are used in the experiment. Nine are from cloud, seven are from mobile, and eight are from grid. Abstract papers are of format .PDF, .DOC & .TXT (Fig. 8) and located in a source folder of local file system. The

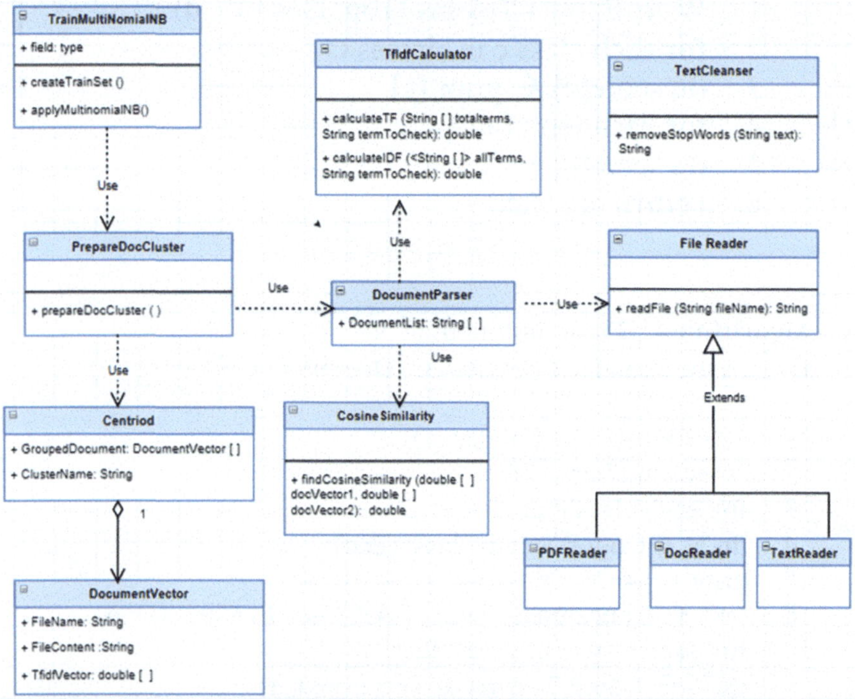

Fig. 7 Class model of proposed method

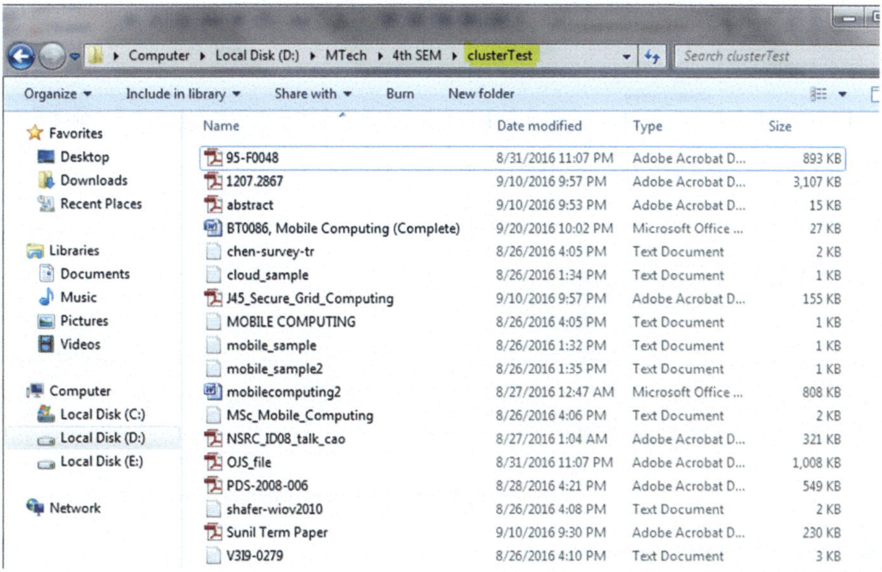

Fig. 8 Source folder of abstract papers

cluster model is also tested for clustering medical surgery report on different subject for example hernia, appendix & gallbladder.

6.2 Experimental Results

Figure 8 shows the source folder where initially all the documents are kept. Source path is browsed with a user interfaces shown in Fig. 9. When user hits on 'Create Cluster' button, documents are processed and progress percentage is shown on screen (Fig. 10). The keyword extraction process is applied to all the abstracts and key words extracted, are "cloud", "grid" & "mobile" i.e. the subject of the abstract paper. Figure 11 describes how the cluster folders are created for each subject t i.e. mobile computing, cloud computing and grid computing. Each cluster contains documents related to the subject of that cluster. Thus documents of initial source folder are organized by cluster. Once the initial cluster set is created, new documents are classified with that cluster model.

For second use case, clusters are created for each subject of surgery report i.e. hernia, appendix & gallbladder. Reports are placed inside the related cluster.

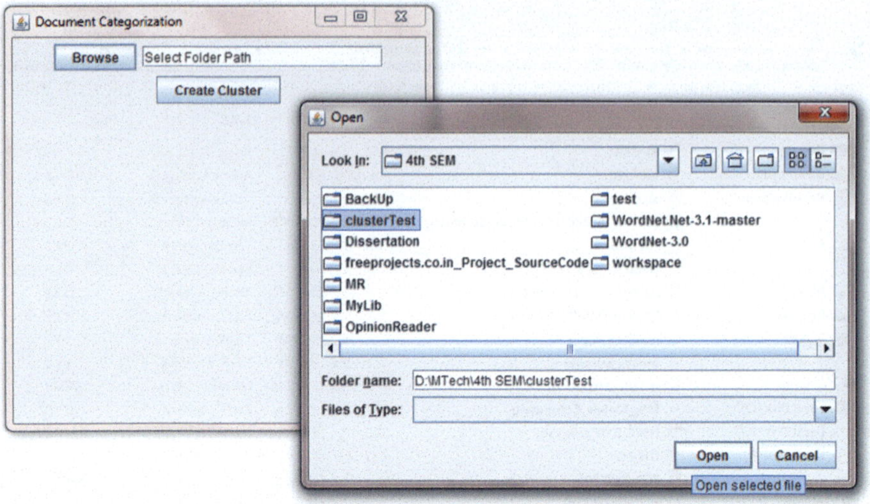

Fig. 9 User interface of browsing source folder

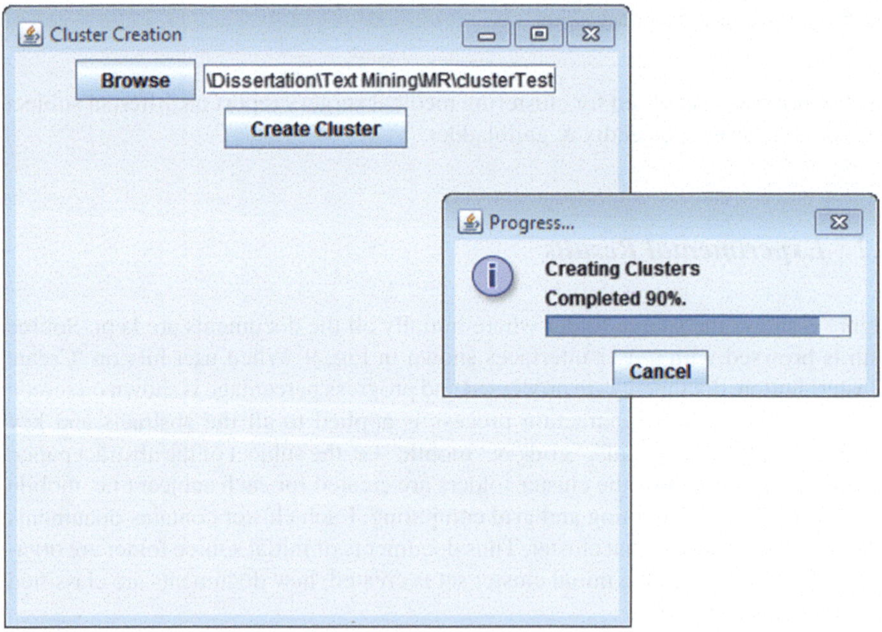

Fig. 10 User interface of progress percentage

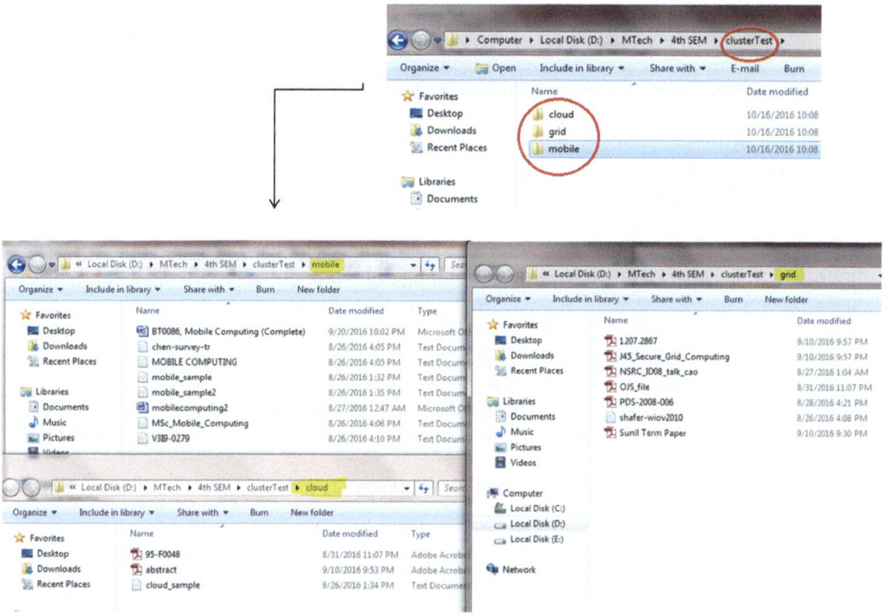

Fig. 11 Organized documents by clusters

7 Conclusion

In this work a method of automatic document categorization is described with an improved technique of automatic creation of training sets using keyword lists of each category. This text learning method can be used to reduce cost of manually labeling huge volume of unlabeled documents to create the training set. This method can also be used in areas where domain of text documents is unknown such as in medical communities, automatic topic detection is often required based on medical domain specific features like symptoms, drug used diagnosis results etc. There are a number of suggestions to extend this work. One enhancement is to re-run the clustering algorithm on the whole data set, initial set of data plus data that comes in subsequent request, incrementally. This will run as an independent process weekly or monthly schedule which will help to enhance the model. As a result, the model accuracy will increase and it can support documents from wider categories. Clustering quality of the model, can be further improved by analyzing the meta data information in the documents (e.g. title, headings, tags), semantic structure of the sentences in document etc. Search engine capability can be introduced here which will return relevant document against a query.

References

1. Ikonomakis, M., Kotsiantis, S., & Tampakas, V. (2005). Text classification using machine learning techniques. *WSEAS Transactions on Computers, 4*(8), 966–974.
2. Purohit, A., Atre, D., Jaswani, P., & Asawara, P. (2015). Text classification in data mining. *International Journal of Scientific and Research Publications, 5*(6), 1–7.
3. Morariu, D. I., Cretulescu, R. G., & Breazu, M.: *Feature selection in document classification.* https://pdfs.semanticscholar.org/.
4. http://www.codeproject.com/Articles/822379/Text-Mining-and-its-Business-Applications.
5. https://www.datanovia.com/en/lessons/determining-the-optimal-number-of-clusters-3-must-know-methods/.
6. Liu, Y. C., Liu, M., Wang, X. L. (2012). *Application of self-organizing maps in text clustering: a review* (vol. 10). https://doi.org/10.5772/50618.
7. https://www.kdnuggets.com/2015/01/text-analysis-101-document-classification.html.
8. Ko, Y., & Seo, J.: Automatic text categorization by unsupervised learning. In: Proceedings of the 18th Conference on Computational Linguistics (vol. 1, pp. 453–459). Association for Computational Linguistics, July 2000.
9. https://www.slideserve.com/nelly/text-mining-overview.

Automated Categorization and Mining Tweets for Disaster Management

Rakhi Patra

Abstract Recent trend has shown that Twitter is an emerging source to monitor disaster events. This article aims at mining tweet messages to facilitate disaster manager for better response in times of emergency. The two main objectives here, are classification and extraction of tweets. In this work machine learning is applied on tweets that are generated during the course of a disaster, to categorize them into different stages of that disaster and extract the metadata. An initial effort has spent for manually labeling 650 tweet messages into predefined categories according to disaster phases to build the classification model. Thereafter, the trained model is used to automatically categorize the new subsequent tweets in a supervised way of machine learning. The five classes or categories are considered in this work are as preparedness, response, impact, recovery and other. Once the classification is done, next objective is to mine informations from the categorized tweet messages such as impacted locations, hardest hit areas, volunteer relief organization who are working on, whether that message is a complain, if any disease has spread etc. The result generated after classification and extraction, helps emergency manager so that they can take action quickly and efficiently in the impacted location.

Keywords Twitter · Text Mining · Data Mining · Disaster Coordination · Disaster Relief

1 Introduction

Since past few years, uses of social media service has been growing rapidly as a standard medium for faster communication, especially in the age where people are always connected on mobiles unlike traditional newspaper & TV. This is the reason of rapidly growing huge amount of data. There are several popular social media such

R. Patra (✉)
Department of Software Engineering, Birla Institute of Technology and Science, Pilani 333031, Rajasthan, India
e-mail: rakhi.patra@gmail.com

S. K. Das et al. (eds.), *Machine Learning Algorithms for Industrial Applications*, Studies in Computational Intelligence 907,
https://doi.org/10.1007/978-3-030-50641-4_3

as Twitter, Face book, YouTube, LinkedIn, etc. Each media directly or indirectly connected with wired or wireless network in term of communication. Sometimes, they are also connected with wireless ad hoc network which is an infrastructure-less network that is used for temporary operation in disaster management [1–3]. This network is based on some intelligent technique such as particle swarm optimization, artificial neural network, fuzzy logic, genetic algorithm, etc. [4–6]. Twitter is a real time micro blogging network which provides a platform where users can post their view and information's called tweets. In times of disasters like cyclone, earthquake Twitter plays a major role in rapid spreading of messages by posting early alerts, warnings, live updates of emergency situations on-the-ground like magnitude of the disaster, loss, impact & damage severity. People tend to use social media for several reasons to check on family and friends, seek support, gather news on daily updates on weather, safe locations and food & water supplies, send donations etc. Thus Twitter can help to track natural disasters in real time. It notify emergency to areas that need urgent aid.

In several countries in Asia, public, government agencies, Non-Governmental Organizations (NGOs), and ad hoc volunteer groups use Twitter to share, exchange and collaborate critical information and organize relief efforts in times of crisis. Twitter improves relief and rescue operation by retrieving actionable data from tweet messages. However, manual examination of millions of unstructured tweet posts is a big challenge in time of crisis. Purpose of this study is to apply supervised machine learning technique to understand & analyze the tweets, posted in the time span of a disaster and classify them according to disaster stages. Classified tweets help emergency managers to understand the transition between stages that save their time to organize information, and set priorities and activities during the course of a disaster. Tweet mining is the next step to identify the hardest hit areas, people in the most need, to understand challenges, damages & impacts so that emergency manager, rescue workers can streamline activities accordingly.

This paper presents a tweet analysis framework that is built with IBM Natural Language Classifier (NLC) & Natural Language Understanding (NLU) Application programming interface (API). With this framework, relevant tweets are categorized and mined from the raw tweet text. The subsequent sections of this paper are structured as follows. Section 2 is a general review of the research on using social media in a disaster, followed by the Sect. 3 describing the methodology for preparing and mining tweets for disaster analysis. Section 4 focuses on how to apply the classifier on new tweets and validating the results. The paper is concluded with a discussion of the issues, challenges and future research directions of using social media data for disaster analysis and study in Sect. 5.

2 Related Works

Social Media has redefined and revolutionized the communication in time of crisis. Recently, in many studies social media data are processed for damage assessment, analysis people's response, and coordination of relief efforts.

Regarding the contribution of social media in flooding management, Bala et al. [8] described how tweet post are processed for polarity classification and irony detection, to facilitate disaster managers for better response operations. In their work, they emphasize on tweets with negative polarity that have additional probability to contain informations concerning dangers or emergency things within the context of a natural disaster. They used KNN machine learning algorithm for sentiment analysis. In their study final outcomes are presented in form of word cloud derived from tweet posts, bar graph of location wise tweets, and the polarity score of sentiment on disaster response. They validated the result on 2015 Chennai flood data.

For wildfire risk management, Athanasis et al. [9], proposed a fire behavior prediction system by processing tweet messages & their geographic location. In their proposed methodology, they used Apache Kafka open source message queue storage to process tweet stream data and Hadoop a distributed big data search and analytics engine. For identifying messages containing relevant to the incident information, they used keyword based filtration of Twitter messages that is common practice in the analysis of Twitter messages. They built a REST API to query inside the Elastic Search big data store, to exclude the off-topic messages and visualize the meaningful messages through the web-based GIS visualization platform.

The work of Nair [10] mainly focused on the 2015 Chennai Flood in India. Here, tweets regarding flood were retrieved with the help of the hash tag #chennai flood along with metadata attributes such as Twitter ID, language in which text is tweeted, date at which it is created, count of likes that the tweet has obtained, whether the text is retweeted or not, number of times the text is retweeted, date at which Twitter user created his account, number of tweets that the user has tweeted, count of tweets that the user liked the most, number of people who are following the user, number of people that the user follows etc. The tweets with meta data attributes were given as input to Weka, a free data mining analytical tool for classification, clustering, data preprocessing and visualization. Tweets collected have been classified using machine learning algorithms such as Decision trees, Random Forests and Naive Bayes. Performances of all the three algorithms are compared in terms of Precision, Recall and F-measure. The tweets obtained mainly fall into five categories that include (i) need for help, (ii) relief Measures, (iii) express gratitude, (iv) complaints and (v) others. This study also identifies the most influential users of Chennai flood on the basis of count of tweets, retweets and the followers that each user has.

The work of Huang and Xiao [11] selected Hurricane Sandy, which struck the Northwestern US on 29 October 2012, as a case study. Tweet messages are processed here along with the attributes like metadata, such as the timestamp of posting, geo-tag (location), and author profile information, which includes author location, profile description, number of tweets, number of followers and friends, etc. Their work also focused on classification of tweets according to disaster phases. Before training process, tweet text are preprocessed by removing all non-words, and tokenized with Apache Lucerne, an open source information retrieval software library. In their methodology, for classification, they selected the logistic regression which outperforms compared to other algorithms including K-Nearest Neighbors (KNN), naïve Bayes, and logistic regression which come with Apache Mahout, an open source machine-learning package.

The intent of this work here is to highlight the role to twitter in real time disaster management. This work focuses on tweets regarding the cyclone.

In this work the tweets are categorized according to disaster stages. Compared to similar studies, the added value is by mining tweets to find insights such as impacted location, company or organization, if disease spreads in affected area etc.

3 Methodology

Here the proposed framework consists of two main parts: classification and extraction (Fig. 1). As the core technology of this framework, IBM Watson Natural Language Classifier (NLC) and Watson Natural Language Understanding Service (NLU) are used to leverage the advanced natural language processing techniques to create a powerful tweet classifier. NLC helps user to classify short text inputs into predefined categories, at scale. IBM NLC combines various advanced machine learning techniques to provide the highest accuracy possible, without requiring a lot of training data. Whereas NLU can analyse text and extract meta-data from unstructured tweet text such as concepts, entities, keywords, categories, sentiments, emotions, relations, semantic roles. A cloud-based Watson Service, NLU comes with a simple to use API framework. In the solution, classifier and extractor are combined and delivered as a cloud based service with simple to use API. Technologies used here to build the solution model, are Python (version 3.6.5) and Pyquery (version 1.4.0) which integrate IBM NLU and NLC.

3.1 Training Data

In this work, cyclone Fani, which hit the state Odisha in India on 1st May 2019, is selected to create the training model. Tweets with hash tag 'cyclonefani' are examined. Cyclone Fani was formed since 26 April 2019 and dissipated 5 May 2019. In order to understand the type of information that has been spread in Twitter before, during and after the disaster, tweets which were posted in the time span of 26th April to 26th May 2019, are retrieved with the help of the hashtag #cyclonefani. Tweet messages are retrieved from the Twitter Source by utilizing the Twitter API with predefined has tag for a specified date range. Tweets are then dumped into a text file.

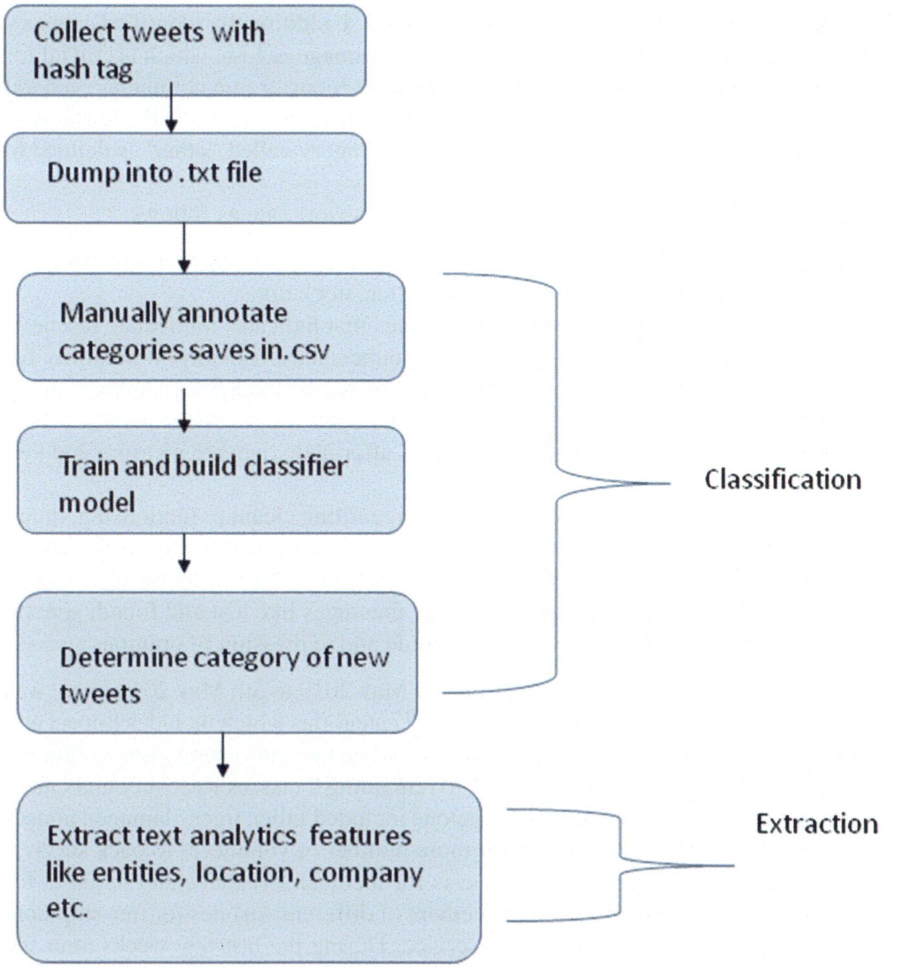

Fig. 1 Architectural components of the proposed methodology

3.2 Tweet Categories

A total of 1,000 tweets were collected for this work. To determine the categories of tweets obtained, each message is manually examined based on specific keywords relevant to the context of disaster phase that is a common practice in the analysis of Twitter messages. To decide the disaster phase I refer official government procedures [12, 13]. Government disaster management process has defined disasters into four stages (a) mitigation (prevention and risk reduction), (b) preparedness, (c) response (immediate rescue) and (d) recovery (build-back better). However, mitigation is not considered in this work. Because tweets of this category are less in count as it concerns the long-term measures to prevent future disasters or minimize their effects.

This paper primarily focuses on other three phases. I additionally identified category Impact, that focuses tweets on impact, hit areas, damage & loss, which is crucial for disaster response. There are many tweet posts that report ironic comments, general comments, expressing gratitude which don't seem to be significantly relevant in the context situational awareness. Therefore, a category called "other" is defined to describe such type of messages.

The five classes or categories considered in this work, are as follows:

i. **Preparedness:** This category contains tweets regarding early alerts, warning, emergency plans, such as evacuation, shelter, stock up.
ii. **Response:** This category includes messages that indicates immediate rescue & relief operations such as, any kind of announcement from airport authority for free shipping of relief aids such as supply of foods, medicines, doctors. If any kind of disease spreads like fever, diarrhea.
iii. **Impact:** This category includes messages aftermath, damage measure and loss etc.
iv. **Recovery:** This category contains tweets regarding cleanup, fundraising, donation, rebuild, restore of transportation, communication, cellular network, school, bank, work, water, power etc.
v. **Others:** This category includes all other messages like lost and found, general comments, messages that express gratitude and expression of opinions.

During the time of disaster that is from 1st May 2019 to 5th May 2019, there was increased number of Twitter messages on the 2 categories which includes Impact and Relief measures. After the disaster that is, once when the wind speed started reducing tweets obtained were more concentrated to remaining 2 classes Recovery and Other. Messages posted during the period of cyclone included fallen trees, damaged houses in various areas and roads; requests for more number of volunteers to pack survival kits that include basic necessities; requests for medical assistance and requests for food, water supplies and boats, announcement of different Airlines for free shipment of aids, announcement of free mobile service. During the last few weeks after the disaster the tweet messages include several repairs and rehabilitation work undertaken by the government and non-government organizations; cleaning the roads and the city; bringing back power supply, restoration of telecommunication, transportation, fund raising, donation, deployment of mobile ATMs, rebuilding Odisha; For tweets of category preparedness, I have collected tweets from 26th April which is 5 days before the cyclone actually hit the Odisha coast in Puri on 1st May. Tweets of this category includes emergency plans, stock up of goods, kits, foods, cancellations of trains flights, shelter after evacuation, event tracking, weather updates etc.

3.3 Train the Classifier Model

To get the tweets ready for training process, set of collected tweets are manually annotated and saved with a .CSV file. During the initial annotation process, I notice that most of the tweets are fallen into others category, and some categories contain only a very small number of tweets. In Figs. 2 and 3 training details are shown. Total 650 example tweets are collected in order to feed the classifier model. Out of total, 162 are from 'Preparedness', 164 are for 'Recover', 'Response' has 65, 'Impact' has 66 and remaining 198 tweets are belongs to 'Other'. Sampling set is prepared by ensuring that each category should have enough tweets to build a successful classification model for the predefined categories. After a market research on ready tool available for text classification, decision was made here to use IBM Watson NLC to meet the classification goals which has shown extremely well performance for several cases by accelerating the rate of data annotation. In the NLC model builder, classes are predefined and text examples are added. NLC accepts the example data in a .CSV file format to train the model. How to use NLC is described in detail in [14–16]. Once the category of each tweet is determined, the initial text file of collected tweets, is saved as a .csv format containing the tweet text with the disaster phase the message depicts, separated by a comma as {tweet-text, category}. In this work IBM watson NLC is accessed through IBM online cloud portal shown in Figs. 2 and 3. The .CSV file is then uploaded to the portal to feed the NLC as an input to train the classifier model. Once training model has been created successfully it is used predict the category of the new tweets.

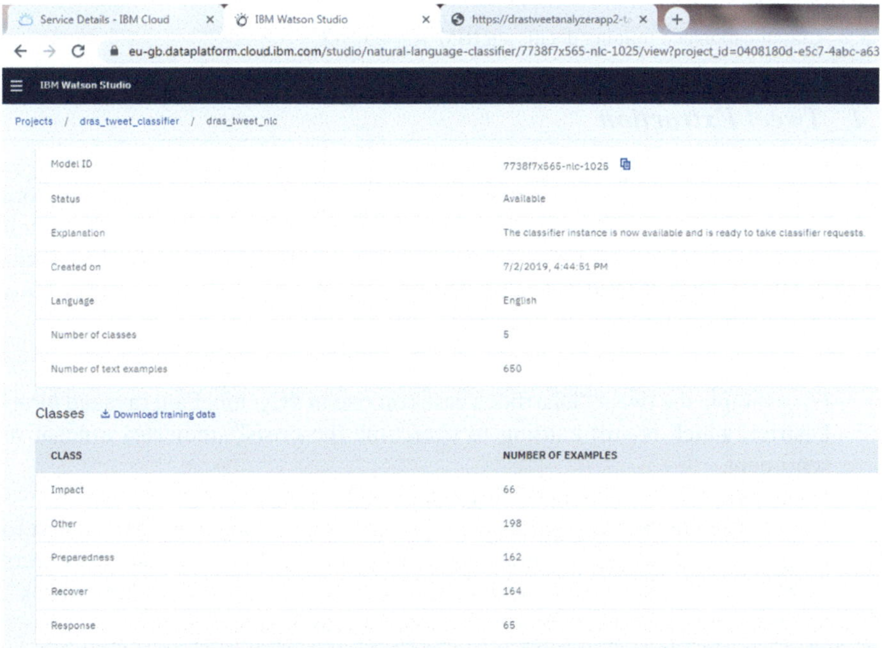

Fig. 2 Tweets distribution in IBM Watson NLC

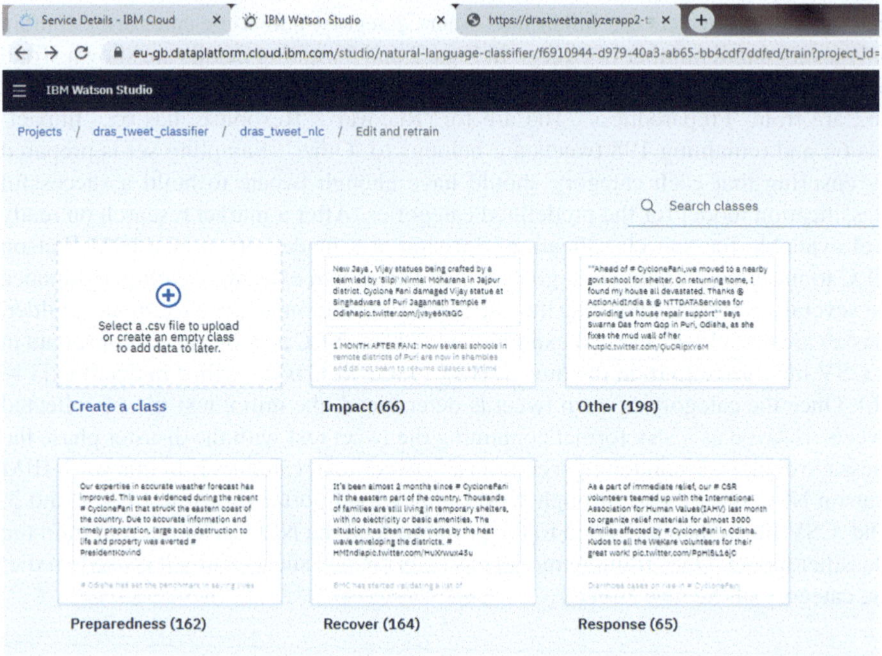

Fig. 3 Classification and training details sin IBM Watson NLC

3.4 Tweet Extraction

After classification, tweets are further mined to retrieve useful information from tweet text for disaster analysis. Here tweets are processed by IBM cloud's natural language understanding (NLU) in order to extract text analytics features like sentiment, location, organization/company, topic etc. Relevant features in context of this works are discussed below with examples.

(i) **Sentiment:** Negative sentiments are considered as complaints.
For example the tweet "Diarrhoea cases on rise in #CycloneFani ravaged #Puri District, which is still battling to overcome the crisis", indicates a negative sentiment.

(ii) **Location:** Any location entity used in tweet post.
For example in the tweet message "Cyclonic storm #Fani now 670 km east to Potuvil. Cloudy skies over most parts of #SriLanka. Rains or thundershowers in Western, Southern, Sabaragamuwa, North-Western and Central provinces", locations are identified here are, SriLanka, Potuvil, Sabaragamuwa.

(iii) **Organization/Company:** If the tweet post contains name of the volunteer, NGO organization. For example, "Odisha Fire Service" is identified as an organization in the tweet message "Odisha Fire Service supplied drinking water to people in Bhubaneswar. This was the first time such a relief operation was carried out by the Fire Service. #CycloneFani."

(iv) **Topic:** It indicates high-level concepts in the content. For example, a tweet message "Diarrhoea cases on rise in #CycloneFani ravaged #Puri District, which is still battling to overcome the crisis", returns the concept "/health and fitness/disease".

4 Results and Discussion

The classifier has been validated on tweets collected with hashtag #CycloneVayu, #CycloneTitli & # NepalEarthquake. Hash tags and date range are captured as inputs though a web based UI (Fig. 4).

Cyclone Vayu was a severe tropical cyclonic storm hit Gujarat in India from the Arabian Sea coast, brought heavy rains and winds that caused moderate damage during June 2019. Cyclone Titli left a trail of destruction in Odisha-Andhra Pradesh coast in India after making landfall on October 6 2018 as a very severe cyclonic storm with wind speeds of 130–140 kmph.

Nepal earthquake of 2015 caused tremendous damage and loss. The severe earthquake occurred near the city of Kathmandu in central Nepal on 25th April 2015, with a magnitude of 7.8Mw.

The consumer API is used to get out the classified and mined tweeter messages. Classification results are generated as expected. Figure 5 describes how the result can be visualized in JSON format through a web browser based visualization platform.

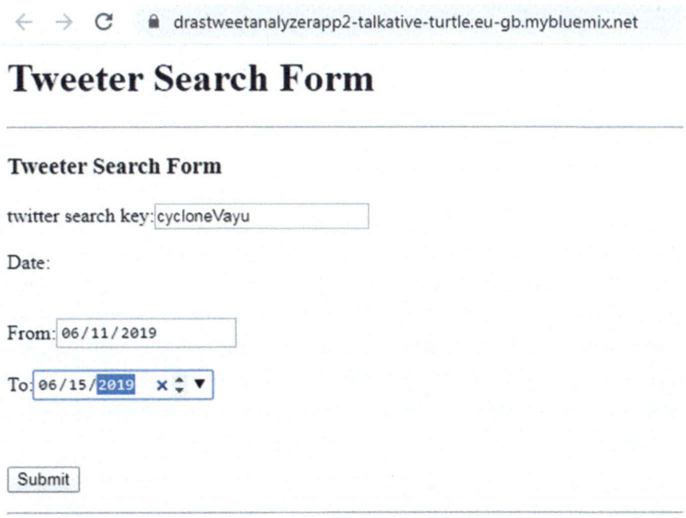

Fig. 4 UI for inputs of hashtag & date range

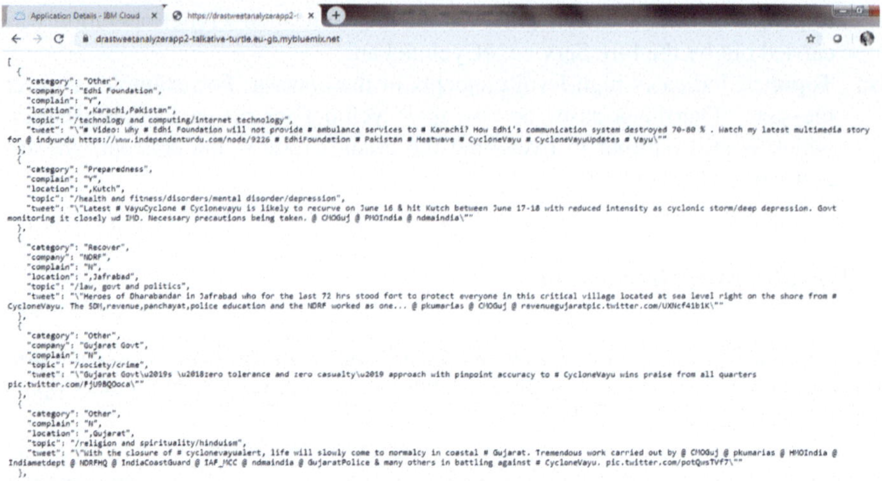

Fig. 5 JSON based visualization of tweet classification result

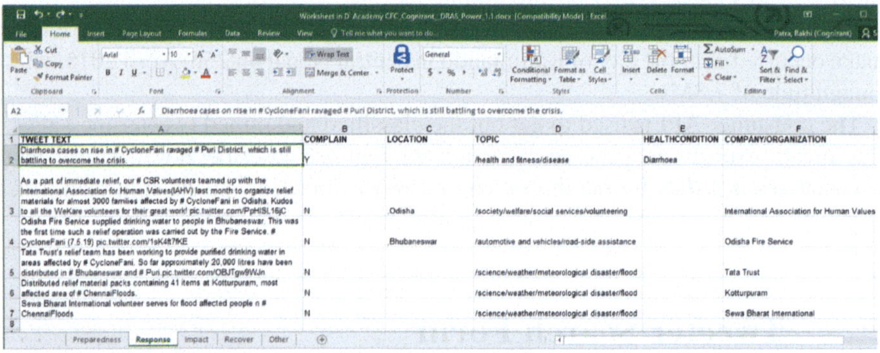

Fig. 6 Excel based visualization tweet classification results

JSON response got for #CycloneVayu for the date range 06/10/2019 (mm/dd/yyyy) to 06/15/2019 (mm/dd/yyyy) is shown in Table 1.

There is also an option available to download the result in excel format. Excel has five sheets according to disaster stages. Each sheet contains tweets of that category. To sum up the result, excel has fields such as if the message is complaint, impacted location, company or organization working on, if any disease spread etc. Figure 6 describes how the final result is presented in excel format after classification and extraction of tweets.

Table 1 JSON response


```
[
  {
    "category": "Other",
    "company": "Edhi Foundation",
"complain": "Y",
    "location": ",Karachi,Pakistan",
"topic": "/technology and computing/internet technolo-
gy",
    "tweet": "\"# Video: Why # Edhi Foundation will
not provide # ambulance services to # Karachi? How
Edhi's communication system destroyed 70-80 % . Watch
my latest multimedia story for @ indyurdu
https://www.independenturdu.com/node/9226 #
EdhiFoundation # Pakistan # Heatwave # CycloneVayu #
CycloneVayuUpdates # Vayu\""
  },
  {
    "category": "Preparedness",
    "complain": "Y",
    "location": ",Kutch",
    "topic": "/health and fitness/disorders/mental
disorder/depression",
    "tweet": "\"Latest # VayuCyclone # Cyclonevayu is
likely to recurve on June 16 & hit Kutch between June
17-18 with reduced intensity as cyclonic storm/deep
depression. Govt monitoring it closely wd IMD. Neces-
sary precautions being taken. @ CMOGuj @ PMOIndia @
ndmaindia\""
  },
  {
    "category": "Recover",
    "company": "NDRF",
    "complain": "N",
    "location": ",Jafrabad",
    "topic": "/law, govt and politics",
    "tweet": "\"Heroes of Dharabandar in Jafrabad who
for the last 72 hrs stood fort to protect everyone in
this critical village located at sea level right on
the shore from # CycloneVayu. The
SDM,revenue,panchayat,police education and the NDRF
worked as one... @ pkumarias @ CMOGuj @
revenuegujaratpic.twitter.com/UXNcf41b1K\""
  },
```


(continued)

Table 1 (continued)


```
  {
    "category": "Other",
    "company": "Gujarat Govt",
    "complain": "N",
    "topic": "/society/crime",
    "tweet": "\"Gujarat Govt\u2019s \u2018zero toler-
ance and zero casualty\u2019 approach with pinpoint
accuracy to # CycloneVayu wins praise from all quar-
ters pic.twitter.com/FjU9BQOoca\""
  },
  {
    "category": "Other",
    "complain": "N",
    "location": ",Gujarat",
    "topic": "/religion and spirituality/hinduism",
    "tweet": "\"With the closure of #
cyclonevayualert, life will slowly come to normalcy in
coastal # Gujarat. Tremendous work carried out by @
CMOGuj @ pkumarias @ HMOIndia @ Indiametdept @ NDRFHQ
@ IndiaCoastGuard @ IAF_MCC @ ndmaindia @
GujaratPolice & many others in battling against #
CycloneVayu. pic.twitter.com/potQwsTVf7\""
  },
  {
    "category": "Other",
    "complain": "N",
    "location": ",Gujarat",
    "topic": "/science/weather/meteorological disas-
ter/hurricane",
    "tweet": "\"A sense of gratitude to PM Shri @
narendramodi , HM Shri @ AmitShah , CM Shri @
vijayrupanibjp and team and people of # Gujarat to
Team up for slowing down the adverse effects of cy-
clone and making all the possible precautions ready
and available to fight against it. # CycloneVayu\""
  },
  {
    "category": "Other",
    "company": "Central Govt",
    "complain": "N",
    "location": ",Gujarat",
    "topic": "/law, govt and politics/government/state
and local government",
    "tweet": " \"THANKS on behalf of Gujarat people,
```


(continued)

Table 1 (continued)


```
the State Government expresses gratitude towards @
PMOIndia @ HMOIndia entire Central Govt @ Indiametdept
@ NDRFHQ @ IndiaCoastGuard @ IAF_MCC @ ndmaindia @
GujaratPolice & all those who extended their helping
hand in our battle against # CycloneVayu\""
  },
  {
    "category": "Other",
    "company": "All Rescue Forces",
    "complain": "N",
    "topic": "/law, govt and politics",
    "tweet": "\"Expressing deep gratitude to People,
Various community organisations & NGOs , All Rescue
Forces, GOI, IMD, NDMA , Media for all their assis-
tance and cooperation for making it Zero Casualty #
VayuCyclone # cyclonevayu\""
  },
  {
    "category": "Preparedness",
    "complain": "N",
    "topic": "/science/weather",
    "tweet": "\"Hon CM @ vijayrupanibjp declares Clo-
sure of Alert for # VayuCyclone # cyclonevayu . Shift-
ed persons to return back to their homes. Cashdoles
would be paid for a duration of 3 days.
Shools/Colleges to become open from tomorrow. ST ser-
vices becoming normal.\""
  },
  {
    "category": "Recover",
    "complain": "Y",
    "location": ",Gujarat,Gujarat",
    "topic": "/law, govt and politics",
    "tweet": "\"Gujarat Chief Minister Vijay Rupani:
It has become clear that # CycloneVayu will not hit
Gujarat, the state is safe now. Government has decided
to call back all senior ministers & officials that
were sent to tackle the situation in 10 areas that
were expected to be affected.
pic.twitter.com/Um7TNa5gJz\""
  }
]
```

5 Conclusion

The focus here was to leverage tweets for disaster management. This paper presents classifier & mining model that could be used to automatically classify tweets of natural hazards, especially related to cyclone. The model could help, support real-time disaster management and analysis by monitoring subsequent events while tweets are streaming, and mining useful information. In this work, I used Cyclone Fani data to train the classifier and Cyclone Vayu and Cyclone Titli to validate the classifier.

Additional informations from tweets like originator of tweets, user type (Individuals, Celebrities, Journalists, and News organizations, Government or NGOs), date & time of tweets could also be processed. IBM NLU service can be leveraged to enhance the feature to identify topic of a tweets text more accurately. Our goal is to run the tweet classifier program as a continuous process during a disaster outbreak which will capture whenever a new tweet is posted. Future plan is to extend the functionalities by adding damage severity assessment by processing the images posted with tweet text messages.

References

1. Das, S. K., & Tripathi, S. (2018). Adaptive and intelligent energy efficient routing for transparent heterogeneous ad-hoc network by fusion of game theory and linear programming. *Applied Intelligence, 48*(7), 1825–1845.
2. Das, S. K., & Tripathi, S. (2017). Energy efficient routing formation technique for hybrid ad hoc network using fusion of artificial intelligence techniques. *International Journal of Communication Systems, 30*(16), e3340, 1–16.
3. Das, S. K., & Tripathi, S. (2019). Energy efficient routing formation algorithm for hybrid ad-hoc network: A geometric programming approach. *Peer-to-Peer Networking and Applications, 12*(1), 102–128.
4. Chatterjee, S., Hore, S., Dey, N., Chakraborty, S., & Ashour, A. S. (2017). Dengue fever classification using gene expression data: A PSO based artificial neural network approach. In *Proceedings of the 5th International Conference on Frontiers in Intelligent Computing: Theory and Applications* (pp. 331–341). Singapore: Springer.
5. Jagatheesan, K., Anand, B., Samanta, S., Dey, N., Ashour, A. S., & Balas, V. E. (2017). Particle swarm optimisation-based parameters optimisation of PID controller for load frequency control of multi-area reheat thermal power systems. *International Journal of Advanced Intelligence Paradigms, 9*(5–6), 464–489.
6. Dey, N., Ashour, A. S., Beagum, S., Pistola, D. S., Gospodinov, M., Gospodinova, E. P., et al. (2015). Parameter optimization for local polynomial approximation based intersection confidence interval filter using genetic algorithm: An application for brain MRI image de-noising. *Journal of Imaging, 1*(1), 60–84.
7. Karaa, W. B. A., Ashour, A. S., Sassi, D. B., Roy, P., Kausar, N., & Dey, N. (2016). Medline text mining: An enhancement genetic algorithm based approach for document clustering. In *Applications of intelligent optimization in biology and medicine* (pp. 267–287). Cham: Springer.
8. Bala, M. M., Navya, K., & Shruthilaya, P. (2017). Text mining on real time Twitter data for disaster response. *International Journal of Civil Engineering and Technology, 8*(8), 20–29.

9. Athanasis, N., Themistocleous, M., Kalabokidis, K., Papakonstantinou, A., Soulakellis, N., & Palaiologou, P. (2018). The emergence of social media for natural disasters management: A big data perspective. *International Archives of the Photogrammetry, Remote Sensing and Spatial Information Sciences, 42*(3), W4.
10. Nair, M. R., Ramya, G. R., & Sivakumar, P. B. (2017). Usage and analysis of Twitter during 2015 Chennai flood towards disaster management. *Procedia Computer Science, 115*, 350–358.
11. Huang, Q., & Xiao, Y. (2015). Geographic situational awareness: Mining tweets for disaster preparedness, emergency response, impact, and recovery. *ISPRS International Journal of Geo-Information, 4*(3), 1549–1568.
12. https://ndma.gov.in/images/policyplan/dmplan/National%20Disaster%20Management%20Plan%20May%202016.pdf.
13. https://www.undp.org/content/dam/india/docs/disaster_management_in_india.pdf.
14. http://www.redbooks.ibm.com/redbooks/pdfs/sg248391.pdf.
15. https://cloud.ibm.com/docs/services/natural-language-classifier.
16. https://dataplatform.cloud.ibm.com/docs/content/wsj/analyze-data/nlc-overview.html.

Sentiment Analysis in Airline Data: Customer Rating Based Recommendation Prediction Using WEKA

Praphula Kumar Jain and Rajendra Pamula

Abstract Customer all over the world gives the ratings and reviews for the services they are using. Internet web sites (IWS) are a good platform for the customer to share their views regarding the offered services. These reviews and ratings are very useful for both customer and service provider, business-related purposes as it increases sales and is also useful for customers as it acts as a source of information. IWS makes it easy to share and access the data, but the presence of a huge amount of data makes it difficult to analyze, so the machine learning (ML) technique is developed to analysis, prediction, and recommendations. In this chapter, we have collected ratings on air transport management given by customers from different sites. There are ratings on seat comfort, cabin staff, food beverage, inflight entertainment and many more, which is further combined to give the overall rating through which recommendation is done. We have used different ML techniques to find out the overall sentiments generated by the customers on different service aspects and give the most suitable recommendation to customers. This helps customers as in travelers in decision making based on service type. We have basically compared Random tree and Decision tree ML techniques for recommendation prediction. In this chapter, we have been used WEKA as a tool to apply these ML techniques. Finally, the accuracy of the result is calculated using precision, recall, and F-scores.

Keywords Machine Learning · Sentiment Analysis · Prediction · Airline Recommendation · Classification

---

P. K. Jain (✉) · R. Pamula
Department of Computer Science and Engineering, Indian Institute of Technology (ISM) Dhanbad, Dhanbad 826004, JH, India
e-mail: praphulajn1@gmail.com

R. Pamula
e-mail: rajendrapamula@gmail.com

S. K. Das et al. (eds.), *Machine Learning Algorithms for Industrial Applications*, Studies in Computational Intelligence 907,
https://doi.org/10.1007/978-3-030-50641-4_4

1 Introduction

Information presents on social media carries an, feeling, sentiment, and desires. These feelings show the positivity, negativity and neutrality of the reviewers. Such types of feelings or emotional datasets are presents on many online platforms such as Twitter, Facebook, and Instagram etc. This online platform contains information from many sources of airline trips, restaurants, tourism planners, movies, lections, and hospitals etc. All the feedback related to these contains a piece of hidden information that shows the comfort/discomfort in the associate area. Hence, this is a great opportunity for analyzing this type of information to find out the associated feeling in these reviews. This type of emotional pattern extraction from the reviews data can help us to distinguish the feelings of the reviewers in the associated area and the causes behind it.

There are many modes of travel nationally or internationally but airline travel is a comfortable mode for distance journeys [1]. There are so many airline service providers are present in the world. This is a very competitive world that motivates airline service providers to draw customers intention. However, a customer considering many points choosing any flight for his or her travel. These points may be ticket cost, time of travel, luggage allowed, number of stoppages, and previous customers feedback at the last but not least. Therefore, all airline service providers are considering these points to improve their services and in-flight comfort for drawing customers' intentions.

This is a need of understanding customer's requirements and comfort level i.e. customer gratification inside the flight. Therefore, customer feedback is a great concern for any airline services provider company. There are many ways to get the customers feedback, the most convenient and easiest way is the collection of feedback inside the airline using feedback forms. There are many disadvantages of this method like customers did not fill feedback properly, inappropriate questionnaires or reviewers biasing on a fixed number of parameters. Another approach for the collection of customers' feedback is an online platform such as websites or mobile Apps of the service provider. After completion of the travel, a message can be sent to the customer associated with the link of feedback request. But guarantees of success this approach is limited. Another approach is to send the message on the customers mobile and request them for rate services (1 to 5 where 1 means poor and 5 shows excellent) on the mentioned parameters. The more easiest and convenient approach for a customer to post their feedback, as they wish. Therefore, the easiest way is to express their feedback is social media in place of filling any feedback form. There are many social media platforms are available where customers can share their feedback freely on issues they noticed during the travel. Twitter [2] is a freely available worldwide platform for customers' feedback. Customer feedback twitter information is a useful source for sentiment analysis.

There are many factors that affect customer emotions inside the air travel. These factors can be loss of baggage, flight delay, cost of travel, cabin crew behavior, quality of food seat comfort etc. All these factors may be contribute to positive or negative emotions. Also, If there is a regular negative impression feedback for a service provider, then it can be a economaly negative impression for that company. Therefore, it is very much important to understand the reason behind the negative feedback so that the associated company may take necessary action timely. There are so many airline service providers are providing services for their customers to travel across the globe [3]. Therefore, we can consider so many travelers are traveling daily in these airlines. In addition, on the social media platform airline reviews rising day by day. Therefore, it is difficult to handle these many reviews manually and extract the customers feeling inside the tweet. Therefore, it is a need of tools and techniques that will be capable to handle such a huge amount of reviews and may extrat emotion inside it.

Machine learning techniques are capable to analyze the huge amount of tweet data and to build a highly accurate prediction or classification model. In this chapter, machine learning techniques are used to develop a prediction model for customer generated airline feedback data. This chapter opted to build a binary classification model for two types of sentiments. i.e positive and negative. We have selected only positive and negative sentiment because neutral sentiment does not associate with any information regarding the services whether it was good or not. Neutral sentiment may be associated with the kind customer or he did not take feedback seriously. Also, it is not very much important for the service provider company because they are interested in only services that are "good or bad". Due to that only positive and negative sentiment have been considered for further analysis in this chapter. We have basically compared tree base ML techniques for recommendation prediction. In this chapter, we have been used WEKA as a tool to apply these ML techniques. Finally, the accuracy of the result is calculated using precision, recall, and F-scores.

This chapter is organized as follows "literature review" in this section different theoretical and methodological points related to the prediction of airline sentiment have been presented. In "methodology" section, a details description of the related dataset and machine learning techniques are given. "Results and Discussion" section, the result analysis presented with detail description. The chapter concluded in the last "conclusion" section.

2 Literature Review

Sentiment analysis is a recent area of research and it is also an important approach to identify the feeling or emotion or sentiment from the qualitative and quantitative customer-generated feedback data i.e. online movie reviews, airline rating, and reviews, twitter data etc. The airline feedback dataset usually contains information for the used services. There are many sources of airline feedback data such as airlinequality.com, twitter.com, tripadvisor.com. This data is very useful in extracting emotion or sentiment using machine learning techniques and provides help to understand the customers' intentions about the services regarding their comfort. There are so many related articles are available in the field of sentiment analysis [4–6] From the literature it has been found that there are two types of sentiment analysis techniques are available Lexicon-based techniques [7] and machine learning techniques [9]. In Lexicon-based techniques sentiment lexicon has been considered for sentiment analysis. Set of predefined and inbuilt terms, idioms and phrases are called as sentiment lexicon. These phrases or idioms are developed in regards to opinion finder, dictionaries, and ontologies [10]. Lexicon based approaches are of two types first one is a dictionary based approach and the second is a corpus-based approach [8].

Online user-generated feedback about airline travel, hotels, and tourism service provider is a source of information for the upcoming travelers [11]. From the observation, It has been found there are a million travelers who have referred to online reviews before their travel [12]. From that many travelers, online review affects the 84% visitors prior to making their reservation decision [13]. The author [14] points out that consumer word of mouth from the other customer affects traveler decision making. The author [15] studied and found that online reviews are more reliable, up to date, and enjoyable information for the inferred customer.

Additionally, the Author [16] and [17] suggested that online user-generated reviews are very important for both customer and service providers. Likewise, [18] point out that online word of mouth can be helpful in brand building, product quality improvement, and product development for the respective service provider. A recent study shows many advantages of online reviews in product or sales. For instance, the author [19] finds out the positive effects of online reviews on book sales from different websites such as Amazon.com, Barnesandnoble.com etc. and also examined that this is a greater influence of word of mouth. The author [20] formed a panel for data analysis for movie box office revenue data, and findout that there was not any sign of the impact on movie box office revenue meanwhile online reviews have a greater impact on box office sales.

Different authors used different data mining techniques for sentiment analysis. The Author [21] uses naive base classifiers for twitter sentiment analysis, they classify tweets in positive, negative, and neutral using R and Rapid Minor tools. They show that naive base classifier performs better. The author [22] implemented a comparative study on different US-based airline companies' data. They have used decision tree, Gaussian Naive Base, SVM, K-Nearest neighbors, Logistic Regression, and Adaboot. In this implementation, 80% of data is used for training a model and 20% data used for the testing model. Results show that Adaboost, Random forest, and SVM outperforms. However, the authors show that the addition of more tweets in the implementation performs improvements in the results.

3 Methodology

3.1 Dataset Description and Data Visualization

Internet is the biggest platform for the recommendation, reviews and rating gives us the way to do the sentimental analysis of customers all over the world which is very

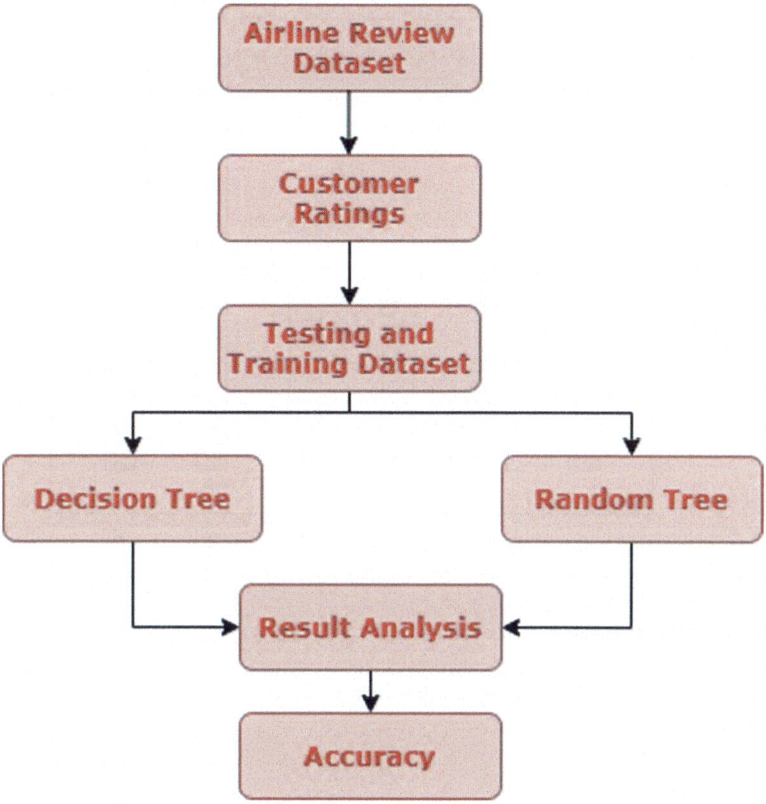

Fig. 1 Flow diagram

useful in case of business and also for the customers. Hence here we have collected the ratings of the customers analyze the rating and then recommend it further to the customers wanting to use the same facilities in the near future. In this study, we collected the ratings on inflight services. Inflight services are the services available to the aircraft passengers during a flight. The data we have collected contains 2000 rows of different service ratings of different customers over the world. Figure 1 represents the flow diagram for the applied methodology and Table 1 shows the different attribute with their rating and average rating. Data visualization represented in Fig. 2–7, where Fig. 2 shows the seat comfort rating with their counts. Cabin staff rating with its counts represented in Fig. 3. Food beverage ratings with their counts shown in the Fig. 4. Figure 5 represents the inflight entertainments ratings with its count. Value of money rating with its counts shown in the Fig. 6. At the last Fig. 7 represents Overall rating with their counts. We divided the data randomly in 60–40 to use 60% for training and 40% for testing.

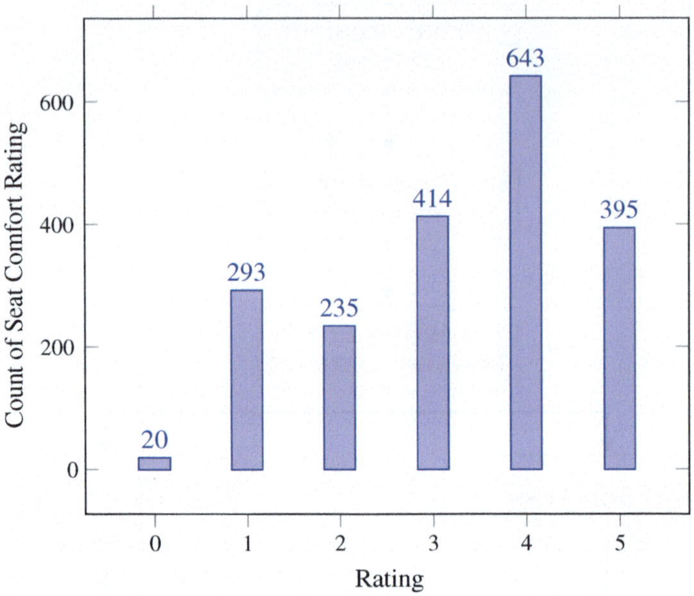

Fig. 2 Seat comfort rating with their counts

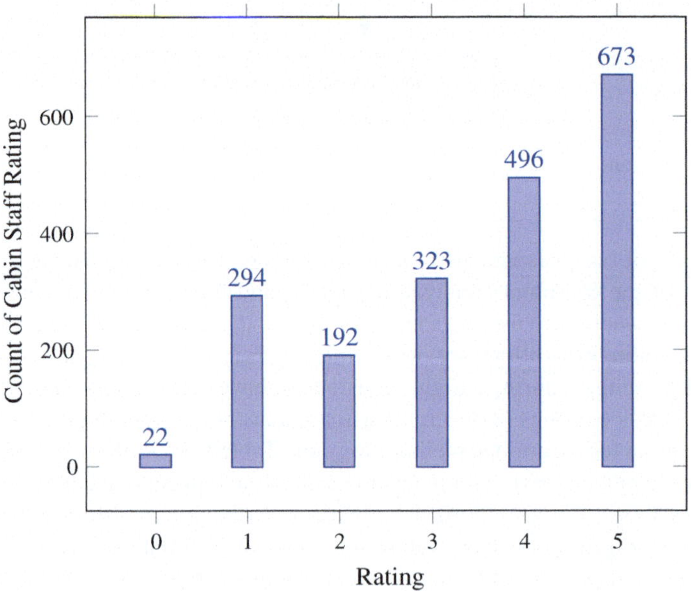

Fig. 3 Cabin staff rating with their counts

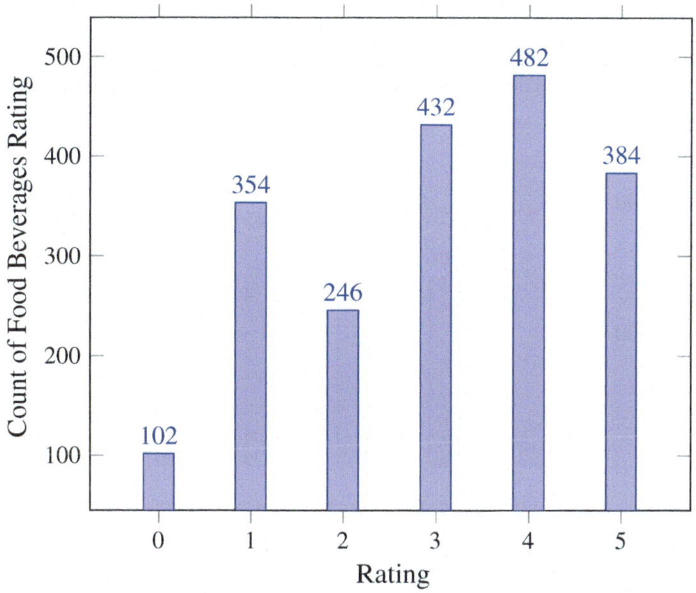

Fig. 4 Food beverages rating with their counts

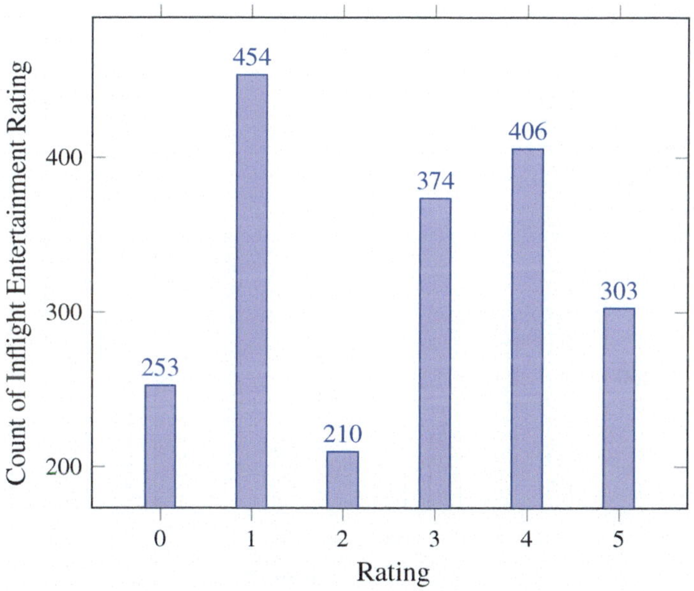

Fig. 5 Inflight entertainment rating with their counts

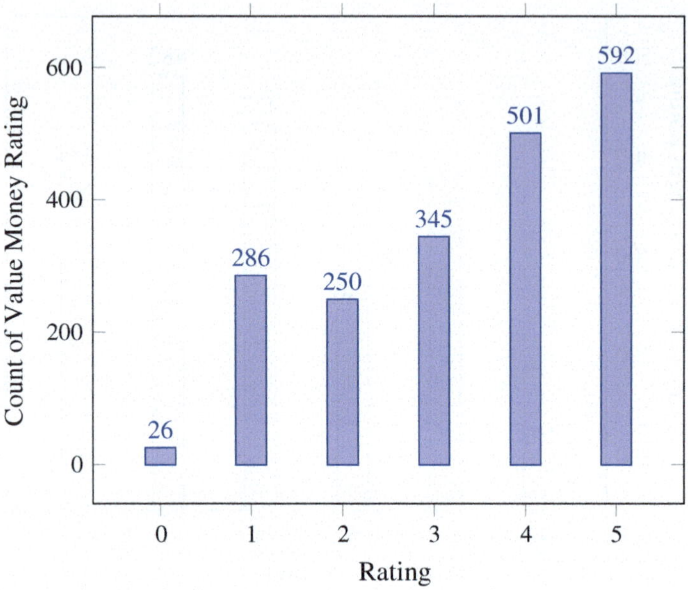

Fig. 6 Value money rating with their counts

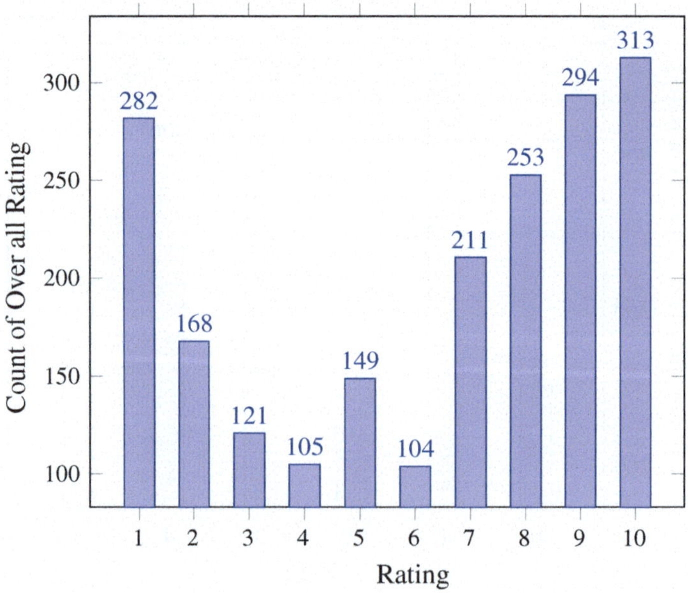

Fig. 7 Over all rating with their counts

Table 1 Attributes summery used in analysis

Variable	Rating	Avg. rating
Overall rating	1–10	6
Seat comfort rating	1–5	4
Cabin staff rating	1–5	3
Inflight entertainment rating	1–5	3
Food beverages rating	1–5	4
Value money rating	1–5	3

3.2 Environment Used for Analysis

Weka stands for Waikato Environment for Knowledge Analysis was developed in Waikato, New Zealand. It is a tool for machine learning written in Java. This software is free. It has a license under the GNU General Public License. It can run on different platforms such as IA-32, x8-64, Java SE. the operating systems of WEKA are Windows, OS X, Linux.

The main purposes of using WEKA is usually to make prediction or assumptions, mostly data file is present in WEKA, it has popular algorithms such as logistic regression, decision tree, neural network, support vector machine, Bayes algorithms, and many others. It is very much useful for a person who has not coded for a while. The algorithms in WEKA can either be applied directly to a dataset or called from your own Java code.

The features of WEKA include machine learning, data mining, preprocessing, classification, regression, clustering, association rules, attribute selection, experiments, workflow, and visualization.

4 Result and Discussion

4.1 Random Tree

Leo Breiman and Adele Cutler were the people to inaugurate Random Tree. It chooses k no of attributes. These attributes are usually randomly chosen for each node. It is a supervised classifier. In this, the node is split using the best set of data. Bagging idea is used in a random tree algorithm to use random sets of data for the making of the decision tree. The prognosticator at the node is chosen randomly. When we put together all the tree prognosticators we make it into a forest. This algorithm can function in both classification and regression. The classifier takes the input feature vector classifies it with every other tree in the forest and then gives the output as the class label like the ones receiving the maximum number of votes. The regression

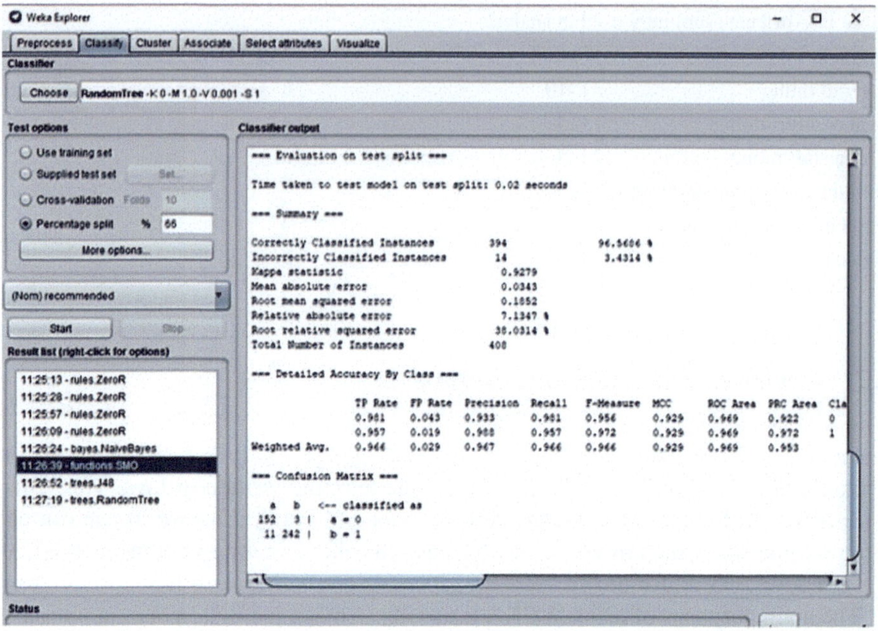

Fig. 8 Result evalution using Random tree

when used gives the average of the responses on the trees present in the forest (6). It is seen in the RandomTree that it basically joins more than one algorithm present in machine learning that is combining a single model tree with Random Forest ideas.

The model tree has a leaf that decides its own local space and contains a linear model that is placed on the local space of each leaf. On the other hand, the Random Forest improves the quality as is present in a model tree as in the decision tree. The randomness increases the quality of the algorithm. It also brings diversity. It is brought into consideration by two methods- First Bagging is used in which training data is replaced with the single tree and Secondly while the tree is grown we need not always take the best split we can always consider the random split and then take the best split into consideration and do the computation. Using a Random tree, we get 94.54% accuracy shown in Fig. 8.

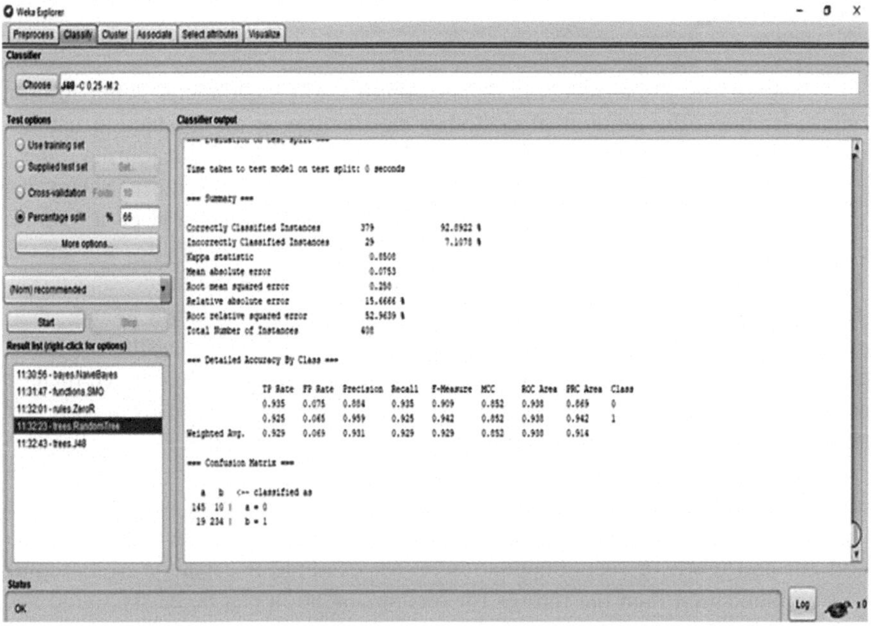

Fig. 9 Result evalution using Decision tree

4.2 Decision Tree

It is also known as the J48 classifier in WEKA. It basically is a structural or graphical representation based on any decision. Such a name is given to a decision tree because it has a single root node at the beginning and ends with multiple leaf nodes. It is a kind of algorithm that contains conditional control statements. It explains both the details of its inputs and outputs. The data in this kind of tree is repeatedly broken down on the basis of some needed parameters. Many Decision Trees makes Random Tree. It is the most powerful algorithm used for prediction and classification.

The most important advantages of decision trees are that it makes the easiest algorithm that is easy to understand by common people, the classification in decision tree can be done easily without doing much of computation and is able to handle both continuous and categorical variables, it is useful to predict which field is important for the computation. Using Decision tree, we get 92.83% of accuracy shown in Fig. 9.

5 Conclusion

In this chapter, customers feedback uses to predict and recommend better inflight services. It shows the amount of customer satisfaction or dissatisfaction using a particular service that is provided by the flight during their inflight journey. There are different services like cabin_staff, food_beverages, inflight_entertainment and others. Ratings are taken for each service particularly and the bar graph is drawn for each inflight service to represents the different ratings. We have divided the dataset in training and testing dataset, training dataset used for train the model and model tested with the help of testing dataset. Customers' recommendation as one of its attributes, basis on that positive and negative recommendation is classified by using different data mining techniques in the WEKA tool. In this work, tree based method have been applied and compared for the customer sentiment analysis. From the above implimentation it can be concluded that random tree outperforms.

Furthermore, these recommendation are used in predictions of customer sentiment. In case of any future to be done, this can be very much helpful for both the company and the customer as the ratings for a particular service can be easily found out and used for recommendation as for example if a person is searching for recommendation then the ratings for recommendation can be easily found and recommendation can be done of that particular flight.

References

1. Flight is best mode of travel. https://www.flightnetwork.com/blog/10-reasons-flying-still-best-way-travel/.s. Accessed 06 May 2019.
2. Twitter. https://en.wikipedia.org/wiki/Twitter. Accessed 23 Jan 2018.
3. Number of flights in a day. https://www.quora.com/How-many-airplanes-fly-each-day-in-the-world. Accessed 25 Jan 2018.
4. Agarwal, A., et al. (2011). Sentiment analysis of Twitter data. In *Proceedings of the Workshop on Language in Social Media (LSM 2011)*.
5. Liu, B., et al. (2012). Sentiment analysis and opinion mining. *Synthesis Lectures on Human Language Technologies*, *5*(1), 1–167.
6. Liu, B., & Zhang, L. (2012). A survey of opinion mining and sentiment analysis. In *Mining text data* (pp. 415–463). Boston: Springer.
7. Taboada, M., et al. (2011). Lexicon-based methods for sentiment analysis. *Computational Linguistics*, *37*(2), 267–307.
8. Pak, A., & Paroubek, P. (2010). Twitter as a corpus for sentiment analysis and opinion mining. In *LREc* (vol. 10).
9. Agarwal, B., & Mittal, N. (2016). Machine learning approach for sentiment analysis. In *Prominent feature extraction for sentiment analysis* (pp. 21–45). Cham: Springer.
10. Bhonde, R., et al. (2015). Sentiment analysis based on dictionary approach. *International Journal of Emerging Engineering Research and Technology*, *3*(1), 51–55.
11. Pan, B., MacLaurin, T., & Crotts, J. C. (2007). Travel blogs and the implications for destination marketing. *Journal of Travel Research*, *46*(1), 35–45.
12. Tripadvisor.com. (2006). Fact Sheet. http://www.tripadvisor.com/pages/factsheet/html. Accessed 20 July 2008.

13. Travelindustrywire.com. (2007). Travel Reviews Consumers are Changing your Brand and Reputation Online. http://www.travelindustrywire.com/article29359.html. Accessed 20 July 2008.
14. Goldenberg, J., Libai, B., & Muller, E. (2001). Talk of the network: A complex systems look at the underlying process of word-of-mouth. *Marketing Letters, 12*(3), 211–223.
15. Gretzel, U., & Yoo, K. (2008). Use and impact of online travel reviews. In P. O. Connor, W. Hopken, & U. Gretzel (Eds.), *Information and communication technologies in tourism 2008* (pp. 35–46). New York: Springer.
16. Zhu, F., & Zhang, X. (2006). The influence of online consumer reviews on the demand for experience goods: The case of video games. In *Proceedings of Twenty-Seventh International Conference on Information Systems (ICIS), Milwaukee, USA* (pp. 367–382).
17. Cheung, C. M. Y., Shek, S. P. W., & Sia, C. L. (2004). Virtual community of consumers: Why people are willing to contribute. In *Proceedings of the 8th Asia-Pacific Conference on Information Systems, Shanghai, China* (pp. 2100–2107).
18. Dellarocas, C. (2003). The digitization of word-of-mouth: Promise and challenges of online feedback mechanisms. *Management Science, 49*(10), 1407–1424.
19. Chevlier, J. A., & Mayzlin, D. (2006). The effect of word-of-mouth on sales: Online book reviews. *Journal of Marketing Research, 43*(3), 345–354.
20. Duan, W., Gu, B., & Whinston, A. B. (2008). Do online reviews matter? An empirical investigation of panel data. *Decision Support Systems, 45*(4), 1007–1016.
21. Dutta, D. D., Sharma, S., Natani, S., Khare, N., & Singh, B. (2017). Sentiment analysis for airline Twitter data. In *IOP Conference Series: Materials Science and Engineering* (vol. 263, no. 4)
22. Rane, A., & Kumar, A. (2018). Sentiment classification system of Twitter data for US airline service analysis. In *Proceedings of the 42nd IEEE Computer Software and Applications Conference, COMPSAC 2018, Tokyo, Japan* (pp. 769–773).

Computer Vision

Image Inpainting for Irregular Holes Using Extreme Learning Machine

Debanand Kanhar and Raunak Chandak

Abstract Image inpainting is usually framed as a constrained image generation problem. It is a method that helps to reconstruct the lost or deteriorated parts of images as well as video. The main focus in image inpainting technique is how precisely to generate the corrupted pixels in an image. Traditional inpainting algorithms are unfortunately not well adapted to handle such corruptions as they rely on image processing techniques that cannot properly infer missing information when the corrupted holes are too large. In this paper, a single pass method of inpaintng method is used which does not require any back propagation for training purpose, hence saving time for training the model. The objective of the proposed model is to reconstruct large continuous regions of missing or deteriorated parts of an image. In this paper, role of Extreme Machine Learning is a single pass neural network model, where we train our model based on region surrounding the corrupted region. Each image is divided into two sections: the missing part that we try to reconstruct, and the context. The network does not require any region to be in some defined shape as it can also work on arbitrary regions. Final evaluation is based on the average element-wise L2 distance between the corrupted image and the original image for the regions which is to be regenerate. We have also used PSNR value for comparison between original image and the image which we reconstruct.

Keywords Image Inpainting · Deep Learning · Image Processing · Extreme Learning Machine

D. Kanhar · R. Chandak (✉)
School of Computer Science and Engineering, National Institute of Science and Technology (Autonomous), Institute Park, Pallur Hills, Berhampur 761008, Odisha, India
e-mail: raunak.chk@gmail.com

D. Kanhar
e-mail: devanand@nist.edu

S. K. Das et al. (eds.), *Machine Learning Algorithms for Industrial Applications*, Studies in Computational Intelligence 907,
https://doi.org/10.1007/978-3-030-50641-4_5

1 Introduction

Image Inpainting is state-of-art technique which is used to reconstruct corrupted regions in an image or video. For example, if we an image is too old or damaged due to any reason, then it is possible to regenerate the lost parts to some extent after applying some inpainting algorithm. What it does is it tries to generate the corrupted regions using surrounding region, similar patches, and textural data or training a bunch of data for learning process.

Feed forward neural networks have been used in multiple fields due to their ability to approximate complex nonlinear functions from input data and to handle cases which are difficult to handle by classic conventional parametric techniques [1, 2]. However learning time for neural networks is quite high. Extreme Learning Machines (ELM) are essentially feed forward neural networks which can be used for various purpose such as regression, classification and sparse approximation [3, 4]. ELM belongs for Single-hidden Layer Feed forward network. It can be used for function approximation using finite set of data and with any nonlinear activation function.

In any feed forward neural network, the weights have to be tuned by training. Traditionally gradient descent algorithm or some of its variation of its variations for training. However these are very slow due to incorrect learning rate or its tendency to be stuck in local minima. Also as training is required for neural network it would take time and not the best choice if one wants to work with few images and has no requirement of that in future. Traditional algorithms which do not require training for inpainting might be an alternative option but it has been seen that

I. They are able to cope up with regular shaped corrupted region and might not perform that well if the region is has no particular shape.
II. Considers that restoration can be done by finding similar patches [5] without considering surrounding regions which carry some info about overall image.

One alternative to this problem could be ELM which can provide better generalization error and is could be 1000 times faster than traditional back propagation algorithm. ELM consists of three layers shown as:

(a) **Input layer:** This layer passes the input to the next layer. Number of nodes in this layer matches the independent feature in the training set.
(b) **Hidden layer:** This layer takes from input layer and applies activation layer to it. After doing the operation it is then passed to the final layer.
(c) **Output layer:** This layer sums the multiplication of input from hidden layer and set of weights which is to be determined during training of the model. This layer can have one or more nodes based on number of nodes required.

The proposed is based on soft computing [6] approach where General Regression Neural Network (GRNN) is used as a soft computing method for image inpainting for irregular holes. There are several soft computing methods are available, among them fuzzy logic, genetic algorithm, particle sward optimization, etc. play an important role [7–13]. These are rapidly used in several application.

The remaining of the paper is organize as follows. Section 2 describe some existing works as related works. The proposed method describe in the Sect. 3. Section 4 describe result and discussion part. Finally, conclusion described in Sect. 5.

2 Related Works

In last few years, several works have been proposed. Some of them described as follows. Ashour et al. [5] proposed cuckoo search based medical image enhancement method. They used Computed Tomography image for experiment and compared the Cuckoo Search method with PSO based method. Samanta et al. [14] applied firefly algorithm to optimize the parameter of Log Transform for enhancement of low contrast aerial images taken by Autonomous Mini Unmanned Aerial Vehicle. Analysis of enhanced images had shown the robustness of the CS based technique. Choudhury et al. [15] introduced the segmentation of microscopic image using quantum inspired evolutionary algorithm. QIEA was used to select the optimum threshold for segmentation microscopic rat hippocampus image. A brief survey was done by Nandi et al. [16] various application of Principal component analysis in medical image processing. Pradhan et al. [17] have proposed another novel multi-level CBIR system to improve retrieval accuracy. Here, the authors have introduced a 3-level image filtering approach to filter out most of the irrelevant images from the image database. Thus, the final image retrieval extracts most of the similar images from the reduced database. To perform the 3-level image filtration task, the authors have used tetrolet transform, edge-joint histograms, and color-channel correlation histograms based image features. In 2019, Pradhan et al. [18] have introduced an image reordering based CBIR system. In their scheme, the authors have used a covariance matrix and Eigenvalue analysis to compute the principal texture direction of the image. Subsequently, they have used the principle texture direction to reorder the input image. Finally, a color edge and directional texture features have been computed to perform the final image retrieval task. Further in 2019, Pradhan et al. [19] have introduced another CBIR system that works on structural and co-relational image features. In this approach, the authors have computed the rotational invariant directional motif patterns to perform image retrieval. In 2019, Majhi et al. [20] have suggested a secure CBIR approach to deal with image integrity and modification problems. In this scheme, the authors have utilized the Itti Koch saliency approach to extract the salient part of the image. Next, they have computed the color and texture features from the salient and some statistical features from the non-salient regions to perform the image retrieval. Finally, they used some effective cryptographic techniques to tackle security issues. Further in 2019, Pradhan et al. [21] have suggested another CBIR approach to enhance retrieval efficiency. In their scheme, the authors have separated the texture dominated and color dominated regions of the image. Subsequently, they have extracted the DT-CWT based texture features from the texture region and semantic annular histogram-based color features from the color regions only. They have also suggested a weighted similarity matching approach to perform image retrieval. Recently, in 2020, Pradhan et al. [22] have suggested an image fusion scheme for image retrieval and medical imaging. In this scheme, the authors have introduced an adaptive weight map and gray-level edge maps based image fusion approach. In contrast to this method, privacy preserving is the other effective and secure approach in the secure image retrieval. In [23], a secure image retrieve scheme is proposed which retrieve images from the corresponding image repositories using encrypted feature vector. Quantized HSV color space histogram and texture features are exploited to build the feature vector which are encrypted by XORed operation with its corresponding sliced biplanes to preserve the distance measure. Finally, image are retrieved from the randomly permuted encrypted feature vector with all security constraints. In a similar approach, for better feature and security aspects, annular distribution density structure descriptor (ADDSD) based feature is proposed in [24] to retrieve the images using encrypted which retrieves the relevant images efficiently without revealing image visual content. In an image there are different features which establish the representative background and foreground regions. In this regard, saliency is a well-known approach and to exploit this concept, an integrated foreground and background based feature with suitable cryptographic security scheme is proposed in [20]. In this scheme a highly secure encrypted feature communication between the user and data owner are illustrated and retrieved images are watermarked to enrich the identity of data owner. This highly secure mechanism ensures that the retrieved images are shared or distributed only to the authorized user to enhance the privacy of the retrieval system. In [25], the authors proposed a segmentation algorithm for fingerprint image where block-based statistics and morphological filtering. The method segments an image in two stages- (i) coarse segmentation using statistical methods and (ii) fine segmentation using morphological filters. In [26], the authors proposed technique developed a concept of fingerprint segmentation based on pixel-wise approach where 2nd order moment had been used to find the Region of Interest. In [27], the authors designed a minutia detection approach from direct gray-scale fingerprint image by using Hit-or-Miss transformation. Instead of using binary image, the proposed method was applied on raw gray-scale image which reduces the computational complexity. In [28], the authors have designed another fingerprint segmentation scheme where the thresholds are generated adaptively. To achieve the more refined segmentation, this article employed extra round of filtering. In [29], the proposed a scheme of fingerprint matching was developed by the authors where templates had been generated using clustering of minutia points. The proposed method was designed for fast and secure process of fingerprint biometric authentication.

3 The Proposed Method

To explain in short terms of what ELM does is that at first it initializes the weights and bias between input layer and hidden layer. Then using the values from hidden layer and the training output it estimates the weights between them using Moore-Penrose

Table 1 Example input and corresponding output

Input	Output
1	4
3	7
5	9
7	10

pseudo-inverse under the criterion of least-squares method. This simple structure of the model enables it to be trained in such short period of time given in Table 1.

Let value to be calculated for 4 and let number of nodes for hidden layer, n = 4 and no activation function is applied at hidden layer. For n = 4 given in Eq. 1 for input x.

$$\text{Input x} = \begin{bmatrix} 1 \\ 3 \\ 5 \\ 7 \end{bmatrix} \tag{1}$$

Let us randomly initialize the weight matrix between input layer and hidden layer. So the values of w and b are shown in Eqs. (2) and (3). And value of h1 shown in Eq. (4).

$$\text{w} = [2.009 \quad -1.13 \quad 0.416 \quad 1.441] \tag{2}$$

$$\text{b} = [1.06 \quad -1.467 \quad -0.885 \quad 1.629] \tag{3}$$

$$\text{h1} = \text{x.w} + \text{b} \tag{4}$$

$$= \begin{bmatrix} 1 \\ 3 \\ 5 \\ 7 \end{bmatrix} \cdot [2.009 \quad -1.13 \quad 0.416 \quad 1.441] + [1.06 \quad -1.467 \quad -0.885 \quad 1.629]$$

$$= \begin{bmatrix} 3.068 & -2.596 & -0.469 & 3.070 \\ 7.088 & -4.852 & 0.363 & 5.953 \\ 11.107 & -7.109 & 1.196 & 8.836 \\ 15.127 & -9.365 & 2.028 & 11.719 \end{bmatrix}$$

The calculation between hidden layer and output layer shown as given:

As output y = h1.T, where T is set of weights between hidden and output layer and needs to be calculated, therefore $\text{T} = \text{h1}^{-1}.\text{y}$ and the values of y and T shown in Eqs. (5) and (6).

$$y = \begin{bmatrix} 4 \\ 7 \\ 9 \\ 10 \end{bmatrix} \tag{5}$$

$$T = \begin{bmatrix} 3.068 & -2.596 & -0.469 & 3.070 \\ 7.088 & -4.852 & 0.363 & 5.953 \\ 11.107 & -7.109 & 1.196 & 8.836 \\ 15.127 & -9.365 & 2.028 & 11.719 \end{bmatrix}^{-1} \cdot \begin{bmatrix} 4 \\ 7 \\ 9 \\ 10 \end{bmatrix}$$

$$= \begin{bmatrix} -0.292 \\ -0.866 \\ -1.364 \\ 0.817 \end{bmatrix} \tag{6}$$

The calculating value for x = 4, using calculated value of every parameter find y′ shown in Eq. (7).

$$y' = (x.w + b).T = 7.492 \tag{7}$$

Similarly, try with n = 5 and values of w, b, h1, and T shown in Eqs. (8) and (11).

$$w = [-1.225 \quad -1.274 \quad 0.581 \quad -1.535 \quad 2.627] \tag{8}$$

$$b = [0.668 \quad 2.123 \quad -0.685 \quad -0.554 \quad -0.504] \tag{9}$$

$$h1 = x.w + b \tag{10}$$

$$= \begin{bmatrix} 1 \\ 3 \\ 5 \\ 7 \end{bmatrix} [-1.225 \quad -1.274 \quad 0.581 \quad -1.535 \quad 2.627] + [0.668 \quad 2.123 \quad -0.685 \quad -0.554 \quad -0.504]$$

$$= \begin{bmatrix} -0.557 & 0.849 & -0.104 & -2.088 & 2.123 \\ -3.006 & -1.699 & 1.059 & -5.157 & 7.378 \\ -5.456 & -4.246 & 2.222 & -8.227 & 12.632 \\ -7.906 & -6.793 & 3.384 & -11.296 & 17.887 \end{bmatrix}$$

$$T = h1^{-1}.y = \begin{bmatrix} 0.115 \\ 1.349 \\ -0.371 \\ -1.054 \\ 0.554 \end{bmatrix} \tag{11}$$

Therefore for x = 4, and y′ shown in Eq. (12).

$$y' = (x.w + b).T = 7.586 \tag{12}$$

For n = 3, values of w, b, h1, T and y′ are shown in Eqs. (13) and (17).

$$w = [-0.004 \quad 0.706 \quad -1.427] \tag{13}$$

$$b = [1.354 \quad 0.674 \quad 1.415] \tag{14}$$

$$h1 = x.w + b = \begin{bmatrix} 1.350 & 1.38 & -0.012 \\ 1.334 & 4.204 & -5.719 \\ 1.325 & 5.617 & -8.572 \end{bmatrix} \tag{15}$$

$$T = h1^{-1}.y = \begin{bmatrix} 1.66 \\ 1.637 \\ 0.104 \end{bmatrix} \tag{16}$$

$$y' = (x.w + b).T = 7.56 \tag{17}$$

Iterating over multiple values of n and select the most appropriate model using least sum of squares method. The proposed algorithm shown in Algorithm 1.

Algorithm 1: The proposed ELM for image inpainting.

Step 1: Identify the corrupted region in the image.
Step 2: Create a binary mask with same dimensions as the image and where 0 represents the corrupted region and 1 represents known regions.
Step 3: Determine number of parameters to be used or iterate over some parameters and choose with the least amount of error.
Step 4: Identify regions having corrupted pixel.
Step 5: For each region apply dilution and erosion properties in the binary mask to get the outermost layer of the region and store as area_to_be_filled.
Step 6: Similarly use dilution and erosion to get nearest layer with known values surrounding the region.
Step 7: Create training set with for all ∀ $y]$ as input and output y=g(x,y) which is the pixel value at that location.
Step 8: Create randomly initialized weights and bias with mean 0 and standard deviation of 1.
Step 9: Calculate value for first hidden layer and then apply an activation function to it (h1).
Step 10: Using h1 and output Y determine T.
Step 11: Use T such that $y] \in$ areas_to_be_filled, calculate the predicted value g'(x,y) and update it in the image.
Step 12: Update the corresponding region in the mask also.
Step 13: Goto **Step 4**.

4 Result and Discussion

The proposed method simulated in the Python by given system configuration shown in Table 2. The simulation performed in 30 iteration which is depicted by few iteration such as iteration 0, iteration 10, iteration 20, and iteration 30. Figure 1 shows corrupted image and its corresponding mask in iteration 0.

Figure 1(a) shows the original image which is corrupted. Corrupted regions are represented by black patch. This is the region which needs to be remade by inpainting.

Figure 1(b) shows the corrupted region in the image is represented in binary format. Here, 0 indicates all the corrupted region and 1 denotes undamaged region.

Figure 2 shows gradual decrease of corrupted regions for iteration 10. Figure 2(a) shows gradual decrease in area occupied by corrupted regions as the model slowly

Table 2 Simulation parameters

Software/Hardware	Specification
Windows OS	8.1
MS Office	Office 16.0
Python	3.6
Processor	Intel i5
Speed	2.4 GHz–3.8 GHz
RAM	8 GB
Hard Disk	1 TB

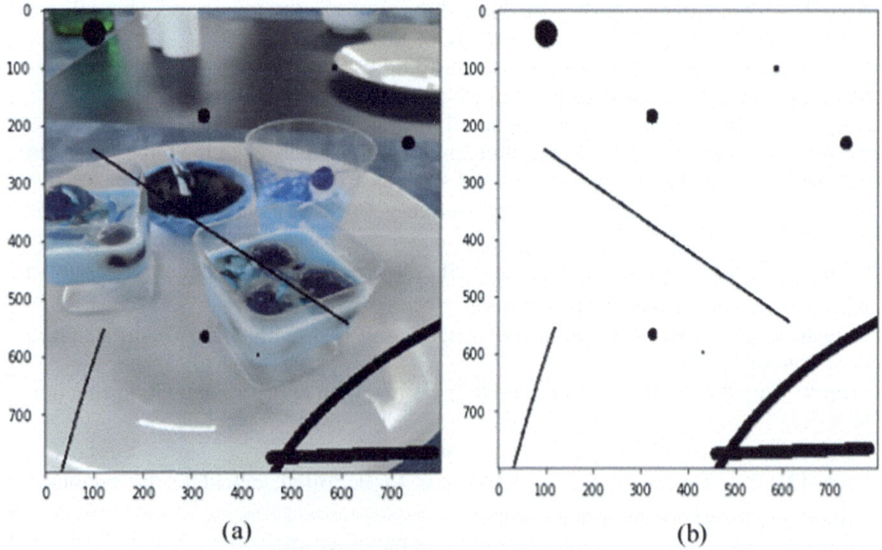

Fig. 1 The corrupted image and its corresponding image

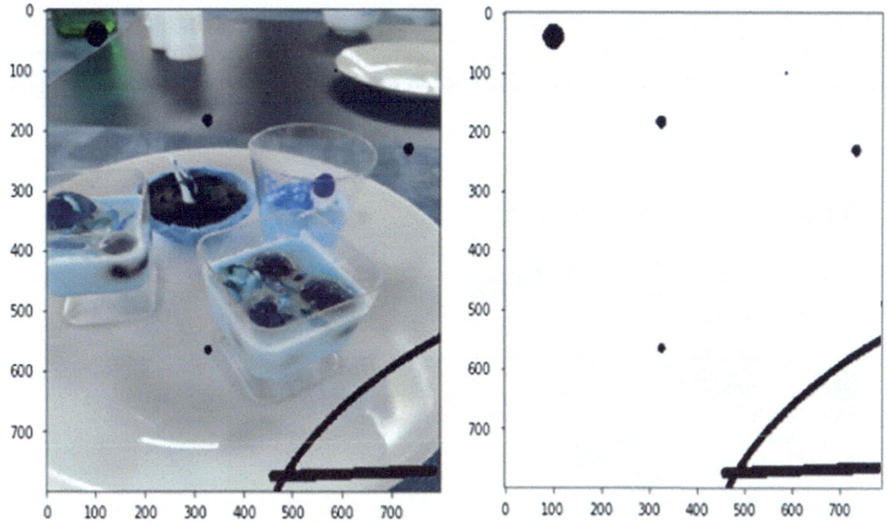

(a) (b)

Fig. 2 Gradual decrease of corrupted regions

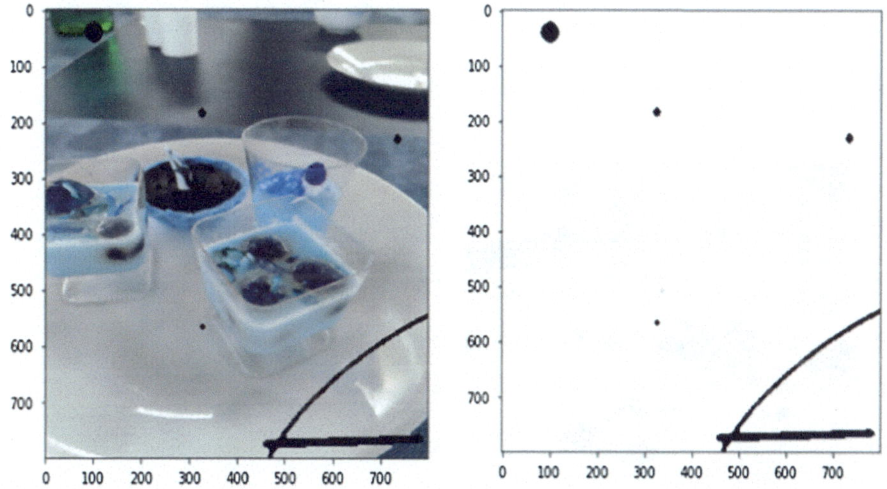

(a) (b)

Fig. 3 Images in iteration 20

starts to fill required regions. Figure 2(b) shows corresponding mask showing the areas which are required to be filled and which are already filled up. Comparing with last mask, it be seen that the corrupted regions are slowly decreasing.

Figure 3 shows the images for iteration 20. Figure 3(a) shows the status of image after current iteration. More regions have been filled up. Slowly moving towards completion. Figure 3(b) shows the status of the mask after current iteration. Changes made in masks are quite visible now.

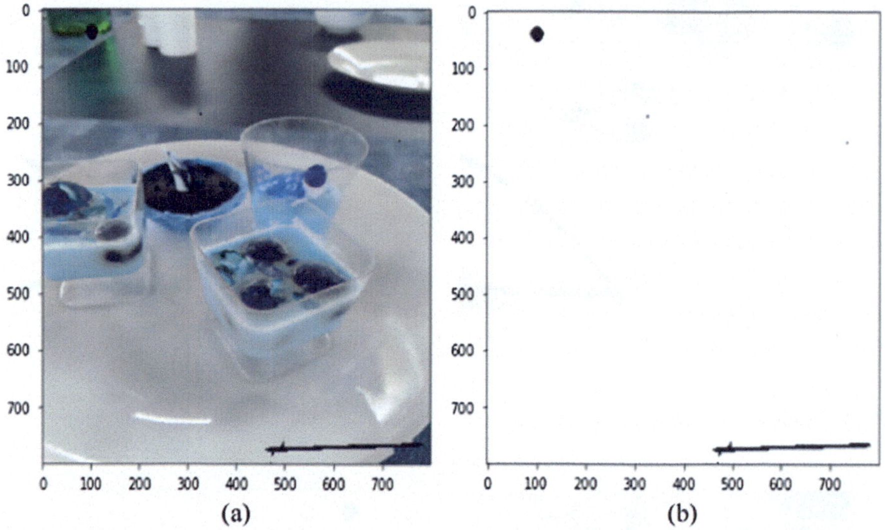

Fig. 4 Images where almost all the areas are covered

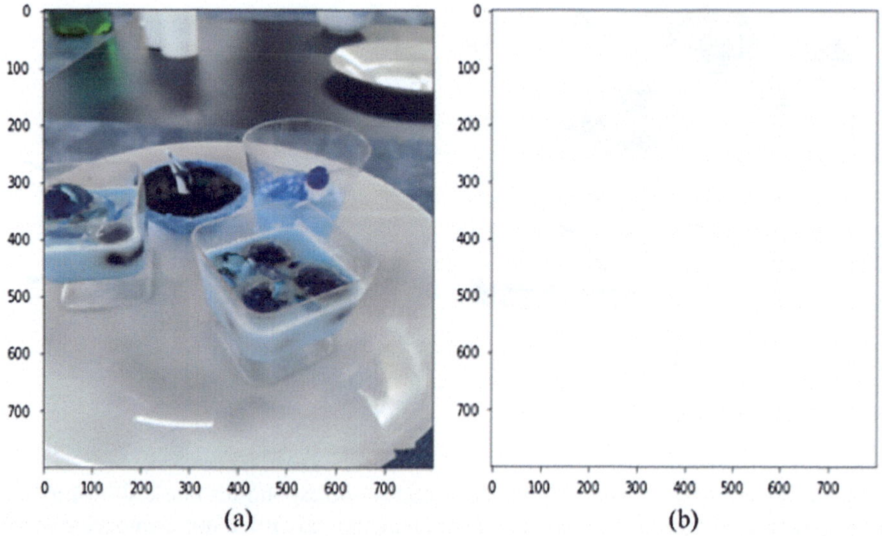

Fig. 5 Images in final iteration

Figure 4 shows the images where almost all the areas are covered in iteration 30. In Fig. 4(a), the task is almost completed. Nearly all the regions have been filled up. Very few corrupted region is left to be computed. In Fig. 4(b), very thin line or spots of black region can be seen which suggests that very few corrupted regions are now left.

Figure 5 shows the final iteration, where all the holes have been covered and the modified image is ready. Figure 5(a) shows result after final iteration. The model gives an inpainted image after it tries to fill up all the corrupted regions. Figure 5(b) shows final mask after completion of the algorithm. As compared with mask in other iteration(s), it can be seen that there are no corrupted regions (denoted by 0) left which has to be redeemed.

5 Conclusion

We tried to present a method which takes very less time for inpainting images and produce images while maintaining a high degree of quality. Step by step analysis of how the method steadily approaches the solution and finally produces the results. At last 2 methods to calculate error has been used for evaluating our results. If very less number of images has to be inpainted, then instead of training a model which would require huge amount of time, this method can be applied to gain results in a managable amount of time. As research is ongoing in this particular field, better algorithm could be invented which would generate results which provide much better results while taking considerably less time.

References

1. Zhang, A., Zhou, H., Li, X., & Zhu, W. (2019). Fast and robust learning in spiking feed-forward neural networks based on intrinsic plasticity mechanism. *Neurocomputing, 365,* 102–112.
2. Zhang, C., Zhang, X., & Zhang, Y. (2018). Dynamic properties of feed-forward neural networks and application in contrast enhancement for image. *Chaos, Solitons & Fractals, 114,* 281–290.
3. Shukla, S., & Raghuwanshi, B. S. (2019). Online sequential class-specific extreme learning machine for binary imbalanced learning. *Neural Networks, 119,* 235–248.
4. Chen, J., Zeng, Y., Li, Y., & Huang, G. B. (2019). Unsupervised feature selection based extreme learning machine for clustering. *Neurocomputing., 386,* 198–207.
5. Ashour, A. S., Samanta, S., Dey, N., Kausar, N., Abdessalemkaraa, W. B., & Hassanien, A. E. (2015). Computed tomography image enhancement using cuckoo search: A log transform based approach. *Journal of Signal and Information Processing, 6*(03), 244.
6. Das, S. K., Kumar, A., Das, B., & Burnwal, A. P. (2013). On soft computing techniques in various areas. *Computer Science and Information Technology, 3,* 59.
7. Das, S. K., & Tripathi, S. (2019). Energy efficient routing formation algorithm for hybrid ad-hoc network: A geometric programming approach. *Peer-to-Peer Networking and Applications, 12*(1), 102–128.

8. Das, S. K., & Tripathi, S. (2018). Adaptive and intelligent energy efficient routing for transparent heterogeneous ad-hoc network by fusion of game theory and linear programming. *Applied Intelligence, 48*(7), 1825–1845.
9. Das, S. K., & Tripathi, S. (2017). Energy efficient routing formation technique for hybrid ad hoc network using fusion of artificial intelligence techniques. *International Journal of Communication Systems, 30*(16), e3340, 1–16.
10. Chatterjee, S., Hore, S., Dey, N., Chakraborty, S., & Ashour, A. S. (2017). Dengue fever classification using gene expression data: A PSO based artificial neural network approach. In *Proceedings of the 5th International Conference on Frontiers in Intelligent Computing: Theory and Applications* (pp. 331–341). Singapore: Springer.
11. Jagatheesan, K., Anand, B., Samanta, S., Dey, N., Ashour, A. S., & Balas, V. E. (2017). Particle swarm optimisation-based parameters optimisation of PID controller for load frequency control of multi-area reheat thermal power systems. *International Journal of Advanced Intelligence Paradigms, 9*(5–6), 464–489.
12. Dey, N., Ashour, A. S., Beagum, S., Pistola, D. S., Gospodinov, M., Gospodinova, E. P., et al. (2015). Parameter optimization for local polynomial approximation based intersection confidence interval filter using genetic algorithm: An application for brain MRI image de-noising. *Journal of Imaging, 1*(1), 60–84.
13. Karaa, W. B. A., Ashour, A. S., Sassi, D. B., Roy, P., Kausar, N., & Dey, N. (2016). Medline text mining: an enhancement genetic algorithm based approach for document clustering. In *Applications of intelligent optimization in biology and medicine* (pp. 267–287). Cham: Springer.
14. Samanta, S., Mukherjee, A., Ashour, A. S., Dey, N., Tavares, J. M. R., Abdessalem Karâa, W. B., et al. (2018). Log transform based optimal image enhancement using firefly algorithm for autonomous mini unmanned aerial vehicle: An application of aerial photography. *International Journal of Image and Graphics, 18*(04), 1850019.
15. Choudhury, A., Samanta, S., Dey, N., Ashour, A. S., Bălas-Timar, D., Gospodinov, M., et al. (2015). Microscopic image segmentation using quantum inspired evolutionary algorithm. *Journal of Advanced Microscopy Research, 10*(3), 164–173.
16. Nandi, D., Ashour, A. S., Samanta, S., Chakraborty, S., Salem, M. A., & Dey, N. (2015). Principal component analysis in medical image processing: a study. *International Journal of Image Mining, 1*(1), 65–86.
17. Pradhan, J., Kumar, S., Pal, A. K., & Banka, H. (2018). A hierarchical CBIR framework using adaptive tetrolet transform and novel histograms from color and shape features. *Digital Signal Processing, 82,* 258–281.
18. Pradhan, J., Pal, A. K., & Banka, H. (2019). Principal texture direction based block level image reordering and use of color edge features for application of object based image retrieval. *Multimedia Tools and Applications, 78*(2), 1685–1717.
19. Pradhan, J., Ajad, A., Pal, A. K., & Banka, H. (2019). Multi-level colored directional motif histograms for content-based image retrieval. *The Visual Computer* 1–22. https://doi.org/10.1007/s00371-019-01773-9
20. Majhi, M., Pradhan, J., & Pal, A. K. (2019, March). An efficient content based image retrieval scheme with preserving the security of images. In *2019 6th International Conference on Signal Processing and Integrated Networks (SPIN)* (pp. 874–879). IEEE.
21. Pradhan, J., Kumar, S., Pal, A., & Banka, H. (2019). Texture and color region separation based image retrieval using probability annular histogram and weighted similarity matching scheme. *IET Image Processing, 14*(7), 1303–1315.
22. Pradhan, J., Raj, A., Pal, A. K., & Banka, H. (2020). Multi-scale image fusion scheme based on gray-level edge maps and adaptive weight maps. In *Proceedings of 3rd International Conference on Computer Vision and Image Processing* (pp. 445–459). Singapore: Springer.
23. Majhi, M., & Maheshkar, S. (2016, December). Privacy preserving in CBIR using color and texture features. In *2016 Fourth International Conference on Parallel, Distributed and Grid Computing (PDGC)* (pp. 276–281). IEEE.

24. Majhi, M., & Maheshkar, S. (2018). Privacy preserving for annular distribution density structure descriptor in CBIR using bit-plane randomization encryption. In *Proceedings of 2nd International Conference on Computer Vision & Image Processing* (pp. 159–171). Singapore: Springer.
25. Das, D., & Mukhopadhyay, S. (2015). Fingerprint image segmentation using block-based statistics and morphological filtering. *Arabian Journal for Science and Engineering, 40*(11), 3161–3171.
26. Das, D., & Mukhopadhyay, S. (2015). A pixel based segmentation scheme for fingerprint images. In *Information systems design and intelligent applications* (pp. 439–448). New Delhi: Springer.
27. Das, D. (2020). A minutia detection approach from direct gray-scale fingerprint image using hit-or-miss transformation. In *Computational intelligence in pattern recognition* (pp. 195–206). Singapore: Springer.
28. Das, D. (2018, October). A fingerprint segmentation scheme based on adaptive threshold estimation. In *2018 11th International Congress on Image and Signal Processing, Biomedical Engineering and Informatics (CISP-BMEI)* (pp. 1–6). IEEE.
29. Das, D., Mukhopadhyay, S., & Biswas, G. P. (2016, August). Cluster based template generation for fast and secure fingerprint matching. In *Proceedings of the International Conference on Advances in Information Communication Technology & Computing* (p. 24). ACM.

OCR Using Computer Vision and Machine Learning

Ashish Ranjan, Varun Nagesh Jolly Behera, and Motahar Reza

Abstract Being a well-researched area, optical character recognition or OCR has seen many advancements. Many state-of-the-art algorithms have been developed that can be used for the purpose of OCR but extracting text from images containing tables while preserving the structure of the table still remains a challenging task. Here, we present an efficient and highly scalable parallel architecture to segment input images containing tabular data with and without borders into cells and reconstruct the tabular data while preserving the tabular format. The performance improvement thus made can be used to ease the tedious task of digitizing tabular data in bulk. The same architecture can be used for regular OCR applications to improve performance if the data is in huge quantities.

Keywords Optical Character Recognition · Computer Vision · Machine Learning · Image Processing · Data Extraction

1 Introduction

We often take things we have, for granted. Vision is one such thing. We don't pay much attention to it unless we have a problem. We have our brains which give us the ability to interpret what we see. For a computer, this is very hard. Vision for a human being is facilitated by eyes that have developed through evolution over millions of years. For a computer, this can somewhat be achieved by the use of a camera. It gives the computer the ability to see.

For long, the vision of a camera was left to a person for analysis. But, with the advent of state-of-the-art Machine Learning techniques, the interpretation part was

A. Ranjan (✉) · V. N. J. Behera · M. Reza
National Institute of Science and Technology, Berhampur, Berhampur, India
e-mail: ashishranjan6282@gmail.com

V. N. J. Behera
e-mail: varunberaalego@gmail.com

M. Reza
e-mail: reza@nist.edu

S. K. Das et al. (eds.), *Machine Learning Algorithms for Industrial Applications*, Studies in Computational Intelligence 907,
https://doi.org/10.1007/978-3-030-50641-4_6

also entrusted to the computer. Optical Character Recognition is the simplest of examples where this is implemented. Optical Character Recognition or OCR is a well-researched area in the field of Machine Learning. An OCR software takes an image of text as input and converts it into digital text. There are several motivations for converting an image of text into digital text as we will find out later.

A major challenge in today's fast-paced world is the rate at which data is sourced. One of the reasons for this is the unavailability of older data in digital format. It requires manual labor to enter the data on a spreadsheet or a digital document. If the data has a specific structure, let's say it is tabular in format then it takes much more time. Image of a document takes up much more storage space than the same document in digital format.

A generic OCR software uses a Neural Network model to recognize characters based on previous training. The model is first trained using thousands of samples of each possible character so that it actually "learns" how that specific character looks. It should not confuse between the character '1' and the character '7'. Only when the accuracy of the predictive model is high enough to faithfully recognize each character, it is allowed to give an opinion on an actual test case.

The aforementioned problem of tabular data can be solved by dividing the image into cells and converting each cell individually. Even then, this takes a lot of time. This can be somewhat solved by parallelism during the analysis of cells over multiple threads. Still, there are many roadblocks in the progress of not just OCR technology but Machine Learning as a whole.

2 Goals and Objectives

The aim is to develop a system that can recognize text from scanned images that can either be printed or handwritten text. Another objective of this system is to recognize text from images of tabular data and maintain the tabular structure of the input image, then store it in a higher-level format such as CSV format or a spreadsheet. We need to take care of many factors for the best results. These factors do not concern the end-user of the system, but internally, these need to be addressed. Some of the factors are discussed as follows.

1. **Input image quality:** The input image can be low resolution, contain noise, have inappropriate and uneven lighting conditions during image acquisition. The human eye can understand and account for the lower quality image because we have millions of years of evolution helping us recognize patterns easily, but for a computer, a lower quality image may seem unrecognizable. The problem can be tackled by pre-processing the image to compensate for the low-quality image.
2. **Text language:** The input text can be in any language. So, the OCR system may support multiple languages. The number of supported languages depend on the training dataset that we will use during the machine learning process. The training can only be done on multiple languages if there is adequate data on which training can be done.

3. **Support for handwritten data:** The system may support the recognition of handwritten data, but only if the machine learning model is trained on handwritten data as well as typed data. The types of handwritten training data limit the types of handwritings the system can recognize.
4. **Presence of borders in tabular data:** If the input image contains tabular data, the presence of borders in the tabular data can help us easily identify the whole table as well as each cell of the table. If there are no borders available, then human assistance may be required to locate the table in the image. The cells of the table can be identified by the space between each text block.
5. **Speed:** The system should be fast enough to do its task in a small amount of time, especially in the case of tabular data, where the entire process of recognizing characters needs to be repeated for each cell. If the system is slower than a human who transcribes the text in the image in digital form, then the whole purpose of the system is defeated.

3 Image Acquisition

The OCR system requires the input to be an image which contains the text that needs to be extracted. The resolution and overall quality of the image plays an important role in the overall process. This image has to be acquired through some means. Some of the means of image acquisition are discussed below.

1. **Scanned document:** The document can be scanned and stored as an image.
2. **Photographs:** The document or text to be recognized can be photographed using a mobile phone or a camera.
3. **Digital images and art:** An image that is digitally generated and contains text can be recognized by the OCR system.
4. **Screenshots:** Sometimes, applications on mobile phones and computers alike contain text that cannot be selected and copied. These text can be recognized if a screenshot of the screen is taken when the text is shown.

The specifications of the image acquisition device should be high enough that all the details in the image are preserved. Things such as focus on text, less noise and proper lighting should be taken care of. Images can be interpreted as 2D arrays with multiple channels representing color, i.e., there are that many number of 2D arrays as there are supported colors. Each element of the 2D array of each channel stores the intensity of that pixel in that channel. The combination of corresponding pixel from each channel gives the unique overall color of the pixel (See Fig. 1). The dimensions of the 2D array is the resolution of the image. Generally, image processing applications are done using grayscale images which has only one channel. This makes the processing efficient.

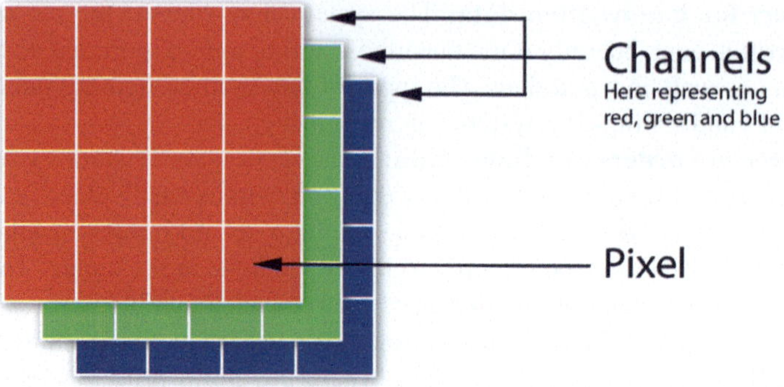

Fig. 1 2D representation of image

4 Dataset

Selection of appropriate dataset is crucial to building a good OCR system. The set of characters in the dataset define what characters the final product is able to recognize. Hence, it is obvious that the dataset used to train the model must contain the characters and symbols from languages that need to be included in the the recognition system. The NIST Database [1], published by the National Institute of Standards and Technology contains handwriting of 3600 writers and includes over 800,000 characters. A subset of the original NIST Database, called the MNIST Database [2] which contains 60,000 examples of handwritten digits.

The input to the OCR system may contain mathematical symbols and expressions. For this, the Mathematics Expressions (CROHME) dataset [3] can be used which contains over 10,000 mathematical expressions and over 100 mathematical symbols. In order to handle different languages, dataset for the languages are required. e.g., Devanagari Character Dataset [4] which can used to recognize Devanagari (Hindi) characters. It contains over 1800 samples from 36 character class collected from by 25 native Hindi writers. Other examples include Chinese Characters (HIT-OR3C) Dataset [5], which include over 900,000 images form about 10 news articles and the Arabic Printed Text Image (APTI) Dataset which contains 113,284 words using 10 Arabic fonts.

Another application of the OCR system can be to identify street signs and boards. For this, the datasets that can be used are the Street View Text [6] which is created from Google Street View and contains images of outdoor street signs and boards. And Street View House Numbers (SVHN) [7], which is also from Google Street View and contain 73000 digits of house street numbers.

In order to handle input from uneven and unreliable lighting conditions, the Natural Environment OCR (NEOCR) Dataset [8] can be used which contains 659 images in real world conditions with 5238 annotations of text. Another dataset that can be used for this purpose is the KAIST Scene Text Database [9] which contains 3000 images in different and uneven lighting conditions. The dataset, MSRA Text Detection 500 Database (MSRA-TD500) [10] contains 500 images taken from a pocket camera and includes two types of images, indoor and outdoor images. The indoor images contain signs and plates in indoor conditions while the outdoor images contain billboards.

The collection of these datasets and others such as IAM Online Document Database (IAMonDo-database) [11], IAM On-Line Handwriting Database [12], Stanford OCR, The Chars74K dataset [13] and many more can be used for building a robust OCR system which can handle printed as well as handwritten data in multiple languages under different lighting conditions.

5 Preprocessing

The input images may still contain some imperfections such as noise and uneven lighting even after careful acquisition. These imperfections need to be removed before applying the OCR algorithm. The preprocessing [14] steps to be taken are discussed further.

5.1 Gaussian Blur

It is a image blurring technique that uses a Gaussian function (it also expresses the normal distribution in statistics) for calculating the modified pixel value for each pixel in the image. The formula of a Gaussian function in one dimension is

$$G(x) = \frac{1}{\sqrt{2\pi\sigma^2}} e^{-\frac{x^2}{2\sigma^2}} \tag{1}$$

In two dimensions, the same formula is used but as a product of two Gaussian functions for each dimension.

$$G(x, y) = \frac{1}{\sqrt{2\pi\sigma^2}} e^{-\frac{x^2+y^2}{2\sigma^2}} \tag{2}$$

where x is the distance from the origin in the horizontal axis, y is the distance from the origin in the vertical axis, and σ is the standard deviation of the Gaussian distribution.

Gaussian Blur is used to smooth the image and reduce noise [15]. We can observe the effect of Gaussian Blur in the Fig. 2.

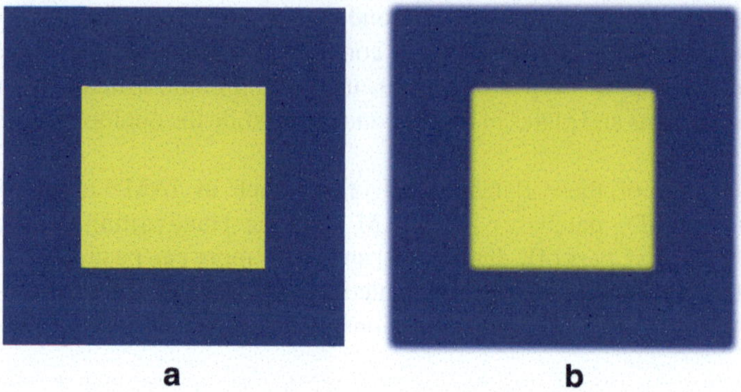

Fig. 2 **a** Original image, **b** Blurred image using Gaussian Blur

5.2 Thresholding and Binarization

The image may have uneven lighting and noise. This can be solved by using thresholding and binarization techniques [15]. Text is ideally written or printed using a color on a contrasting background. In other words, if the text is darker in color, the background is supposed to be in a lighter color (e.g., Black board with chalk writing) and vice versa (e.g., text on a white paper). We can exploit this to our advantage.

5.2.1 Thresholding

We first define a fixed threshold intensity value manually and filter out the text. This means that for a lighter text on a darker background, the intensity of a pixel is above the threshold value, we keep it and discard the rest of the pixels by replacing the intensity by zero and vice versa for a darker text and lighter background. This can also be achieved by just implementing the algorithm for one of the forms (out of dark text light background and light text and dark background) and inverting the image and using the same algorithm if the other form is given as input. If the passing intensity value while thresholding is replaced with the maximum possible intensity value (white) while others are replaced with the lowest value (black), i.e., zero, we have the image in only two colors. This image is called the binarized image and the process is called binarization.

5.2.2 Adaptive Thresholding

This thresholding technique does not take a fixed threshold value for the whole image but the threshold value is calculated for smaller regions so that it has different thresholds for different regions of the image. This is especially helpful when the image contains regions of uneven lighting. The control in this type of thresholding is the block size of neighbouring pixels that are taken into consideration for calculating the threshold value.

5.2.3 Otsu's Method

Otsu's Method [16, 17], named after Nobuyuki Otsu is used to perform automatic image thresholding. It automatically determines the optimum threshold for the image. This threshold is determined by minimizing intra-class intensity variance, or equivalently, by maximizing inter-class variance. It is equivalent to a globally optimal k-means algorithm performed on the intensity histogram. This method gets the best of both simple and adaptive threshold by having a single threshold value for the whole image and remove the unwanted background which may happen in adaptive threshold. Its application is subjective to the input image. We have to use adaptive or Otsu's thresholding, whichever works better on the current image.

Figure 3 shows the differences in outputs of the various thresholding methods. The applicability of these methods is use case based. One may work in some case while the other works in some other case. Generally, adaptive thresholding works best if proper parameters are used.

5.3 Morphological Operations

The thresholded or binarized image may contain unwanted gaps and cracks which can be filled using dilation with an appropriate structuring element. This is especially necessary for table grid reconstruction when the image contains a table which needs to be extracted while preserving the structure.

One way to reconstruct the structure of the table or the grids is to use morphological opening [18, 19] operation with sufficiently large structuring element in both vertical and horizontal directions separately. Morphological opening operation is erosion followed by dilation. This will remove all the text from the image but preserve and enhance all vertical and horizontal lines respectively. We can merge both vertical and horizontal lines to get the grid structure of the table. The process is shown in Fig. 4.

Once the grid is reconstructed, morphological closing operation (dilation followed by erosion) can be done to fill in the remaining gaps or inconsistencies in the structure. There are smarter methods to achieve better results, some of these even involve machine learning [20, 21].

This approach may not work if the table is skewed or the image was taken at an oblique angle. To counter this problem, de-skewing and perspective correction may be required. This is discussed later in Sect. 7.2.

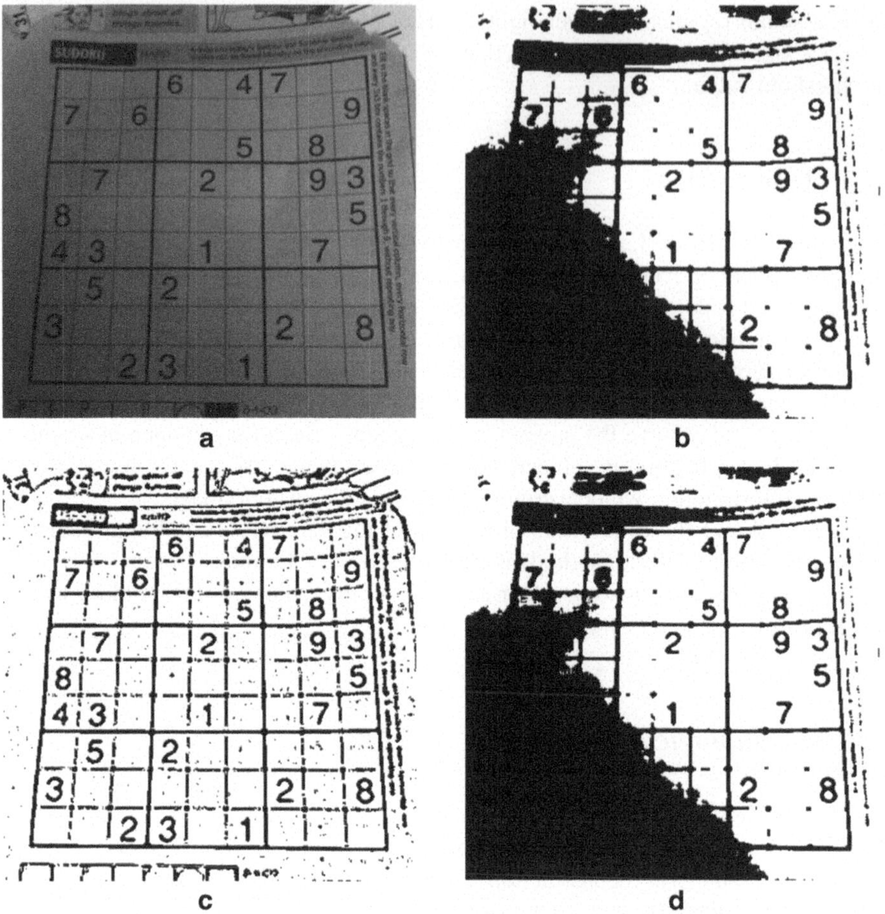

Fig. 3 **a** Original image, **b** Result of manual thresholding with a threshold value of 127, **c** Result of adaptive thresholding, **d** Result of Otsu's Method

6 Page Extraction

The input image may be a scanned document or a photograph. If it contains the whole page, it is necessary to transform it in such a way that the text is properly aligned and the image as a whole seems to be taken at a perfect top down manner as if it was scanned.

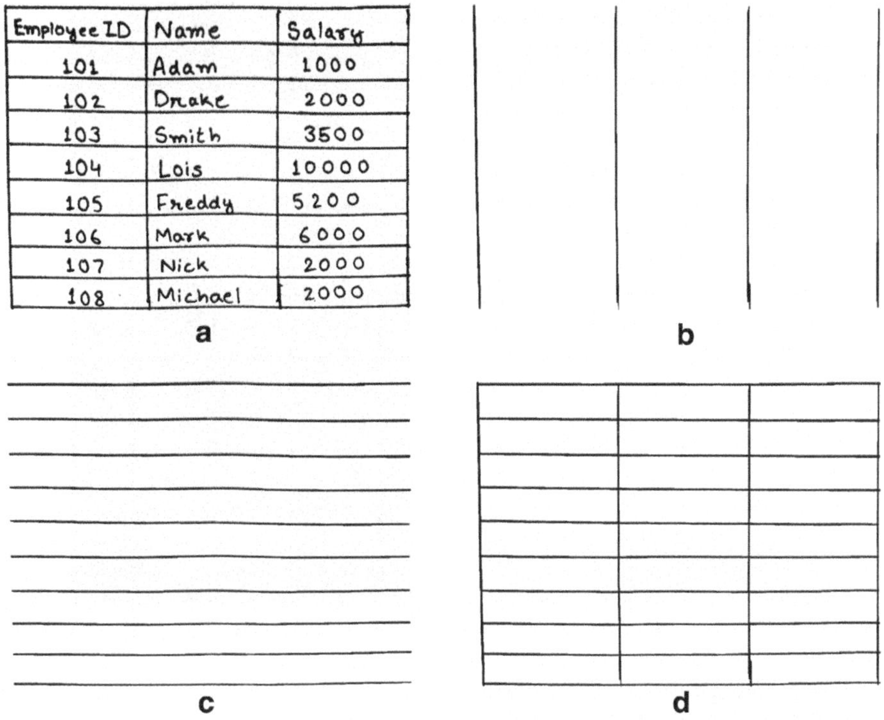

Fig. 4 **a** Original image after thresholding, **b** Result of opening in vertical direction, **c** Result of opening in horizontal direction, **d** Result after combining both results

We do this by detecting the edges using Canny edge detection [22] on the image to get the boundary of the page. Then, we find the largest contour having four sides. We apply perspective warp transform on this contour to stretch it out into a rectangle of suitable size (See Fig. 5). This is further discussed in Sect. 7.2.

7 Table Identification

If the image contains a tabular structure and the borders are visible then it will inherently be rectangular in shape. There are many approaches to table identification but we approach the problem by using contours [23].

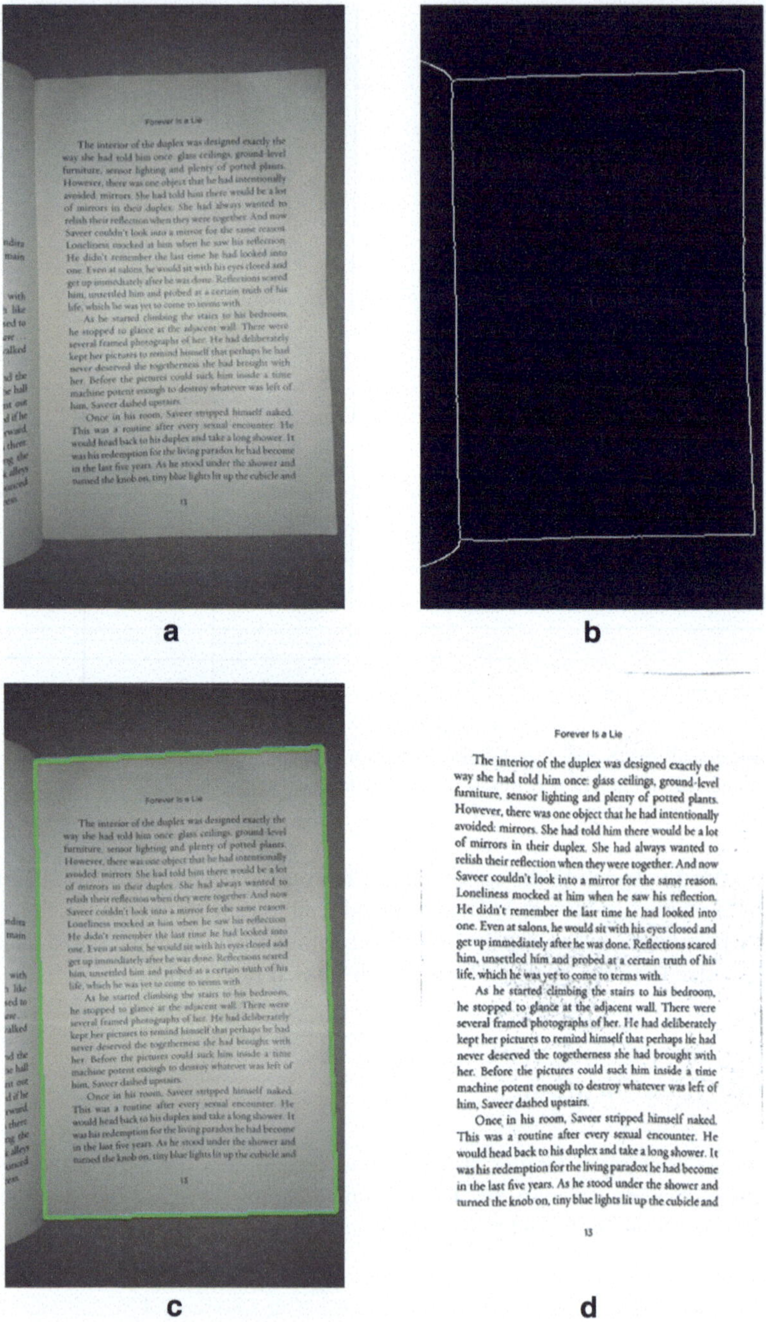

Fig. 5 **a** Original image taken using phone's camera, **b** Canny Edge Detection, **c** Largest contour having four sides, **d** Corrected image by stretching the largest contour having four sides into a rectangle

7.1 Contours

Contours can be explained simply as a curve joining all the continuous points (along the boundary), having same color or intensity. It essentially marks the boundaries of areas of image having similar color.

In a thresholded image, the table grid is either black or white while the background is of the opposite color. This gives a clear boundary for contours to cover the whole area of the table. Many contours may be detected in an image representing each cell of the table, markings and text as well but we are interested in finding the entire table. The largest contour or the contour covering the greatest area is considered to be the table and inherently, any contour inside the table is a cell except for the text.

The text can be ignored by ignoring the smaller contours because text covers only a small area as compared to the cells and the entire table. We only require the entire table and not the text at the moment because the table may need to be properly oriented for further processing.

The process of getting the largest contour can serve the purpose of transforming the image so that it seems like the image was taken at the perfect angle while it was actually taken at an oblique angle to keep the flow of text and the table in the correct direction. The process is shown in Fig. 6.

7.2 Deskewing and Perspective Transformation

The image may have been taken at an oblique angle which can make the table look distorted. This can be solved by locating the four corners of the table contour and applying a perspective warp transform to compensate for the oblique angle at which the image was taken. The perspective transform takes four points on the image as input and stretches them out to a target rectangular dimension. Three of these four points must not be co-linear. See Fig. 5.

If the image is a scanned document, the only problem it may have is that the text is rotated and needs to be adjusted. This can be done by rotating the image in such an angle that the bounding box around the text covers the smallest possible area. OpenCV's [24] minAreaRect function can be helpful in this regard. Figure 7 shows an example.

7.3 Text Detection

The previous method of using contours to identify and locate tables only works on images that contain tables with visible borders. If the table has no visible borders, e.g., a printed receipt but obviously has a tabular structure, then the method of using contours fail.

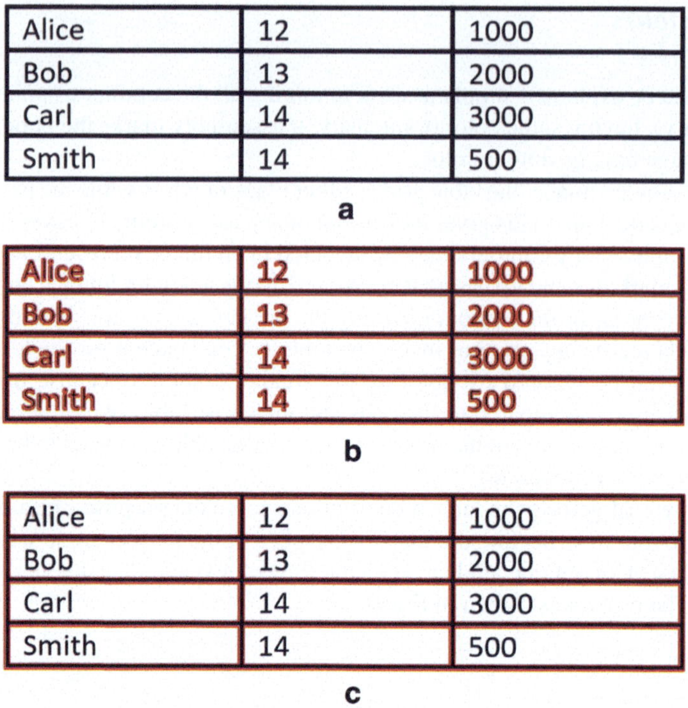

Fig. 6 **a** Original image after thresholding, **b** Red borders represent each contour, **c** Contours with smaller contours are ignored to ignore text (Color figure online)

The Efficient and Accurate Scene Text Detector or EAST [25] is a machine learning approach for text detection. It actually detects where text is present in the image and does not recognize the text itself. Figure 8 shows an output for EAST. It works irrespective of the orientation of the text and returns a bounding box within which text is present. There are many approach to text detection available such as YOLO [26] and SSD [27] which are faster but less accurate. These also work in the similar way.

For an input image with table that has invisible or no borders. The user has to specify that the image actually contains a table. This choice is entirely subjective as there is nearly no signs that can help confirm with certainty that the image is tabular or not. There can be text that seem to show tabular nature but are actually non tabular.

In this case, in order to identify the table, we first identify the cells. For this, we detect the location of text within the image using EAST because, the text in a cell are nearby and can be fit in one bounding box while text in different cells are far apart.

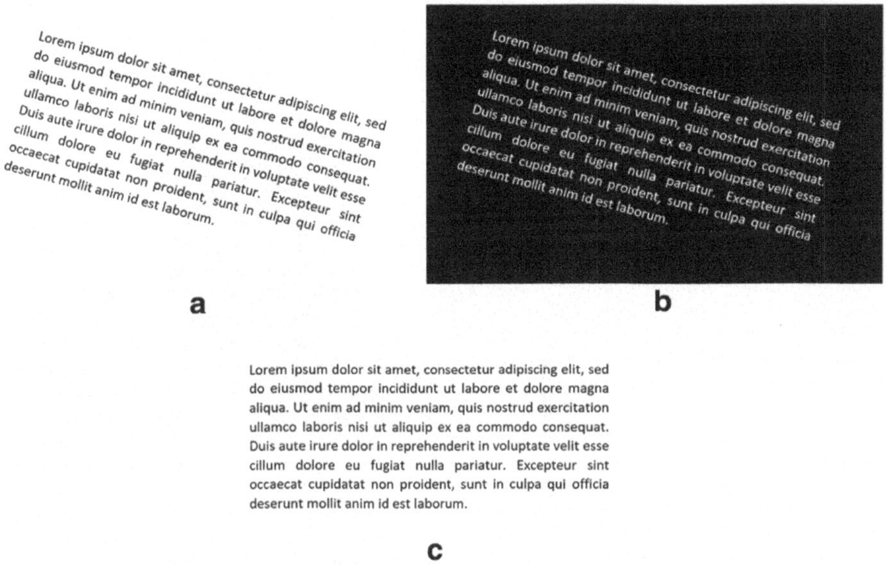

Fig. 7 Deskewing: **a**. Input skewed image, **b**. Binarized image with white pixels for text, **c**. Corrected image

8 Tabular Analysis

The tabular analysis is slightly different for bordered and non bordered tables. For both types of tables, we follow these methods for tabular analysis.

8.1 For Tables with Borders

After the table has been located, we know that all contours inside the table which have sufficiently large area, must be the cells of the table. These cells can be sorted based on the top left corner coordinates of the cells. The cells are sorted based on the y-axis value of the cell's top left corner and if two cells have the same y-axis value, then they are sorted based on the x-axis value. This will generate a list of cells stored in row major wise. Similarly, the list can also be generated in column major wise. The number of rows or columns must be found by counting the number of cells which have the same y-axis or x-axis coordinates respectively. Later, this information can be used to recreate the tabular structure of the input image. The cells hence obtained can be fed to the OCR engine to recognize the text in each cell.

Fig. 8 Text detection

8.2 For Tables Without Borders

As previously established, the cells are identified and the bounding boxes are found. The corners of the corner cells can be used to get the entire table. Similar to the process of tabular analysis for tables with borders, the cells are sorted in row major wise or column major wise. Now, the cells can be fed to the OCR engine. Boxes having similar x-value will be in same column and items having similar y-value will be in same row. This is not a very reliable method and can fail very easily if the text detection does not perform well.

Another approach is by analyzing white pixel columns and rows to draw the missing borders. For vertical borders, We find the first column of pixels where no black pixel was found and the next column where at least one black pixel was found. The mid-point of these two columns of pixel will be the x-value of border separating the text columns. Once the x-value is found out, we can draw a vertical line. This process continues till the whole image is covered. Similarly, horizontal borders can also be drawn. Examples for both approaches are shown in Fig. 9.

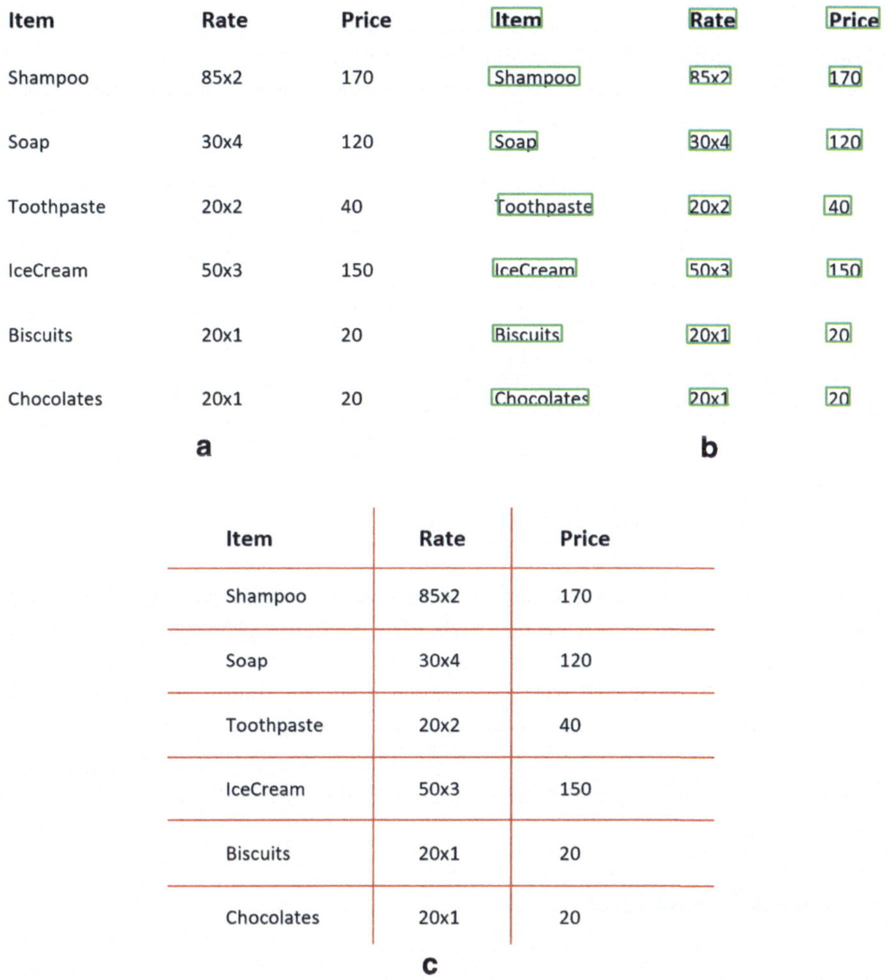

Fig. 9 **a** Original image of table without borders, **b** Text detected where items having similar x-value will be in same column and items having similar y-value will be in same row, **c** Drawing grid by finding the mid-point of continuous white pixels

9 Optical Character Recognition Engine

The Optical Character Recognition Engine or the OCR Engine is an algorithm implementation that takes the preprocessed image and finally returns the text written on it. This is the actual piece of software that recognizes the text. There are many standard deep learning approaches to the problem of text recognition. Some of the most popular ones are YOLO [26], SSD [27], Mask RCNN [28] and Faster RCNN [29]. These architectures are basically object detectors which can be trained for the task of character recognition. Algorithms such as Mask RCNN and Faster RCNN are region based detectors. This means that the algorithm first looks for objects (text in this case) in the image and then classifies the objects (characters in this case). This two step process makes it slow but more accurate.

Algorithms such as YOLO and SSD are Single Shot Detectors look for objects and classify them at the same time. The single step process make them faster but perform worse for smaller objects i.e., text in our case. Any of the previously mentioned dataset can be used for training these models and the trained model can be used to predict or recognize the text in any input image.

There is another solution. Tesseract [30] is an OCR engine which was originally developed by HP and was made open source in 2005 and now its development is sponsored by Google. It supports over 100 languages with great accuracy thanks to the latest version (4.1.0) which is LSTM based.

10 The Problem with Tabular Data

Tabular data needs to be fed into the OCR engine cell wise to maintain the tabular structure. An image may contain hundreds of cells and to process that, the entire OCR procedure needs to be run again and again. This takes considerable time. If an image without a tabular structure is given as input, the output may be produced within seconds. But, if the image contains a table with many cells, it can even take several minutes to produce the output. Speed of processing is the major problem with tabular data. Both tables with and without borders suffer from this problem.

11 Parallel Workflow

The speed of processing can be improved by processing each cell in parallel by using the parallel architecture of the CPU and GPU which are highly efficient in parallelizing small and similar tasks. The concept of multi-threading can be used to parallelize the process and divide the workload among multiple threads. The overall algorithm is summarized in Algorithm 1.

The parallel workflow can further be extended in certain use cases. If there are huge number of images on which OCR has to be done, then multiple images can be processed in parallel to speedup the overall process. The workflow is shown in Fig. 10.

12 Experimental Results

The algorithm was implemented in Python 3.6 programming language using the open source computer vision library, OpenCV and the Tesseract OCR engine. For the experimental evaluation, we used a system with an Intel Core i7 6700HQ processor, with 4 cores and 8 threads. The OCR system was tested with several input data. One of such input data is shown in the Figure which contains 65 rows and 11 columns (It is a sample Microsoft Excel Spreadsheet).

Figure 11 shows the image with the identified cells. The OCR system is working as expected as proper output is produced. Since, the OCR system converts visual information into text, the input can easily produce incorrect result due to the presence of noise and visual artifacts. That is why the OCR system needs manual testing or labeled testing dataset for verification of accuracy.

Multiple test cases for handwritten images was done. One of those images is shown in Fig. 12.

We found that the approach works very well with scanned documents and digital tabular images. The performance decreases as expected, if photographed images or handwritten data is given as input. This is because of the possible noise, camera angle while taking the image, orientation and layout of the document where the image, etc. These lead to unrecognizable characters and structure. The most important factor

Algorithm 1: Algorithm for parallel approach

Input: Image segment containing tabular data
Result: The Generated Table

```
READ Input Image
APPLY Preprocessing
FIND all contours in image
SET number of parallel processes
foreach contour in image do
    DISTRIBUTE contours among workers
    if area of contour = expected cell area then
        USE OCR engine to detect text within cell
        APPEND text, yValue to list L
    end
end
prevY ← ∞
p ← 0
foreach i in 0 to length of L do
    if prevY < L[i].yValue then
        APPEND L[i-p:i] to list T
        p ← 0
    end
    p ← p + 1
    prevY ← L[i].yValue
end
```

Fig. 10 Algorithm workflow

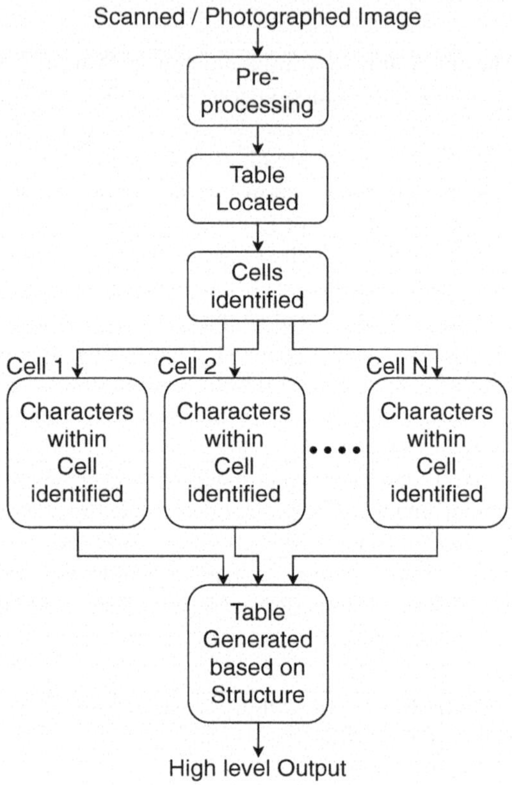

for good accuracy is lighting. If the image is taken in a well lit environment, the preprocessing algorithms can easily enhance the text for the OCR engine to recognize the text.

The algorithm was tested using many such test inputs and the output was found to be accurate in about 90% of the cases with printed text data and 80% of the cases with handwritten data (non cursive). This reduction in accuracy can be attributed to variations in handwritings of people where they write certain characters differently and tendency of certain characters to look similar. For cursive handwriting, text becomes nearly unrecognizable as people may have very unique writing styles and some handwritings are even unrecognizable for people let alone the OCR system. This is a limitation of the OCR engine we use. This means that the OCR system is very reliable and if clean and systematic data is supplied.

Government	Canada	Carretera	None	1618.5	$ 3.00	$ 20.00	$ 32,370.00	$ -	$ 32,370.00	$ 16,185.00
Government	Germany	Carretera	None	1321	$ 3.00	$ 20.00	$ 26,420.00	$ -	$ 26,420.00	$ 13,210.00
Midmarket	France	Carretera	None	2178	$ 3.00	$ 15.00	$ 32,670.00	$ -	$ 32,670.00	$ 21,780.00
Midmarket	Germany	Carretera	None	888	$ 3.00	$ 15.00	$ 13,320.00	$ -	$ 13,320.00	$ 8,880.00
Midmarket	Mexico	Carretera	None	2470	$ 3.00	$ 15.00	$ 37,050.00	$ -	$ 37,050.00	$ 24,700.00
Government	Germany	Carretera	None	1513	$ 3.00	$ 350.00	$ 5,29,550.00	$ -	$ 5,29,550.00	#########
Midmarket	Germany	Montana	None	921	$ 5.00	$ 15.00	$ 13,815.00	$ -	$ 13,815.00	$ 9,210.00
Channel Partners	Canada	Montana	None	2518	$ 5.00	$ 12.00	$ 30,216.00	$ -	$ 30,216.00	$ 7,554.00
Government	France	Montana	None	1899	$ 5.00	$ 20.00	$ 37,980.00	$ -	$ 37,980.00	$ 18,990.00
Channel Partners	Germany	Montana	None	1545	$ 5.00	$ 12.00	$ 18,540.00	$ -	$ 18,540.00	$ 4,635.00
Midmarket	Mexico	Montana	None	2470	$ 5.00	$ 15.00	$ 37,050.00	$ -	$ 37,050.00	$ 24,700.00
Enterprise	Canada	Montana	None	2665.5	$ 5.00	$ 125.00	$ 3,33,187.50	$ -	$ 3,33,187.50	#########
Small Business	Mexico	Montana	None	958	$ 5.00	$ 300.00	$ 2,87,400.00	$ -	$ 2,87,400.00	#########
Government	Germany	Montana	None	2146	$ 5.00	$ 7.00	$ 15,022.00	$ -	$ 15,022.00	$ 10,730.00
Enterprise	Canada	Montana	None	345	$ 5.00	$ 125.00	$ 43,125.00	$ -	$ 43,125.00	$ 41,400.00
Midmarket	United States of America	Montana	None	615	$ 5.00	$ 15.00	$ 9,225.00	$ -	$ 9,225.00	$ 6,150.00
Government	Canada	Paseo	None	292	$ 10.00	$ 20.00	$ 5,840.00	$ -	$ 5,840.00	$ 2,920.00
Midmarket	Mexico	Paseo	None	974	$ 10.00	$ 15.00	$ 14,610.00	$ -	$ 14,610.00	$ 9,740.00
Channel Partners	Canada	Paseo	None	2518	$ 10.00	$ 12.00	$ 30,216.00	$ -	$ 30,216.00	$ 7,554.00
Government	Germany	Paseo	None	1006	$ 10.00	$ 350.00	$ 3,52,100.00	$ -	$ 3,52,100.00	#########
Channel Partners	Germany	Paseo	None	367	$ 10.00	$ 12.00	$ 4,404.00	$ -	$ 4,404.00	$ 1,101.00
Government	Mexico	Paseo	None	883	$ 10.00	$ 7.00	$ 6,181.00	$ -	$ 6,181.00	$ 4,415.00
Midmarket	France	Paseo	None	549	$ 10.00	$ 15.00	$ 8,235.00	$ -	$ 8,235.00	$ 5,490.00
Small Business	Mexico	Paseo	None	788	$ 10.00	$ 300.00	$ 2,36,400.00	$ -	$ 2,36,400.00	#########
Midmarket	Mexico	Paseo	None	2472	$ 10.00	$ 15.00	$ 37,080.00	$ -	$ 37,080.00	$ 24,720.00
Government	United States of America	Paseo	None	1143	$ 10.00	$ 7.00	$ 8,001.00	$ -	$ 8,001.00	$ 5,715.00
Government	Canada	Paseo	None	1725	$ 10.00	$ 350.00	$ 6,03,750.00	$ -	$ 6,03,750.00	#########
Channel Partners	United States of America	Paseo	None	912	$ 10.00	$ 12.00	$ 10,944.00	$ -	$ 10,944.00	$ 2,736.00
Midmarket	Canada	Paseo	None	2152	$ 10.00	$ 15.00	$ 32,280.00	$ -	$ 32,280.00	$ 21,520.00
Government	Canada	Paseo	None	1817	$ 10.00	$ 20.00	$ 36,340.00	$ -	$ 36,340.00	$ 18,170.00
Government	Germany	Paseo	None	1513	$ 10.00	$ 350.00	$ 5,29,550.00	$ -	$ 5,29,550.00	#########
Government	Mexico	Velo	None	1493	$ 120.00	$ 7.00	$ 10,451.00	$ -	$ 10,451.00	$ 7,465.00
Enterprise	France	Velo	None	1804	$ 120.00	$ 125.00	$ 2,25,500.00	$ -	$ 2,25,500.00	#########
Channel Partners	Germany	Velo	None	2161	$ 120.00	$ 12.00	$ 25,932.00	$ -	$ 25,932.00	$ 6,483.00
Government	Germany	Velo	None	1006	$ 120.00	$ 350.00	$ 3,52,100.00	$ -	$ 3,52,100.00	#########
Channel Partners	Germany	Velo	None	1545	$ 120.00	$ 12.00	$ 18,540.00	$ -	$ 18,540.00	$ 4,635.00
Enterprise	United States of America	Velo	None	2821	$ 120.00	$ 125.00	$ 3,52,625.00	$ -	$ 3,52,625.00	#########
Enterprise	Canada	Velo	None	345	$ 120.00	$ 125.00	$ 43,125.00	$ -	$ 43,125.00	$ 41,400.00
Small Business	Canada	VTT	None	2001	$ 250.00	$ 300.00	$ 6,00,300.00	$ -	$ 6,00,300.00	#########
Channel Partners	Germany	VTT	None	2838	$ 250.00	$ 12.00	$ 34,056.00	$ -	$ 34,056.00	$ 8,514.00
Midmarket	France	VTT	None	2178	$ 250.00	$ 15.00	$ 32,670.00	$ -	$ 32,670.00	$ 21,780.00
Midmarket	Germany	VTT	None	888	$ 250.00	$ 15.00	$ 13,320.00	$ -	$ 13,320.00	$ 8,880.00
Government	France	VTT	None	1527	$ 250.00	$ 350.00	$ 5,34,450.00	$ -	$ 5,34,450.00	#########
Small Business	France	VTT	None	2151	$ 250.00	$ 300.00	$ 6,45,300.00	$ -	$ 6,45,300.00	#########
Government	Canada	VTT	None	1817	$ 250.00	$ 20.00	$ 36,340.00	$ -	$ 36,340.00	$ 18,170.00
Government	France	Amarilla	None	2750	$ 260.00	$ 350.00	$ 9,62,500.00	$ -	$ 9,62,500.00	#########
Channel Partners	United States of America	Amarilla	None	1953	$ 260.00	$ 12.00	$ 23,436.00	$ -	$ 23,436.00	$ 5,859.00
Enterprise	Germany	Amarilla	None	4219.5	$ 260.00	$ 125.00	$ 5,27,437.50	$ -	$ 5,27,437.50	#########
Government	France	Amarilla	None	1899	$ 260.00	$ 20.00	$ 37,980.00	$ -	$ 37,980.00	$ 18,990.00
Government	Germany	Amarilla	None	1686	$ 260.00	$ 7.00	$ 11,802.00	$ -	$ 11,802.00	$ 8,430.00
Channel Partners	United States of America	Amarilla	None	2141	$ 260.00	$ 12.00	$ 25,692.00	$ -	$ 25,692.00	$ 6,423.00
Government	United States of America	Amarilla	None	1143	$ 260.00	$ 7.00	$ 8,001.00	$ -	$ 8,001.00	$ 5,715.00
Midmarket	United States of America	Amarilla	None	615	$ 260.00	$ 15.00	$ 9,225.00	$ -	$ 9,225.00	$ 6,150.00
Government	France	Paseo	Low	3945	$ 10.00	$ 7.00	$ 27,615.00	$ 276.15	$ 27,338.85	$ 19,725.00
Midmarket	France	Paseo	Low	2296	$ 10.00	$ 15.00	$ 34,440.00	$ 344.40	$ 34,095.60	$ 22,960.00
Government	France	Paseo	Low	1030	$ 10.00	$ 7.00	$ 7,210.00	$ 72.10	$ 7,137.90	$ 5,150.00
Government	France	Velo	Low	639	$ 120.00	$ 7.00	$ 4,473.00	$ 44.73	$ 4,428.27	$ 3,195.00
Government	Canada	VTT	Low	1326	$ 250.00	$ 7.00	$ 9,282.00	$ 92.82	$ 9,189.18	$ 6,630.00
Channel Partners	United States of America	Carretera	Low	1858	$ 3.00	$ 12.00	$ 22,296.00	$ 222.96	$ 22,073.04	$ 5,574.00
Government	Mexico	Carretera	Low	1210	$ 3.00	$ 350.00	$ 4,23,500.00	$ 4,235.00	$ 4,19,265.00	#########
Government	United States of America	Carretera	Low	2529	$ 3.00	$ 7.00	$ 17,703.00	$ 177.03	$ 17,525.97	$ 12,645.00
Channel Partners	Canada	Carretera	Low	1445	$ 3.00	$ 12.00	$ 17,340.00	$ 173.40	$ 17,166.60	$ 4,335.00
Enterprise	United States of America	Carretera	Low	330	$ 3.00	$ 125.00	$ 41,250.00	$ 412.50	$ 40,837.50	$ 39,600.00
Channel Partners	France	Carretera	Low	2671	$ 3.00	$ 12.00	$ 32,052.00	$ 320.52	$ 31,731.48	$ 8,013.00

Fig. 11 Sample input data

This parallel approach was found to be nearly four times faster than serial approach if the number of cells in the image was huge. The performance metrics can be seen in the Fig. 13. As can be seen, our algorithm scales very well over large numbers of cells within a table, and this high scalability improves efficiency of the data [31] process.

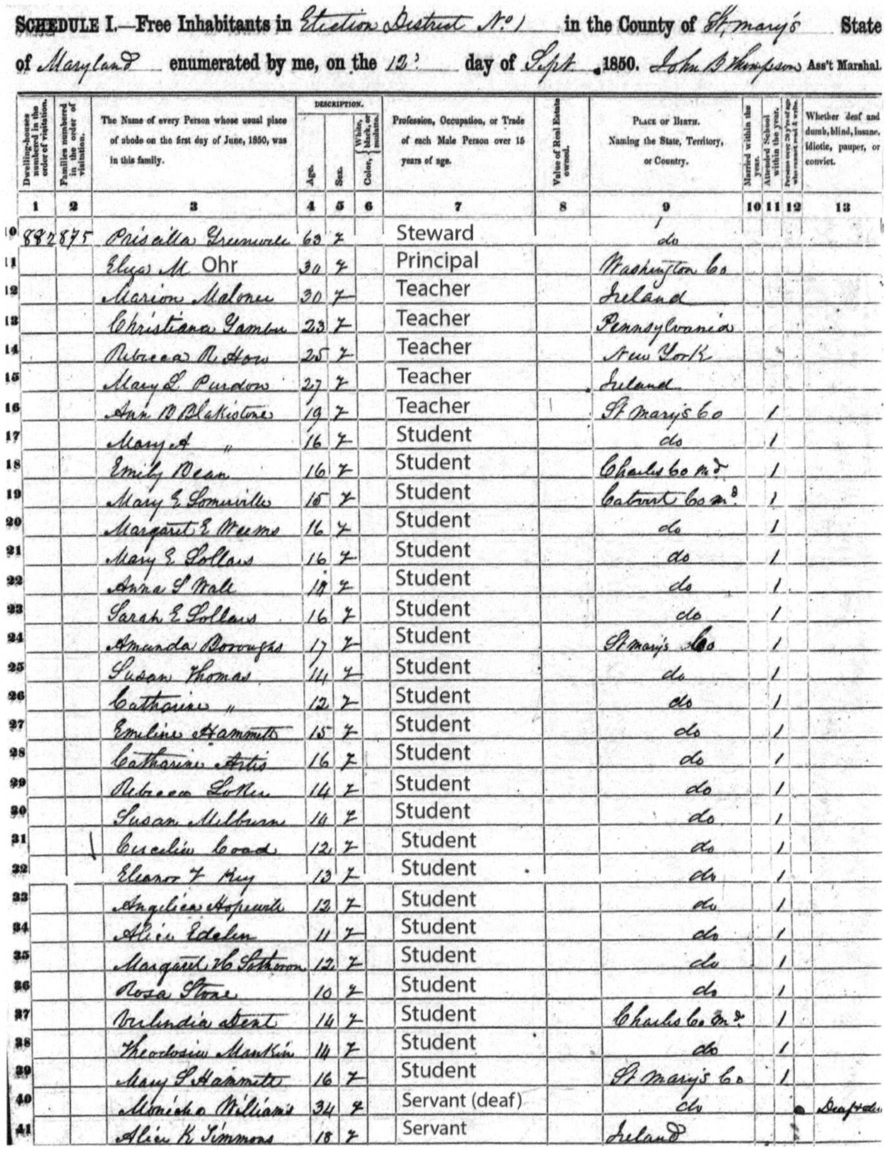

SCHEDULE I.—Free Inhabitants in Election District No 1 in the County of St Mary's State of Maryland enumerated by me, on the 12th day of Sept 1850. John B Thompson Ass't Marshal.

	Dwelling-houses numbered in the order of visitation.	Families numbered in the order of visitation.	The Name of every Person whose usual place of abode on the first day of June, 1850, was in this family.	Age.	Sex.	Color, (White, black, or mulatto.)	Profession, Occupation, or Trade of each Male Person over 15 years of age.	Value of Real Estate owned.	Place of Birth. Naming the State, Territory, or Country.	Married within the year.	Attended School within the year.	Persons over 20 y'rs of age who cannot read & write.	Whether deaf and dumb, blind, insane, idiotic, pauper, or convict.
	1	2	3	4	5	6	7	8	9	10	11	12	13
10	882	875	Priscilla Greenwell	63	F		Steward		do				
11			Eliza M Ohr	30	F		Principal		Washington Co				
12			Marion Maloney	30	F		Teacher		Ireland				
13			Christiana Gamber	23	F		Teacher		Pennsylvania				
14			Rebecca R How	25	F		Teacher		New York				
15			Mary L Purdon	27	F		Teacher		Ireland				
16			Ann B Blakistone	19	F		Teacher		St Mary's Co		1		
17			Mary A "	16	F		Student		do		1		
18			Emily Dean	16	F		Student		Charles Co Md		1		
19			Mary E Somerville	15	F		Student		Calvert Co Md		1		
20			Margaret E Weems	16	F		Student		do		1		
21			Mary E Sollars	16	F		Student		do		1		
22			Anna S Wall	18	F		Student		do		1		
23			Sarah E Sollars	16	F		Student		do		1		
24			Amanda Boroughs	17	F		Student		St Mary's Co		1		
25			Susan Thomas	14	F		Student		do		1		
26			Catharine "	12	F		Student		do		1		
27			Emeline Hammett	15	F		Student		do		1		
28			Catharine Artis	16	F		Student		do		1		
29			Rebecca Loker	14	F		Student		do		1		
30			Susan Milburn	14	F		Student		do		1		
31			Cecelia Coad	12	F		Student		do		1		
32			Eleanor F Key	13	F		Student		do		1		
33			Angelica Hopewell	13	F		Student		do		1		
34			Alice Edelen	11	F		Student		do		1		
35			Margaret H Sothoron	12	F		Student		do		1		
36			Rosa Stone	10	F		Student		do		1		
37			Verlinda Dent	14	F		Student		Charles Co Md		1		
38			Theodosia Mankin	14	F		Student		do		1		
39			Mary S Hammett	16	F		Student		St Mary's Co		1		
40			Monica Williams	34	F		Servant (deaf)		do				Deaf & dumb
41			Alice K Simmons	18	F		Servant		Ireland				

Fig. 12 Handwritten data

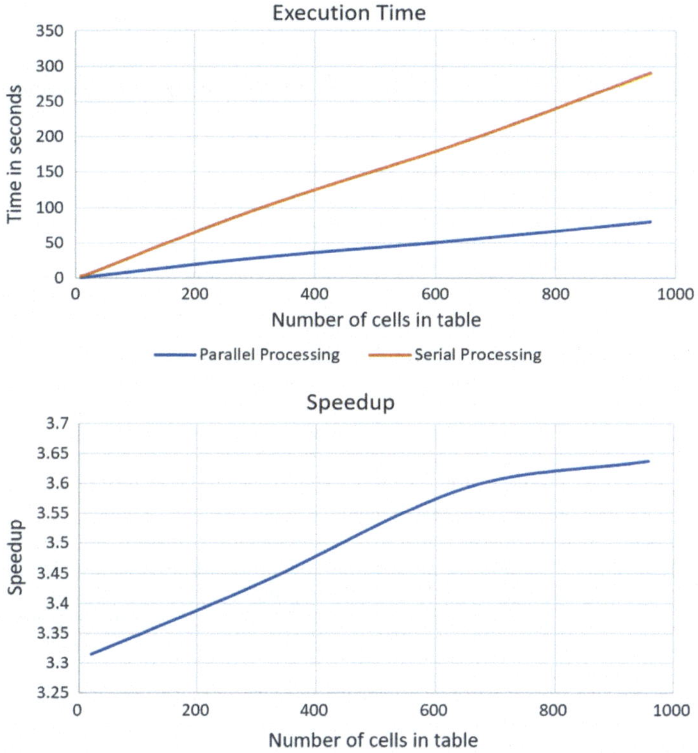

Fig. 13 Performance metrics

13 Conclusion

We studied the fundamentals of OCR, the objectives, requirements and implementation as well. Preprocessing is an important step in the OCR system. It enhances the image using standard image processing techniques such that it becomes easier for the OCR system to recognize the text. If the input image contains tabular data, the processing becomes even more complex if the tabular structure is to be preserved. Contours can help in tabular data extraction if borders of table are visible but fail when the borders are not visible. In that case we need a text detection algorithm to locate the text and assume the size of the cells. The cells are individually fed to the OCR system for text recognition. This process can be slow if huge amounts of tabular data is to be processed. For this, we use the concept of multi-threading to process the cells in parallel to improve performance. We found the results were good and the system is reliable.

References

1. Grother, P. J. (1995). NIST special database 19. Handprinted forms and characters database. National Institute of Standards and Technology.
2. LeCun, Y., Cortes, C., & Burges, C. J. C. (1998). *The MNIST Database of Handwritten Digits* (vol. 10, p. 34). http://yann.lecun.com/exdb/mnist.
3. Mouchere, H., et al. (2011). Crohme2011: competition on recognition of online handwritten mathematical expressions. In: *2011 International Conference on Document Analysis and Recognition*. IEEE.
4. Acharya, S., Pant, A. K., & Gyawali, P. K. (2015). Deep learning based large scale handwritten Devanagari character recognition. In: *9th International Conference on Software, Knowledge, Information Management and Applications (SKIMA)*, p. 2015. IEEE.
5. Zhou, S., Chen, Q., & Wang, X. (2010). HIT-OR3C: an opening recognition corpus for Chinese characters. In: *Proceedings of the 9th IAPR International Workshop on Document Analysis Systems*. ACM.
6. Wang, K., & Belongie, S. (2010, September). Word spotting in the wild. In: *European Conference on Computer Vision* (pp. 591–604). Berlin: Springer.
7. Netzer, Y., Wang, T., Coates, A., Bissacco, A., Wu, B., & Ng, A. Y. (2011). Reading digits in natural images with unsupervised feature learning.
8. Nagy, R., Dicker, A., & Meyer-Wegener, K. (2011, September). NEOCR: a configurable dataset for natural image text recognition. In: *International Workshop on Camera-Based Document Analysis and Recognition* (pp. 150–163). Berlin: Springer.
9. Lee, S., Cho, M. S., Jung, K., & Kim, J. H. (2010, August). Scene text extraction with edge constraint and text collinearity. In: *2010 20th International Conference on Pattern Recognition* (pp. 3983–3986). IEEE.
10. Yao, C., Bai, X., Liu, W., Ma, Y., & Tu, Z. (2012, June). Detecting texts of arbitrary orientations in natural images. In: *2012 IEEE Conference on Computer Vision and Pattern Recognition* (pp. 1083–1090). IEEE.
11. Indermühle, E., Liwicki, M., & Bunke, H. (2010, June). IAMonDo-database: an online handwritten document database with non-uniform contents. In: *Proceedings of the 9th IAPR International Workshop on Document Analysis Systems* (pp. 97–104). ACM.
12. Liwicki, M., & Bunke, H. (2005, August). IAM-OnDB-an on-line English sentence database acquired from handwritten text on a whiteboard. In: *Eighth International Conference on Document Analysis and Recognition (ICDAR 2005)* (pp. 956–961). IEEE.
13. de Campos, T. E., Babu, B. R., Varma, M. (2009, February). Character recognition in natural images. In: *Proceedings of the International Conference on Computer Vision Theory and Applications (VISAPP)*, Lisbon, Portugal.
14. Bieniecki, W., Grabowski, S., Rozenberg, W. (2007, May). Image preprocessing for improving OCR accuracy. In: *2007 International Conference on Perspective Technologies and Methods in MEMS Design* (pp. 75–80). IEEE.
15. Shapiro, L., & Stockman, G. (2001). *Computer Vision*. Upper Saddle River: Prentice Hall.
16. Sezgin, M., & Sankur, B. (2004). Survey over image thresholding techniques and quantitative performance evaluation. *Journal of Electronic Imaging*, *13*(1), 146–166.
17. Otsu, N. (1979). A threshold selection method from gray-level histograms. *IEEE Transactions on Systems, Man, and Cybernetics*, *9*(1), 62–66.
18. Serra, J., & Soille, P. (Eds.). (2012). *Mathematical Morphology and Its Applications to Image Processing* (Vol. 2). Heidelberg: Springer.
19. Srisha, R., & Khan, A. (2013). Morphological Operations for Image Processing: Understanding and its Applications.
20. Bertalmio, M., Sapiro, G., Caselles, V., & Ballester, C. (2000, July). Image inpainting. In: *Proceedings of the 27th Annual Conference on Computer Graphics and Interactive Techniques* (pp. 417–424). ACM Press/Addison-Wesley Publishing Co.

21. Sasaki, K., Iizuka, S., Simo-Serra, E., & Ishikawa, H. (2017). Joint gap detection and inpainting of line drawings. In: *Proceedings of the IEEE Conference on Computer Vision and Pattern Recognition* (pp. 5725–5733).
22. Canny, J. (1986). A computational approach to edge detection. *IEEE Transactions on Pattern Analysis and Machine Intelligence, 6*, 679–698.
23. Maire, M. R. (2009). *Contour Detection and Image Segmentation*. Berkeley: University of California.
24. Bradski, G., & Kaehler, A. (2008). Learning OpenCV: Computer Vision with the OpenCV Library. O'Reilly Media, Inc.
25. Zhou, X., et al. (2017). EAST: an efficient and accurate scene text detector. In: *Proceedings of the IEEE Conference on Computer Vision and Pattern Recognition* (pp. 5551–5560).
26. Redmon, J., Divvala, S., Girshick, R., & Farhadi, A. (2016). You only look once: unified, real-time object detection. In: *Proceedings of the IEEE Conference on Computer Vision and Pattern Recognition* (pp. 779–788).
27. Liu, W., et al. (2016, October). SSD: single shot multibox detector. In: *European Conference on Computer Vision* (pp. 21–37). Cham: Springer.
28. He, K., Gkioxari, G., Dollár, P., & Girshick, R. (2017). Mask R-CNN. In: *Proceedings of the IEEE International Conference on Computer Vision* (pp. 2961–2969).
29. Ren, S., He, K., Girshick, R., & Sun, J. (2015). Faster R-CNN: towards real-time object detection with region proposal networks. In: *Advances in Neural Information Processing Systems* (pp. 91–99).
30. Smith, R. (2007, September). An overview of the Tesseract OCR engine. In: *Ninth International Conference on Document Analysis and Recognition (ICDAR 2007)* (vol. 2, pp. 629–633). IEEE.
31. Karaa, W. B. A., Ashour, A. S., Sassi, D. B., Roy, P., Kausar, N., & Dey, N. (2016). Medline text mining: an enhancement genetic algorithm based approach for document clustering. In: *Applications of Intelligent Optimization in Biology and Medicine* (pp. 267–287). Cham: Springer.

Few Shot Learning for Medical Imaging

Jai Kotia, Adit Kotwal, Rishika Bharti, and Ramchandra Mangrulkar

Abstract While deep learning systems have provided breakthroughs in several tasks in the medical domain, they are still limited by the problem of dependency on the availability of training data. To counter this limitation, there is active research ongoing in few shot learning. Few shot learning algorithms aim to overcome the data dependency by exploiting the information available from a very small amount of data. In medical imaging, due to the rare occurrence of some diseases, there is often a limitation on the available data, as a result, to which the success of few shot learning algorithms can prove to be a significant advancement. In this chapter, the background and working of few shot learning algorithms are explained. The problem statement for few shot classification and segmentation is described. There is then a detailed study of the problems faced in medical imaging related to the availability of limited data. After establishing context, the recent advances in the application of few shot learning to medical imaging tasks such as classification and segmentation are explored. The results of these applications are examined with a discussion on its future scope.

Keywords Few Shot Learning · Medical Imaging · Meta Learning · Image Classification · Image Segmentation

---

J. Kotia and A. Kotwal—Both authors have contributed equally to this chapter.

J. Kotia (✉) · A. Kotwal · R. Bharti · R. Mangrulkar
Dwarkadas J. Sanghvi College of Engineering, Vile Parle, Mumbai, India
e-mail: jaikotia10@gmail.com

A. Kotwal
e-mail: adit.kotwal29@gmail.com

R. Bharti
e-mail: bhartirishika@gmail.com

R. Mangrulkar
e-mail: ramchandra.mangrulkar@djsce.ac.in

S. K. Das et al. (eds.), *Machine Learning Algorithms for Industrial Applications*, Studies in Computational Intelligence 907,
https://doi.org/10.1007/978-3-030-50641-4_7

1 Introduction

Advancements in the field of deep learning systems have been beneficial to a lot of other disciplines beyond computer science. One of the most promising applications for these deep learning systems is in the field of biomedicine. Deep learning methods have successfully been used for various tasks in biomedicine such as prediction, early detection of diseases and especially in biomedical image analysis [1]. In [2], the authors demonstrated a deep learning algorithm surpassing human level accuracy in a classification task. Medical image analysis can be quite complex even for trained professionals as it requires intricate understanding and study of details in the images. It would take someone years of experience to become an expert at a task such as identifying the class of brain tumor or segmenting and annotating human organs. In such a scenario, the introduction of deep learning systems is of great assistance as it can learn at a much faster pace from data and form complex internal representations. This ability of the model to represent data according to the objective (for example, classification) allows the model to achieve high levels of accuracy, often competent with human accuracy.

In light of such success, the application of deep learning systems for medical image analysis has seen a rapid rise [3]. But there are still certain drawbacks that need to be addressed. One of the biggest reliance of deep learning systems is on the amount of available labeled data and class imbalance [4, 5]. The models require a large amount of data to train on. Often, the success and accuracy of models is largely dependent upon the size of the training data. While this might not be a problem for some use cases, in medical image analysis, there is a range of rare medical conditions for which there is insufficient data available. It is difficult to acquire images for certain rare diseases due to which, deep learning image analysis for such tasks does not yield good results. Even if an extensive effort is put into acquiring images for all conditions, often it is not exactly a feasible and viable option. A natural method to counter this problem is to reduce the dependency of deep learning systems on data. But this is not that easy and requires an approach similar to how humans can infer from fewer data.

Few shot learning aims to represent the limited amount of available data such that it can generalize entire classes of data. This involves identifying the key features that distinguish the classes. While convolutional neural networks are successful in such representations given a large data set, they tend to be heavily dependent on the data that it trains on. This dependency is what few shot learning overcomes by focusing on a general representation, rather than specific. In subsequent sections of the chapter, there will be an introduction and explanation of few shot learning algorithms. The different approaches to solving this problem are presented and once there is enough understanding of the few shot learning algorithm, its use cases in medical imaging are discussed.

As suggested by the name, few shot learning is a technique used in the training of a model by providing it with very small amounts of data. This practice is different from the norm which generally uses large quantities of data in facilitating model training for better accuracy in prediction. The learning technique proposed in the chapter is stated as follows: A model is subjected to a sample that belongs to a previously unseen class. The model also has access to a support set S which comprises of n examples from k different unseen classes. The ability of the model to perform few-shot learning is measured by whether or not the model is able to determine which support set the sample query is associated with.

The chapter is organized into four main parts. First, the introduction and related works are explained. This introduces the reader with sufficient background for this chapter. The next part elaborates on the few shot learning problem, including problem statement definitions for classification and segmentation tasks. The other subsections in this part describe the various few shot learning algorithms and their working. This part also briefly delineates the problems faced in medical imaging that are pertinent to few shot learning. The third part of this chapter provides an overview of some of the applications of few shot learning that have been applied to medical images, with discussions about their performance. The final part of the chapter discusses the future scope of few shot learning in medical imaging and an overview of the chapter is presented in the conclusion.

2 Related Work

Currently, few shot learning algorithms are a very active research area with encouraging improvements in performance. In one of the first works on few shot learning, Miller et al. [6] used shared densities on transforms to derive a one shot classifier. The seminal work in few shot learning was authored by Li et al. [7, 8], which used a Bayesian approach for few shot image classification. Lake et al. [9] proposed the use of a generative model to learn visual concepts. Over the last few years, there has been a rapid rise in interest in few shot learning and as a result, several new techniques have been proposed. Koch et al. [10] proposed the use of Siamese neural networks for few shot learning, that functioned on the principle of ranking similarity between inputs. Santoro et al. [11] adopted the use of memory-augmented neural networks for one shot learning task. The matching networks were proposed by Vinyals et al. [11] that significantly improved state of the art in one shot learning by using an attention mechanism. Ravi and Larochelle [12] introduced a LSTM based meta learning for few shot learning. In [13], Finn et al. suggested the fine tuning of model parameters such that the method is transferable to any model that is trained with gradient descent. Snell et al. [14] proposed a prototype based representation for few shot learning. Relational networks introduced by Sung et al. [15] performs few shot classification by computing relation scores between query images and the input images. In [16], the focus is especially on training algorithms to remember rare occurrences, something that will be directly helpful for image analysis of rare medical conditions. For few shot image segmentation, Shaban et al. [17] suggested a successful initial approach that required semantic supervision. Rakelly et al. suggested the use of conditional [18] and guided [19] networks for few shot image segmentation.

In 2019 itself there were multiple new approaches that were introduced for few shot learning such as Variational Prototyping-Encoder [20], Differentiable Convex

Optimization [21], Edge-labeling Graph Neural Network [22], Task-Agnostic Meta-Learning [23], Meta-Transfer Learning [24], Category Traversal [25], Class-Agnostic Segmentation Networks [26], Saliency-guided Hallucination of Samples [27], Image Deformation Meta-Networks [28], use of multiple semantics [29] and Task-Aware Feature Embeddings [30].

For few shot learning classification of medical images, Puch et al. [31] have used Deep Triplet Networks for recognizing brain modalities. Further, Kim et al. [32] used few shot learning for Glaucoma diagnosis with encouraging performance. In [33], meta learning is used to classify the condition of diabetic retinopathy.

Few shot medical image segmentation was performed by Ronneberger et al. [34], using the U-Net with promising results. Lahiani1 et al. [35] segmented condition of tissue using color deconvolution deep neural networks. In [36], the authors used *squeeze & excite* blocks to segment volumetric images. Incorporating the use of Generative Adversarial Networks [37], in [38] 3D volumes of multiple modalities are segmented. Zhao et al. [39] present data augmented networks, that synthesize more samples for training and performing one shot biomedical image segmentation.

3 Few Shot Learning

The success of several deep learning systems is dependent on the amount of data available. This is a significant drawback that limits its application in cases where there are insufficient amounts of training data available to train the model on. In light of this problem, there has been a lot of promising research in few shot learning algorithms. Few shot learning algorithms, as the name suggests, are primarily tasked with providing accurate results with a limited amount of data. The few shot learning algorithms are expected to represent small amounts of data in such a way, that it can generalize to represent a much broader range of data.

As discussed in [40] the few shot learning problem can be compared to semi-supervised learning, imbalanced learning, transfer learning and meta learning.

In the following subsections, there is a complete explanation of the few shot learning problem and architecture.

3.1 Problem Definition

Few shot learning refers to the availability of a limited amount of data, from which the model is expected to make accurate inference for tasks. The problem of few shot learning is often also referred to as low shot learning. In a similar vein, the one shot learning problem explicitly states that only one training sample is available. The approach for this method is similar to that of few shot learning and thus it is included within the scope of this chapter.

The few shot learning problem definition needs to be expressed for two applications with different goals. First, the few shot classification problem is defined followed by the few shot segmentation problem.

3.1.1 Few Shot Classification

In an image classification problem, the system is given a set of images as input and is expected to classify them into respective classes. The data is usually labeled, with each label indicating the class the image belongs to. Computer systems have become very good at this task, with models displaying very high accuracy. In few shot classification though, the model is expected to provide high accuracy with the support of very few input samples. This task becomes harder with less data, as the model has very few samples of each class from which to extract defining features. In classification, the models focus on identifying features of images, that make it distinct to the class which it belongs to. With a larger sample, this task becomes easy to generalize a set of identical features, but with less data, it is challenging to make such inferences. For few shot learning, image classification is one of the primary focus areas, as due to its general success, it should be relatively easier to replicate with lesser data. There are already very advanced models that have been developed specifically for classification and thus it provides a good starting base for few shot algorithms to experiment.

The few shot classification problem is generally of the form K-shot N-way. Here K refers to the number of training samples available for the model to learn from and N refers to the number of classes that the sample may belong to.

Each time the model is given input data to classify, it is referred to as a sample task T. To illustrate, consider an example 5-shot 2-way classification problem. Here, the model will train on 5 training samples and will have to label each sample as belonging to one of two classes (say class A or class B).

The elements in the training data can be represented using the following notation,

$$S_{\mathrm{T}} = \text{Support set} \tag{1}$$

$$Q_{\mathrm{T}} = \text{Query set} \tag{2}$$

The distribution of the set of tasks is denoted as $P(t)$. Further, the joint distribution of T_{i} from $P(t)$ is denoted as,

$$P^{T_i}_{X,Y}(x, y) \tag{3}$$

Here T_{i} is to predict y given x. For each task T_{i},

$$T = (S_{\mathrm{T}}, Q_{\mathrm{T}}) \tag{4}$$

where,

$$S_{\mathrm{T}} = S^S_T \cup S^U_T \tag{5}$$

Table 1 Data distribution based on the type of learning

Learning method	Samples	Labels
Supervised	$S_T^S, S_T^S \neq \emptyset$	$S_T^U, S_T^U = \emptyset$
Semi-supervised	$S_T^S, S_T^S \neq \emptyset$	$S_T^U, S_T^U \neq \emptyset$
Unsupervised	$S_T^S, S_T^S = \emptyset$	$S_T^U, S_T^U \neq \emptyset$

Thus the objective can be written as,

$$\text{Given:} S_{\text{T}}, \text{Minimize:} Q_{\text{T}}. \tag{6}$$

Now that the objective for the few shot classification problem is defined, the type of learning is taken into consideration to further expand on the problem definition. This is delineated in Table 1.

3.1.2 Few Shot Segmentation

In few shot segmentation problem, for a given set of inputs, the model has to accurately segment the image into constituent parts and label them. The segment annotations are provided with each image and they are stored as a pair. Computer vision is laden with image segmentation tasks making it very frequent and important in the domain. Image segmentation refers to the process of partitioning a digital image into multiple sections so as to simplify the analysis of the image by processing the important parts with useful information. The human brain is very efficient in focusing on important details of the image fed by the eye and deriving information from it. Consider the example of a binary classification task to identify the images containing cats and those containing dogs. Humans can identify the animal in the pictures in a matter of seconds. But even for a small test set, machine learning a deep learning algorithms require large amounts of labeled samples and training time to make a meaningful inference. However, recent advances in deep learning and associated fields have enabled technology to advance in computer vision.

The few shot learning task requires a labeled subset of the data for support. This comes in use for supervising the task at hand. Few shot prediction is then performed on a new class label or query which had not been encountered in the supervised data. A few shot learning task can be parameterized by two variables and defined as K-way, S-shot learning problem. This works well for classification tasks but in the case of segmentation, additional considerations are required. To the aforementioned set-up, pixel dimensions for each image are also added up in order to account for the total number of labeled support pixels in each image and the total count of the images that are labeled and used for supervising the task. The labeled pixels per image is denoted mathematically by P and the total number labeled support images by S. Then, in various simulations, different values of P and S are chosen and the (P, S) shot learning is performed.

Usually, the focus is given to smaller values of P for the task because they are relatively easier to find and their only requirement is another annotator to specify the point of interest.

The few shot learning segmentation task can be represented in the form of the input-output pair denoted by (τ_i, Y_i). This has been sampled from a class distribution ϕ. The inputs given to the few shot learning tasks are as follows,

$$\tau = (x_1, L_1), ...(x_s, L_s), (\bar{x_1}, ...\bar{x_Q}) \tag{7}$$

$$L_s = (p_j, l_j) : j\epsilon 1....P, l\epsilon 1....K \tag{8}$$

Here S denotes the total number of labeled support images. The labeled images are represented as x_s. Q denotes the unlabeled set of images represented by $\bar{x}_q$. The labels are represented by point-label pairs (*p*, *l*). The task outputs are represented as $Y = (y_1, ..., y_Q)$ where $y_q = (p_j, l_j) : j\epsilon\bar{x_q}$.

In image segmentation using few shot learning, every task is considered to give a binary result. Each task has its own positive output and the negative, also known as the complement is the background of the image segmentation task. The particular image segmentation tasks in which only a single image is to be segmented like video object segmentation and interactive segmentation generally choose the binary set-up for problem definition. Although this is the case, even higher order tasks can follow the binary image segmentation model.

3.2 *Image Augmentation*

Several medical imaging systems incorporate augmentation of images in order to increase the training data. As a result, image augmentation plays a very important role in few shot learning algorithms. If similar images can be synthesized, it essentially increases the amount of training data available for the model.

For successful image augmentation, the initial data set chosen must be of very high quality. It should also be able to generalize well to the rest of the class. This can ensure that by making minute changes to the initial set of images, a larger set of augmented images can be derived that would be classified as belonging to the same class. The augmentation can be performed by applying simple transformations such as rotation and scaling. More complex augmentation methods involve mapping the spatial transformations of images [39].

3.3 Meta Learning

Few shot learning is closely related to another class of algorithms known as meta learning algorithms. Meta learning algorithms aim to learn to perform a range of tasks, by training on a limited number of tasks. This allows meta learning algorithms to learn to learn on its own [41]. Meta learning is commonly used to tackle the few shot learning problem [42] as both the algorithms aim to generalize representation of tasks using limited input information.

An accurate machine learning model requires large amounts of training data in order to be able to make accurate predictions. As opposed to that, a human can learn to discern between objects at a relatively faster rate with a much smaller number of instances to learn from. For example, a small child can learn to distinguish between a cat and a dog by only observing each of the kind a few times. Meta learning aims to emulate this behavior and incorporates the method of learning new concepts with fewer examples. Some meta learning models aim to obtain a very optimized version of a neural network's architecture while others are focused on finding the right data sets to train the model on.

A good meta learning model is capable of adapting to situations previously not encountered during training [43]. Eventually, the model should be able to accomplish new tasks. Hence the term, *learning to learn*. These tasks intended for completion can range from wide variety of problems which can include supervised, unsupervised or even reinforcement learning. For instance, an agent completing a maze during testing even though it was trained on only straightforward maps of the environment at the time of training is an example of meta learning.

The primary step of meta learning involves the characterization of the type of data [44]. This can be done using statistical measures to gain a better understanding of the underlying meta data or a concept known as land-marking that makes use of a set of simple learners, with inherently different mechanisms, to characterize various data sets where each of the learners can be regarded as an expert. The next step involves mapping the data sets to predictive models. This may include actual handcrafting of the predictive rules or making use of a technique called ranking where a data set is mapped to multiple models before a final satisfactory model is ready. In order to then simulate the 'learning process' is to transfer the meta knowledge across domains. This process of incorporating the meta knowledge into the learning task is known as Inductive transfer.

As shown in Fig. 1, the architecture of meta learning includes a meta-data store and an extractor for the features. Important distinguishing features from test data sets are extracted and stored which are then used later while analyzing the features of a new data set.

In this chapter, the focus will be on the desired task to be a supervised learning problem such as image classification. One way to achieve this is by using a meta learning technique known as neuro-evolution which makes use of evolutionary algorithms to learn neural architectures. This procedure aims to capture the learning process in order to make the network more versatile. According to the Darwinian theory of evolution,

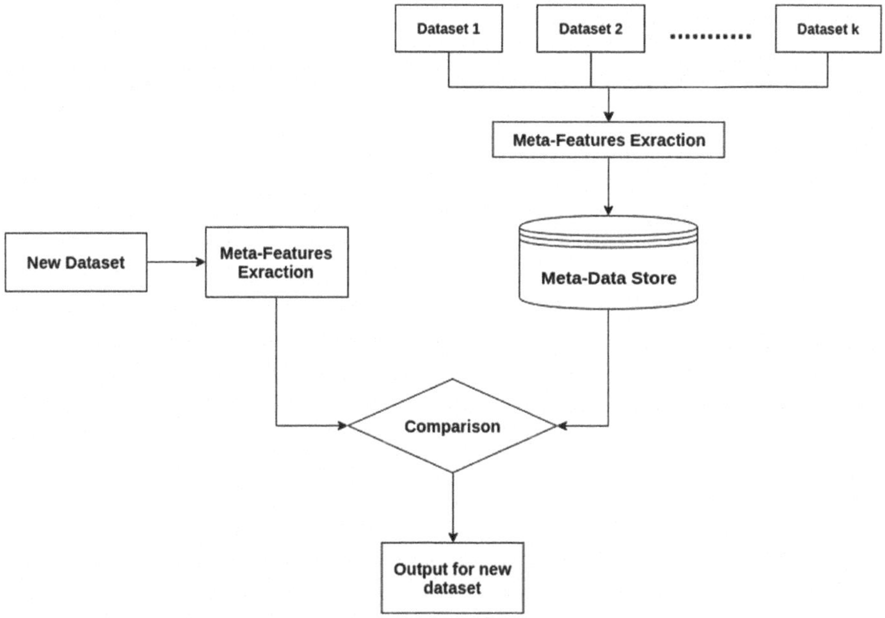

Fig. 1 Meta learning architecture

newer generations of living beings adapt better to their surroundings as compared to their ancestors due to slight variations in their DNA which is basically a blueprint that guides the replication of cells. The biological systems reserve some aspects of the parent generation which helped them survive in the first place but at the same time, they replace the negative features with ones that could enhance the capabilities of future generations. If a similar process can be applied to determine the best algorithm among a pool of several, a model which is most immune to changes in the environment can be obtained. To begin with, several algorithms that have been trained on different data sets are chosen to form part of the initial population. Each algorithm is assigned a score based on a threshold function and the ones with the highest score progress into newer environments (future generations). Some algorithms with a poor score may also be randomly kept in order to avoid getting stuck in the local maxima and can also aid in discovering an even more efficient algorithm, similar to mutation. Based on these permutations and combinations, a strong model is obtained based on the most positive features of the primary models. This evolutionary framework can also be introduced into weight training of the neural network in order to obtain the most optimal set of weights in each neuron and make the network more flexible.

3.3.1 Prototypical Networks

Consider a neural network is trained for a binary classification task. For instance, given an image, check if the image contains a car. This task requires training the network with pictures having a car and pictures without it. But the drawback of this approach is that the network has to be fed several images without a car as well, which would contain a lot of useless features from a broader spectrum of objects that will need processing, resulting in unnecessary use of computing power and time. In order to tackle this problem, only positive classes should be examined instead of taking into account even the negative ones. In this way, the network will be able to detect the most distinguishing features of a car and still produce the desired results of classification.

Prototypical networks [14, 45] work on this exact principle where they only require positive support examples from which they construct a prototype with the most important features that will be comparable to unseen instances in the future. An internal embedding of the whole class is created which is a low dimensional representation of the discrete data. An embedding in a neural network is the summary of the whole class being considered for classification and meaningfully represent it in the transformed space. This means a large number of training images containing a car (possibly thousands) can be taken and each one can be represented using only a single vector. This method of embedding also overcomes the drawbacks of representing categorical variables and can also represent similar entities closer to one another in the vector space. These embeddings can be further improved using neural networks that can minimize the loss on the task.

The prototype of each categorical class can be represented in the form of an M-dimensional vector c_k given by the equation:

$$c_k = \frac{1}{|S_k|} \sum_{(x_i, y_i) \in S_k} f_\phi(x_i) \tag{9}$$

where S_k denotes the positive support examples for the classes labeled 'k', each x_i in R^d is the D-dimensional feature vector and each y_i is the corresponding label.

As shown in Fig. 2, the prototypical network then calculates a centroid of the support classes in question which corresponds to the mean of their embedded instances. Once the network has been trained, the query embedding for the new unseen examples is compared with the internal embedding of the whole class. The attributes of these, after dimensionality reduction, are compared to the embedding vector to check for a similarity measure based on which the instance will receive a positive score for that category.

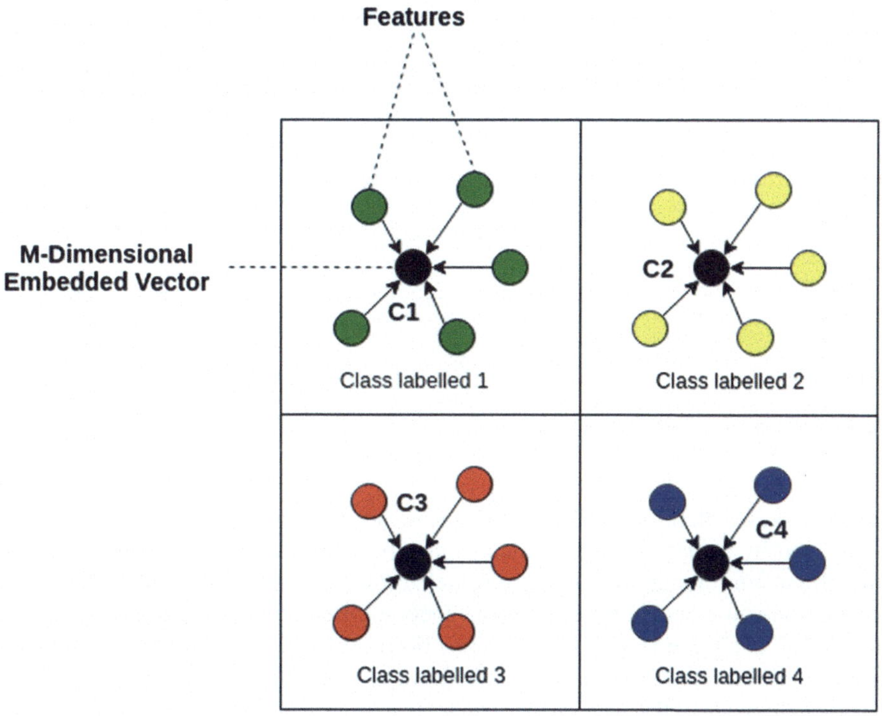

Fig. 2 Embedding in prototypical networks from a given set of features

3.3.2 Matching Networks

Matching networks [11] work on the principle of augmentation with the help of memory. The network attempts to learn the working of a classifier from only a few number of examples. A traditional neural network requires hundreds, possibly even a thousand examples to learn from. While on the other hand, the k-nearest neighbours algorithm requires no training whatsoever. Matching networks attempt to blend these two extremes with the help of an attention mechanism that is used to access the memory.

The network is initially fed a support set depicted as $S = (x_i, y_i)_{i=1}^{k}$ where x_i is the input image and y_i is the corresponding label. A classifier is then defined as $c_s(x)$ on which the support set is mapped. When this network encounters a new example $\hat{x}$, the probability of this belonging to a particular class of the labels needs to be obtained, which is given by $P(\hat{y}|\hat{x}, S)$. Mathematically, the problem can be stated as finding the output category with the highest probability. The estimated output label is computed using the given equation

$$\hat{y} = \sum_{i=1}^{k} a(\hat{x}, x_i) y_i \tag{10}$$

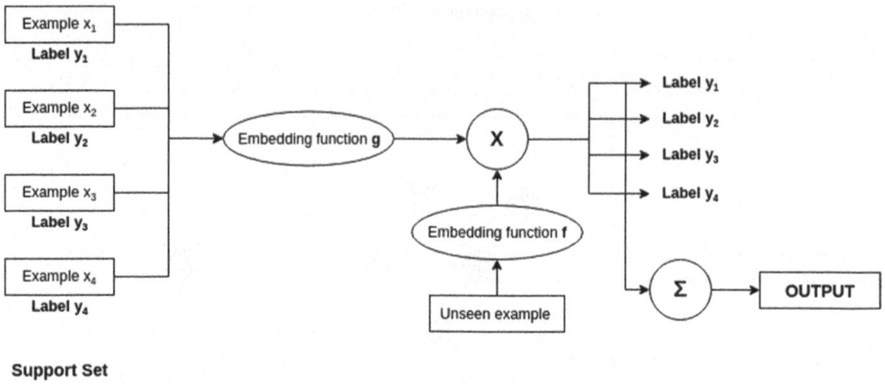

Fig. 3 Architecture of matching nets

where 'a' is the attention mechanism that specifies how comparable is $\hat{x}$ to x_i. This is achieved by computing the cosine distance between the two and then passing it through a softmax function in order to normalize the output between 0 and 1. The attention mechanism makes further use of an encoding function 'g' which is essentially modeled as a bidirectional LSTM. This function considers the full set S in addition to each individual x_i and becomes $g(x_i, S)$ allowing it to embed x_i optimally based on the other elements of the set.

Consider a small batch B to be used while training given only a small number of instances. The matching network then works to minimise the error based on the conditions supplied by the embedding of the support set S by the function g. As shown in Fig. 3, this translates to a form of meta learning where the network learns to learn from a given set to reduce the loss function over a batch.

3.3.3 Relational Networks

Relational networks (RNs) [15] are neural networks that have been augmented with the help of a plugin component that aims to tackle problems based on relational reasoning. The architecture of this network is such that it captures the essence of reasoning. The major role of an RN is to infer the manner in which two input objects are related. The composite function of an RN is given below.

$$RN(O) = f_\phi \sum_{i,j} g_\theta(o_i, o_j) \tag{11}$$

where $O = o_1, o_2, ...o_n$ represents a set of objects which are applied to the function f_ϕ and g_θ.

The function g_θ is used for the calculation of the relation between object pairs. Consider a traditional Multi Layer Perceptron model that would collect all objects from the input set at once and would have to perform dimensionality reduction making the complexity n^2 as opposed to the relational network where it will consider each object pair only once making it a more data efficient network. Another salient feature of these networks is that they are not dependent on the order of the objects present in the input set and also ensures at the same time that the output is order invariant as well. This ensures that the information produced is an accurate representation of the relations between objects.

A relational network does not explicitly require only images or only text. It only requires embeddings in the form of objects as input. The semantics of what the object represents need not be taken into consideration making the model even more versatile.

3.3.4 Model-Agnostic Meta-Learning

MAML stands for Model-Agnostic Meta-Learning [13] which translates to a model that is quite flexible and makes considerations for a wide range of problems. This method aims to achieve swift adaptability as and when more data becomes available. The meta learning phase involves training only a small number of instances in such a manner that it also avoids over-fitting due to a small data set. The final goal of this is the minimization of loss functions for new tasks.

Consider θ to be the model's parameters that have been randomly chosen for a particular task chosen from a support set during the training phase. The model is then trained using a certain number of specimens for that task following which the feedback is generated. Next, the model's performance is evaluated using a test set and its parameters are improved. The fine-tuned parameters for this task are given by:

$$\theta_i^{'} = \theta - \alpha\delta_\theta L_{T_i}(f_\theta) \tag{12}$$

where α is the learning rate, L is the loss function for the task T_i and f_θ is the model

The model is trained batch-wise consisting of a particular number of tasks. Before moving on to the next batch of tasks, the parameters of this model are updated.

4 Medical Imaging

Medical imaging encompasses a critical component of modern day treatments for several diseases. The use of these imaging services is paramount in assessing and documenting the progression of a particular disease leading to its eventual cure. A Computerized Tomography Scan, more commonly known as the CT Scan or the CAT Scan, can provide a three-dimensional view of several cross-sections of the body. These images are constantly used to guide physicians during delicate surgical procedures.

Medical imaging is an area with a wide array of applications within it. The primary problems discussed in this chapter are of image segmentation, where for example an image of the brain is to be segmented, and image classification, where images labeled as different classes are to be identified.

The general idea behind medical image analysis is to measure the most discriminating features of the image. The pixels which make up the entire image represent these features in the form of parameters such as X-ray attenuation, water density, acoustic impedance, electrical activity, etc. Images can be processed manually by humans or automatically with the help of computer techniques.

4.1 Procedure

Images obtained from certain procedures may sometimes diminish the contrast of the image leading to lesser contrast between the features under investigation. The quality of the images may also drop due to the presence of noise that creeps in during the measurement process because of which it is difficult for algorithms to perform well. Several techniques such as histogram equalization, wavelet shrinkage de-noising or speckle reduction can be employed to tackle this problem.

The most essential step in the entire process is segmentation which involves the separation of the structures of interest and can be viewed as a pre-processor for further analysis. It involves partitioning an image into regions that are consistent with respect to certain defining characteristics of a feature. Segmentation can be used to classify image pixels into anatomical or pathological areas in the body. In spite of several algorithms being proposed, there is no standard technique that can generalize the process and produce satisfactory results for all medical applications.

Segmentation can be done on a smaller region of the image (local segmentation) which involves manipulating and analysing fewer pixels or it can be applied to the image as a whole (global segmentation). Edge detection forms an essential part of this procedure which discovers the discontinuity between the objects that form the image. It works on the principle of recognizing large variation in intensity which generally depicts the border between the entities in question.

4.2 Problem Statement

The reason for the skepticism behind adopting image processing as means of diagnosing certain illnesses by the medical community is mainly the occurrence of false positives or negatives, the consequences of which may have a significant impact on the patient's health. Technically speaking, the problem of segmentation is notably higher as anatomical structures, captured from most devices, do not have well defined edges. Generalizing the structure of most organs in the human body is not an easy task as each individual may have a different structure and therefore building a model that can detect and segment such organs becomes an arduous problem. Constructing 3D images from various 2D images of a particular organ also requires expensive equipment with several calculations to be performed. To develop a system that can completely mimic what a radiologist observes and detects in the image is also difficult to emulate. Despite good accuracy being observed in standalone decision systems, there are several related concerns that prevent them from being adopted in real world hospitals or labs. For instance, consider the cost of labeling a cancerous tumor as benign in spite of the model displaying an accuracy greater than 95% on the training set. From the system's point of view, it is just one error out of possibly hundreds. If doctors were to completely rely on this system, they would not prescribe correct treatments for the actual cancerous tumor and the patient may eventually die due to incorrect diagnosis.

5 Few Shot Learning Approach for Medical Imaging

After studying the few shot learning algorithm and understanding the problems in medical imaging, there is sufficient context for discussing research where few shot learning algorithms are applied to medical imaging applications.

5.1 Classification Using Few Shot Learning

In this section, there will be discussion regarding the results obtained by authors that implemented few shot learning networks to classify medical images. Few shot learning approach has primarily used through meta learning algorithms for this task. Each work has been tested on a different category of medical image (two of them related to the eye) and as a result, provides a holistic view of its application.

5.1.1 Glaucoma Diagnosis

In [32], the authors have experimented with the use of few shot learning in order to diagnose cases of glaucoma from images of the fundus of the eye. The network used in this experiment is the matching neural network that was introduced earlier in the chapter. A key aspect in most medical imaging task is maintaining a high resolution image. Not only will it be beneficial for the network to train with more features available at disposal, but also it is requisite to maintain high standards in medical image quality for validation in practical scenarios.

As a result, the authors have used a high resolution CNN along with an attention mechanism. This means that the input data image will be decomposed in the form of an embedding. It acts as a representation of the input image that will help the CNN extract more detailed features. Further, without downscaling or altering the image in any manner, the images were centre-cropped in order to focus more on the distinguishing

Table 2 Accuracy comparison of various predictive models

Neural network	Resolution used	Accuracy
VGG-16	Low	65.2%
Inception ResNet V2	Low	89.5%
AM-CNN (without Aug)	High	88.1%
AM-CNN (with Aug)	High	87.9%

features present between the macula and the optic disc. While the authors did experiment with data augmentation to avoid over-fitting and introduce more data for the network to train on, it is something that may not be accepted as an ethical practice for use in medical diagnosis.

The results of 20-shot learning have been documented to display the highly significant accuracy as opposed to 1, 5 and 10 shot learning making use of twenty positive and twenty negative instances per category. This method is applied to previously existing architectures and the performance is compared with the new CNN making use of an attention mechanism (AM-CNN). The VGG-16 obtained the lowest accuracy even for the 20-shot approach while the Inception ResNet V2 displayed the highest. But the drawback of ResNet is that it requires a large amount of diversity to be introduced in the data which translates to augmentation. The AM-CNN also produced an accuracy higher than the human diagnosis accuracy of about 80% (Table 2).

5.1.2 Diabetic Retinopathy Detection

In [33], the authors have incorporated the use of meta learning in order to detect if a patient has diabetic retinopathy. The conditions are measured in the form of 5 forms of severity, effectively generating an N-shot 5-way classification task for few shot learning. In this short experiment, the Reptile model has been used to perform classification. This model relies on the ability to accurately initialize weights such that they can be transferred efficiently to succeed at similar tasks. It is similar to the MAML model that was discussed earlier in the chapter.

The mini-Imagenet data set was used to initialize the weights learned by the network. This training was performed for the 5-shot 5-way condition. The weights are then transferred and used to fine tune on the target data set.

The models are evaluated based on quadratics weighted kappa due to a large imbalance in the data. This metric is used to assess the amount of agreement between the predictions made by an algorithm and the trusted labels of the same objects. It takes into account the label being predicted "by-chance" based on the ratio of various instances of each category. This method can be applied to data-set with a large number of categories by assigning weights which operates on the supposition that some categories more similar than others and thus the credit for mismatching pairs of these categories should be divided partially.

In this work, only 50% of training data and 50% of validation data is used to implement meta learning due to the presence of a majority class that may inadvertently result in over-fitting the model. This is the optimum amount as a further reduction in the number instances may not produce a very robust output.

5.1.3 Brain Imaging Modality Recognition

With the rise of biomarkers obtained from rare image modalities, the authors in [31] have identified a need to work with fewer data. Biomarkers that are observed in image modalities aid in early diagnosis of diseases and in the treatment of patients. But the scarcity of some of these image modalities presents a crucial task of making the most out of the limited data.

To tackle this problem, the authors have suggested the use of a metric learning algorithm, Deep Triplet Networks. It involves generating an embedding of an anchor, true and false samples. It then learns a distance function that learns the similarities between the three images. Further, in order to obtain an optimized network architecture, the authors used grid search to find the ideal hyper-parameters.

This task involved training the triplet network on 150 slices of data. While this is a seemingly larger amount of data, it takes into account the complexity of the image modalities obtained from the different orthogonal axes. The comparison baseline was a deep convolutional neural network.

5.2 Image Segmentation Using Few Shot Learning

In the following section, there is a study of the work done in using few shot learning for segmenting medical images. Three different works have been included, each covering a different base. The first paper discussed involves segmenting of brain MRI scans. The second paper focuses on the segmentation of neuronal structures in electron microscopic stacks. Finally, the authors of the third paper perform the segmentation of 3D multi modal brain MRI scans.

5.2.1 Segmentation of Brain MRI Scans

A pioneering approach specific to magnetic resonance image (MRI) brain scan segmentation is introduced in [39]. This is used to combat the difficulty associated with labeling biomedical images in terms of expertise and time. This is a one-shot approach that requires only one segmented scan which is correctly labeled. The semi-supervised learning approach is designed as follows. To start with, only one labeled image is taken and along with that multiple unlabeled samples are also present. Using the labeled example, spatial and appearance transforms are applied to it in order to generate a new set of usable images. Through this, the transforms are able to capture the dissimilarities between the data-augmentation generated images and the original unlabeled ones in terms of features such as non-linear deformations and image intensity. The use of these synthesized images is to train a model under supervision.

$y^{(i)}$ denotes the set of biomedical images. To denote the set of references that are labeled and their corresponding segmentation maps by (x, l_x). Transformation τ^x is applied to the labeled images to perform data augmentation. Two transform models, namely, spatial and appearance models are used to acquire the difference between the tagged biomedical images and the unlabeled ones in terms of anatomy and appearance. By using the transformations from these two learned models, labeled data is then created. Using the setup, accurately labeled data is guaranteed due to the fact that both the label map and data volume are generated using the same spatial transformation and this was not the case in former procedures.

To define the spatial and appearance transform models, a voxel-wise displacement field denoted by u and identity function id are defined. Using the aforementioned functions, another function known as the deformation function is used and it is represented by the equation:

$$\phi = id + u \tag{13}$$

The equations that represent the spatial and appearance transformations respectively are given as,

$$\tau_s^{(i)} = x.\phi^{(i)} \tag{14}$$

$$\tau_s^{(i)} = x.\phi^{(i)} \tag{15}$$

Here the "." indicates the application of the transformation function. This paper uses an unsupervised learning approach which is a variation of VoxelMorph in order to learn for reducing the image similarity loss. Using this learning approach for data augmentation, the paper achieves one shot learning for brain MRI scan segmentation. This approach can also be extended to Computed Tomography images because the technique does not use any details that are exclusive to MRI images only.

When tested against the one-shot learning mechanism, the method elaborated in the paper exceeds the performance standards for every example in the test set. This architecture is particularly useful for segmentation in the medicine domain because large amounts of data cannot be labeled manually as a result of limited available time.

5.2.2 Segmentation of Neuronal Structures in Electron Microscopic Stacks

Although convolutional neural networks benefit greatly from large data sets, few shot learning has established an alternative path to high accuracy in image segmentation. As known, the human anatomy poses different modes of variation and as a result of this biomedical images exist with great amounts of variability. Automating the segmentation of these images requires large quantities of data that may not be available for a particular application. The U-Net [34] is a convolutional neural network designed to address this shortage of biomedical data using data augmentation techniques and the productive utilization of available data samples. The design of the U-Net can be split up into two parts- the contracting section and the expansive section. The contracting section comprises of two recurrent unpadded 3×3 convolutions followed by a rectified linear unit and a max pooling operation of size 2×2 and stride 2 for downsampling. By doubling the number of feature channels at every downsampling step the expanding step is established. The expanding path mirrors the contraction path by performing upsampling using 2×2 convolutions along with two 3×3 convolutions and a rectified linear unit. Finally, a fully-connected 1×1 convolution layer is used to map the feature vector to the class labels.

5.2.3 3D Multi-modal Image Segmentation of Brain MRI

Another scenario arises in the medical field with regard to 3D images. Segmentation of medical images is necessary to determine the exact location of certain parts and to correctly identify their precise structures. This, in turn, helps in the accurate diagnosis of diseases down to the finest details and thereby designing an effective treatment plan. The dearth of available labeled data makes the segmentation of these images difficult. Other challenges posed to the task of medical image segmentation include the low interpretability of the scanned images due to the noise points and low contrast between distinguishable regions. To add to that, detailed anatomical constructions are highly irregular in the images and the data generated as a result is varying in many respects. The work elaborated in [38] discusses a semi-supervised approach using adversarial learning which has gained momentum in recent years. A Generative Adversarial Network is trained on both segmented and unlabeled images.

The extant segmentation models such as the U-Net use completely labeled data to train the neural network for segmentation tasks and then use a pixel-based loss function such as cross-entropy for the procedure. However, in the model explained in the paper, the aim is to reduce the requirement of labeled data down to just a few samples. For the same purpose, unlabeled data is substituted in its place and the Generative Adversarial Network is used to create synthetically labeled images. All types of images including the annotated, un-annotated and fake ones generated by the GANs are made a part of the data set for training. A similar approach as used in classification tasks with the GANs is extended to classification. There exist a generator and discriminator along with K+1 classes. The additional class represents the set of fake images that the GAN has generated. Further, this model must also be modified to adjust to the 3D multi-modal images. One such change is the processing of these images in smaller sizes because they have larger computing and memory requirements.

Other changes that were added to the standard model include the weight normalization layer that was added in place of the batch-normalization layer to counter the adverse effects it had on semi-supervised training of the GANs. In addition to that, the rectified linear units were replaced by leaky rectified linear units to allow a small gradient benchmark for the neural units that produce an output below zero and are thereby inactive. Another noteworthy change that was made was the replacement of the max pooling operation with that of average pooling so as to prevent the hindrance during semi-supervised GAN training.

The design of the semi-supervised model that was presented in this paper was tested for results again completely supervised architectures. The testing was done on iSEG-2017 and MRBrains data sets. It was found that even with just a small number of training samples, this method was able to achieve equivalent results like those of the fully-supervised models. Another peculiar achievement mentioned in this paper is the ability of the model to eliminate over-fitting to some extent by being able to identify the actual and generated fake patches using adversarial training. In addition to that, this work is unique because it provides a new perspective to few shot learning without a starting pre-trained network by using the capabilities of the GANs.

5.3 Medical Image Retrieval

A majority of the literature regarding the use of deep learning for medical applications has been centred around the use of convolutional neural networks that are trained under complete supervision through the use of a large labeled data set. Another breakthrough design of a Siamese Convolutional Neural Network has been proposed in [46] which requires only binary image pair information for training. This training through less supervision is then tested on a content based medical imaging system that contains fundus images of diabetic retinopathy cases. After evaluating the results it was deduced that the Siamese Convolutional Neural Network produced results comparable to the other state-of-the-art models and as another benefit, it did not require large training time.

This Siamese Convolutional Neural Network is designed specifically for content-based image retrieval tasks which basically represents the idea of medical personnel identifying the end result of their diagnosis based on similar cases by gathering image related information from the electronic system. There had been limited introduction of deep learning in content based retrieval and it was restricted to specific types of images such as lung CT scans, prostrate MRIs and X-Rays. Another limitation of these techniques was their heavy dependence on good quality images along with ground labelling. To counter this, a Deep Siamese Convolutional Neural Network was designed.

A Deep Siamese Neural Network is a form on a convolutional neural network that is used to pinpoint how two input objects are related and whether they are identical or not. This network comprises of two completely identical sub-networks. In addition, these sub-networks also share weights with each other and that is how the term "Siamese" is coined. A siamese neural network works by learning the significant descriptors from the input which is fed into it and then compare the results generated by the two sub-networks.

Various experiments were conducted to evaluate the performance of this proposed network to see where it stands against the existing networks. It was first tested on a diabetic retinopathy data set of fundus or retinal images. Diabetic retinopathy is a common condition prevalent in developing nations. It is one of the leading causes of blindness and often occurs in patients suffering from long term diabetes mellitus. When not detected on time, it can eventually cause blindness and this fact makes a timely discernment very important. A total of 35,125 fundus images of the retina released by EyePacs, were used for these experiments. For the neural network training, the fundus images were labeled into five categories that signify the acuteness of the diabetic retinopathy from normal to severe.

Further, a series of image processing steps were performed on the images. To do away with the disparities caused by the camera and lighting conditions, firstly all images are brought down to the same size. Now that all images are of the same radius, the local average value of color is subtracted and the median ninety percent of images are retained. The sizes of the images are then uniformly changed to 224×224 pixels. Non-uniform distribution of data is countered by the use of data augmentation techniques. A large portion of the images was concentrated in the class normal and relatively fewer examples in the other classes. Techniques like the Krizhevsky style cropping using a random offset, random horizontal and vertical flipping, rotation through angles 0 to 360° and Gaussian blurring. Finally, combining images from all the classes the train and test split was a seventy to thirty ratio.

5.4 *Medical Text Classification*

There are different categories under which the labels of large data sets used for text classification can come under. This occurs in a multi-labeled data set scenario where many labels can be assigned to the same item. A particular label can be given to a large number of items which makes it's of the frequent type. There are other labels that are given rarely which come in the few shot category and the labels which are never present in the training data which is called the zero-shot category. Most notably in these text classification tasks, the few and zero shot groups are usually left undiscovered due to their scarcity in the training set. The research in [47] targets these labels in an attempt to evaluate the performance of various models on scant labels in the data set. A fine-grained evaluation of these labels in a large size and multi-labeled scenario provided by medical data sets like the MIMIC 2 and 3 which are freely available data sets containing around 60,000 vital signs, demographics, medicines, laboratory assessments and other attributes of the intensive care unit cases are done. The distribution of labels is as follows. A small number of labels have more than 10,000 occurrences. Some 5000 labels are only seen from 1 to 10 times total and 50% of these labels are not encountered during training.

The entire task is split into four major components. Firstly, a comprehensive descriptor of every label in English is required for the label to be predicted. In the next step, the words in each descriptor are used to generate word embeddings for that particular word. Further, these embeddings are averaged in order to create a vector representation for the label associated with the words. Secondly, the labeled vectors are used as attention vectors to find the most important n-grams in the document in terms of the information they contain. Every label leads to a distinct vector representation in the document. Following that, these vectors are passed through a Graph Convolutional Neural Network with two layers. The need for this is to culminate in the hierarchical information stored in the label space. The output vectors of the Graph CNN are then compared with the document's vector and generated predictions through that.

The exploration of infrequent labels through few shot and zero-shot learning techniques through a neural network architecture that uses the hierarchical structure of the has opened up an area of future work in the domain. One such possibility is the use of structured and unstructured information stored in all the articles that are indexed by PubMed which are hierarchically organized. Valuable information can be extracted from even the scarce labels using techniques such as transfer learning and multi-task learning. To delve into further details, this method can be extended to finding whether a particular item in the training data has an infrequent label or not. It can also be used to determine how many infrequent labels it should be annotated with.

6 Future Scope

Few shot learning algorithms have seen a constant evolution in terms of performance lately, as seen in the previous sections. This also implies that these algorithms have not completely developed yet. To help achieve this, defining an effective baseline is very crucial. If the models are tested against suitable criteria, it can help scientists design better algorithms. Chen et al. [48] and Dhillon et al. [49] have made efforts in delineating such baselines for comparison of few shot learning algorithms. Further, there need to be more thorough experiments performed for few shot learning on medical images. While there are some promising results currently, for use in a practical clinical setup, extensive testing and analysis need to be done. In medical applications, the black box architectures of deep networks are not encouraged, although that may change with increasing performance. [50]. Another approach might be to altogether look for machine learning alternatives that are not deep networks [51]. Yet it is established that there is great scope for its use for image classification and segmentation as discussed. Few shot learning can be a great assisting tool for diagnosis and validation of medical images. Future works will only benefit its applications further, with the hope that it can be applied in practice soon.

Magnetic Resonance Imaging (MRI) technology is a radiology technique that uses the concepts of magnetism and radio waves in order to produce images of body tissues and structures. The MRI scanner tube contains a large magnet which creates a strong magnetic field further exposed to radio waves. The MRI technology is so beneficial because they clearly point out the differences between healthy and diseased tissues and structures, better than any other extant technology. A special kind of MRI technology used is the fMRI which stands for Functional Magnetic Resonance Imaging. fMRI techniques are used for brain imaging and they measure the amounts and changes in the blood flow to different parts of the brain, thereby helping is determine the functions associated with each brain part. This technology has gelled achieve great advancement towards the diagnosis of conditions such as Alzheimer's disease, multiple sclerosis, tumors and others. However, a major drawback that MRI imaging suffers from is the presence of noise in the surrounding environment of the region of interest, acquisition of noise from MRI equipment, noise due to background tissue presence, breathing motion of patients, presence of body fat in the area and many such shortcomings [52]. The various types of noise may lead put a limit on the efficiency of MRI scanning techniques and prevent them from achieving their true potential. As the MRI data is loaded with many irregular fluctuations, few shot learning could be used to counter this inherent flaw by eliminating the need for bulk data for automating diagnosis using MRI scans.

7 Conclusion

In this chapter, the authors have presented a thorough examination of the recent trends and advances in few shot learning and how this method can be used to counter the problem of limited data in the medical domain. Specifically, the few shot classification and segmentation tasks are studied. The problem statement for each is defined and explained. Further, there is a detailed discussion on meta learning which translates to the ability of a system to *learn how to learn*. Several prominent few shot learning algorithms are discussed with a detailed analysis of their working.

Within this context, the authors explain the applications of these algorithms for medical imaging. The applications covered are image classification, image segmentation, image retrieval and text classification. There is in depth discussion about specific applications that have been experimented by various authors and their success. Finally, the future scope of these applications is delineated where we discuss some limitations and concerns that need to be addressed. This chapter concludes with the hope it furnishes a comprehensive explanation of the work done so far and provides an impetus for future work.

References

1. Razzak, M. I., Naz, S., & Zaib, A. (2018). Deep learning for medical image processing: overview, challenges and the future. In N. Dey, A. Ashour, & S. Borra (Eds.), *Classification in BioApps* (Vol. 26)., Lecture Notes in Computational Vision and Biomechanics Cham: Springer.
2. He, K., Zhang, X., Ren, S., Sun, J. (2015). Delving deep into rectifiers: surpassing human-level performance on ImageNet classification. In: *Proceedings of the 2015 IEEE International Conference on Computer Vision (ICCV 2015)* (pp. 1026–1034). IEEE Computer Society. https://doi.org/10.1109/ICCV.2015.123
3. Lee, J. G., et al. (2017). Deep learning in medical imaging: general overview. *Korean Journal of Radiology*, *18*(4), 570–584. https://doi.org/10.3348/kjr.2017.18.4.570.
4. Litjens, G., et al. (2017). A survey on deep learning in medical image analysis. *Medical Image Analysis*, *42*, 60–88. https://doi.org/10.1016/j.media.2017.07.005. (ISSN 1361-8415).

5. Shen, D., Wu, G., & Suk, H.-I. (2017). Deep learning in medical image analysis. *Annual Review of Biomedical Engineering*, *19*(1), 221–248. https://doi.org/10.1146/annurev-bioeng-071516-044442.
6. Miller, E. G., Matsakis, N. E., Viola, P. A. (2000). Learning from one example through shared densities on transforms. In *Proceedings IEEE Conference on Computer Vision and Pattern Recognition (CVPR 2000)* (Cat. No.PR00662), Hilton Head Island, SC (Vol. 1, pp. 464–471). https://doi.org/10.1109/CVPR.2000.855856
7. Li, F.-F., Fergus, & Perona (2003). A Bayesian approach to unsupervised one-shot learning of object categories. In: *Proceedings Ninth IEEE International Conference on Computer Vision*, Nice, France (Vol. 2, pp. 1134–1141). https://doi.org/10.1109/ICCV.2003.1238476
8. Li, F.-F., Fergus, R., & Perona, P. (2006). One-shot learning of object categories. *IEEE Transactions on Pattern Analysis and Machine Intelligence*, *28*(4), 594–611. https://doi.org/10.1109/TPAMI.2006.79.
9. Brenden, L., Ruslan, S., Jason, G., Joshua, T. (2011). One shot learning of simple visual concepts.
10. Koch, G. R. (2015). Siamese Neural Networks for One-Shot Image Recognition.
11. Santoro, A., Bartunov, S., Botvinick, M., Wierstra, D., & Lillicrap, T. (2016). One-shot learning with memory-augmented neural networks. In *Proceedings of the 33nd International Conference on Machine Learning*.
12. Ravi, S., & Larochelle, H. (2017). Optimization as a model for few-shot learning. In *ICLR* (2017).
13. Finn, C., Abbeel, P., & Levine, S. (2017). Model-agnostic meta-learning for fast adaptation of deep networks. In *Proceedings of the 34th International Conference on Machine Learning - Volume 70 (ICML 2017)* (pp. 1126–1135). JMLR.org.
14. Snell, J., Swersky, K., Zemel, R. (2017). Prototypical Networks for Few-shot Learning.
15. Sung, F., Yang, Y., Zhang, L., Xiang, T., Torr, P., Hospedales, T. (2018). Learning to compare: relation network for few-shot learning (pp. 1199–1208). https://doi.org/10.1109/CVPR.2018.00131.
16. Kaiser, Ł., Nachum, O., Roy, A., & Bengio, S. (2017). Learning to Remember Rare Events.
17. Shaban, A., Bansal, S., Liu, Z., Essa, I., & Boots, B. (2017, September). One-shot learning for semantic segmentation. In T.K. Kim, S. Zafeiriou, G. Brostow & K. Mikolajczyk (Eds.), *Proceedings of the British Machine Vision Conference (BMVC)* (pp. 167.1–167.13). BMVA Press
18. Rakelly, K., Shelhamer, E., Darrell, T., Efros, A. A., & Levine, S. (2018). Conditional networks for few-shot semantic segmentation. In *ICLR*.
19. Rakelly, K., Shelhamer, E., Darrell, T., Efros, A., & Levine, S. (2018). Few-Shot Segmentation Propagation with Guided Networks.
20. Kim, J., Oh, T., Lee, S., Pan, F., & Kweon, I. (2019). Variational prototyping-encoder: one-shot learning with prototypical images. In *CVPR*.
21. Lee, K., Maji, S., Ravichandran, A., & Soatto, S. (2019). Meta-learning with differentiable convex optimization. In: *CVPR*.
22. Kim, J., Kim, T., Kim, S., & Yoo, C. D. (2019). Edge-labeling graph neural network for few-shot learning. In *2019 IEEE/CVF Conference on Computer Vision and Pattern Recognition (CVPR)* (pp. 11–20), Long Beach, CA, USA. https://doi.org/10.1109/CVPR.2019.00010
23. Jamal, M. A., Qi, G., & Shah, M. (2018). Task-agnostic meta-learning for few-shot learning. In: *CVPR*.
24. Sun, Q., Liu, Y., Chua, T., & Schiele, B. (2018). Meta-transfer learning for few-shot learning. In: *CVPR*.
25. Li, H., Eigen, D., Dodge, S.F., Zeiler, M.D., & Wang, X. (2019). Finding task-relevant features for few-shot learning by category traversal. In: *CVPR*.
26. Zhang, C., Lin, G., Liu, F., Yao, R., & Shen, C. (2019). CANet: class-agnostic segmentation networks with iterative refinement and attentive few-shot learning.
27. Zhang, H., Zhang, J., & Koniusz, P. (2019). Few-Shot Learning via Saliency-Guided Hallucination of Samples (pp. 2765–2774). https://doi.org/10.1109/CVPR.2019.00288.

28. Chen, Z., Fu, Y., Wang, Y., Ma, L., Liu, W., & Hebert, M. (2019). Image deformation meta-networks for one-shot learning. In: *CVPR*.
29. Schwartz, E., Karlinsky, L., Feris, R., Giryes, R., & Bronstein, A. (2019). Baby steps towards few-shot learning with multiple semantics.
30. Wang, X., Yu, F., Wang, R., Darrell, T., & Gonzalez, J. (2019). TAFE-Net: task-aware feature embeddings for low shot learning. In *CVPR*.
31. Puch, S., Sánchez, I., & Rowe, M. (2019). Few-shot learning with deep triplet networks for brain imaging modality recognition. In *DART/MIL3ID@MICCAI*.
32. Kim, M., Zuallaert, J., De Neve, W. (2017). Few-shot learning using a small-sized dataset of high-resolution FUNDUS images for glaucoma diagnosis. In *Proceedings of the 2nd International Workshop on Multimedia for Personal Health and Health Care (MMHealth 2017)* (pp. 89–92). New York: Association for Computing Machinery. https://doi.org/10.1145/3132635.3132650
33. Hu, S., Tomczak, J. (2018) Max Welling: Meta-Learning for Medical Image Classification.
34. Ronneberger, O., Fischer, P., & Brox, T. (2015). U-Net: convolutional networks for biomedical image segmentation. In: N. Navab, J. Hornegger, W. Wells, A. Frangi (eds.), *Medical Image Computing and Computer-Assisted Intervention (MICCAI 2015). Lecture Notes in Computer Science* (vol. 9351). Cham: Springer.
35. Lahiani, A., Gildenblat, J., Klaman, I., Navab, N., & Klaiman, E. (2018). Generalizing multistain immunohistochemistry tissue segmentation using one-shot color deconvolution deep neural networks.
36. Guha Roy, A., Siddiqui, S., Pölsterl, S., Navab, N. & Wachinger, C. (2019). 'Squeeze & Excite' Guided Few-Shot Segmentation of Volumetric Images.
37. Goodfellow, I. J., et al. (2014). Generative adversarial nets. In *Proceedings of the 27th International Conference on Neural Information Processing Systems - Volume 2 (NIPS 2014)* (pp. 2672–2680). Cambridge: MIT Press.
38. Mondal, A., Dolz, J., & Desrosiers, C. (2018). Few-shot 3D Multi-modal Medical Image Segmentation Using Generative Adversarial Learning.
39. Zhao, A., Balakrishnan, G., Durand, F., Guttag, J., & Dalca, A. (2019). Data augmentation using learned transforms for one-shot medical image segmentation.
40. Wang, Y., Yao, Q., Kwok, J. T., & Ni, L. M. (2019). Generalizing from a Few Examples: A Survey on Few-Shot Learning.
41. Thrun, S. (1998). Lifelong learning algorithms. In S. Thrun & L. Pratt (Eds.), *Learning to Learn*. Boston: Springer.
42. Ren, M., et al. (2018). Meta-Learning for Semi-Supervised Few-Shot Classification.
43. Vilalta, R., & Drissi, Y. (2002). A perspective view and survey of meta-learning. *Artificial Intelligence Review*, *18*, 77–95. https://doi.org/10.1023/A:1019956318069.
44. Vilalta, R., Giraud-Carrier, C., Brazdil, P. (2010). Meta-Learning - Concepts and Techniques.
45. Kruspe, A. (2019). One-Way Prototypical Networks. https://doi.org/10.13140/RG.2.2.31516.95367.
46. Chung, Y.-A., & Weng, W.-H. (2017). Learning Deep Representations of Medical Images using Siamese CNNs with Application to Content-Based Image Retrieval.
47. Rios, A., & Kavuluru, R. (2018). Few-shot and zero-shot multi-label learning for structured label spaces. In *Proceedings of the Conference on Empirical Methods in Natural Language Processing. Conference on Empirical Methods in Natural Language Processing* (pp. 3132–3142).
48. Chen, W., Liu, Y., Kira, Z., Wang, Y. F., & Huang, J. (2019). A closer look at few-shot classification. In *International Conference on Learning Representations 2019*. ArXiv, abs/1904.04232.
49. Dhillon, G. S., Chaudhari, P., Ravichandran, A., & Soatto, S. (2019). A baseline for few-shot image classification. In *International Conference on Learning Representations 2020*. ArXiv, abs/1909.02729.
50. London, A. (2019). Artificial intelligence and black-box medical decisions: accuracy versus explainability. *The Hastings Center Report*, *49*, 15–21. https://doi.org/10.1002/hast.973.

51. Yoichi, H. (2019). The right direction needed to develop white-box deep learning in radiology, pathology, and ophthalmology: a short review. *Frontiers in Robotics and AI*, *6*, 24. https://doi.org/10.3389/frobt.2019.00024.
52. Dey, N., et al. (2015). Parameter optimization for local polynomial approximation based intersection confidence interval filter using genetic algorithm: an application for brain MRI image de-noising. *Journal of Imaging*, *1*, 60–84. https://doi.org/10.3390/jimaging1010060.

Hyperspectral Remote Sensing Image Classification Using Active Learning

Vimal K. Shrivastava and Monoj K. Pradhan

Abstract Hyperspectral remote sensing images capture a large number of narrow spectral bands ranging between visible and infrared spectrum. The abundant spectral data provides huge land cover information that helps in accurate classification of land use land cover of earth's surface. However, obtaining labelled training data in hyperspectral images (HSIs) is labour expensive and time consuming. Therefore, designing a classifier that uses fewer labelled samples as possible for classification is highly desirable. Active learning (AL) is a branch of machine learning that finds most uncertain samples in an iterative way from unlabelled dataset resulting relatively smaller training set to achieve adequate classification accuracy. Support vector machine (SVM) has been extensively used as a classifier in AL approach. However, it has high computational complexity. Recently, a non-iterative learning algorithm based on least square solution known as extreme learning machine (ELM), has been integrated in AL framework for HSI classification. It provides a comparable classification accuracy while reducing computation time drastically.

Keywords Hyperspectral Remote Sensing Image · Classification · Active Learning · Query Strategy · Support Vector Machine · Extreme Learning Machine

---

V. K. Shrivastava (✉)
School of Electronics Engineering, Kalinga Institute of Industrial Technology (KIIT), Bhubaneswar, India
e-mail: vimal.shrivastavafet@kiit.ac.in

M. K. Pradhan
Department of Agricultural Statistics and Social Sciences (L), Indira Gandhi Agricultural University, Raipur, India
e-mail: monojpradhan76@gmail.com

S. K. Das et al. (eds.), *Machine Learning Algorithms for Industrial Applications*, Studies in Computational Intelligence 907,
https://doi.org/10.1007/978-3-030-50641-4_8

1 Introduction

The recent advances in variety of space borne and air borne sensors have provided hyperspectral images (HSIs) with large number of spectral bands. These spectral bands provide rich and subtle information with more discriminative ability to identify objects. In the recent era, hyperspectral image (HSI) has been extensively used in many applications such as: (i) crop classification, (ii) soil classification, (iii) urban planning, (iv) vegetation classification, (v) forest monitoring, (vi) land cover classification, etc. Therefore, the classification of HSIs has become an important task in almost all fields such as geosciences, environmental science, mathematics, and computer vision. The fast developing trend of HSI classification lies mainly in three aspects [1]. First, HSI classification has applications in several domains such as agriculture, military object detection and environmental monitoring etc. The second aspect is the development of new computer vision and advanced machine learning techniques namely deep learning, sparse representation and graph model for HSI classification. The third aspect is spatial-spectral classification approaches that provide an opportunity to develop new algorithms.

As per literature, HSIs provide abundant information for classification. But at the same time, there are many challenges while performing the classification task. Some of the challenges are as follows:

(i) High computational cost because of large dimensionality.
(ii) Difficult to train the model efficiently due to limited labelled training samples.
(iii) It is challenging to develop competent HSI classification model due to large variability in spatial and spectral signature.

The efficacy of classification model is highly dependent on training data. However, manual labelling of remotely sensed data is labor expensive, time taking and erroneous. In addition, efficient training of a classifier is difficult with less number of labelled training data. Therefore, selection of limited and suitable training data for classification model is a challenging task that could reduce the labelling and computation cost without sacrificing its performance. Active Learning (AL) is one such approach which is an emerging field in machine learning.

1.1 Objective

The objective of this chapter is to discuss the concept and advantages of AL approach in hyperspectral image (HSI) classification. Further, it discusses three query strategies: (i) random sampling, (ii) multiview based maximum disagreement and (iii) entropy query by bagging. Lastly, the performance analysis of HSI classification with the integration of these query strategies and classifiers such as SVM and ELM has been presented.

2 Active Learning

AL is the process of building a compact and effective training dataset by picking the most informative samples from the unlabelled set iteratively. It is essentially a resampling strategy. It is biased [2], and draws the samples having the most information from the unlabelled set by certain query strategy. It has been considered that interactively constructed training set is optimal and have the most useful information for the model. The prime objective of AL is to train the model with fewer samples as compared to general supervised learning methods to achieve the desired performance. This approach has been generally used in the datasets that have high dimensionality in nature [3–6]. The AL technique consists of mainly six components denoted as: M, Q, S, D_L, D_P and D_U [7]. The dataset is first randomly partitioned into training and test set. Further, the training set is partitioned into a small labelled set (D_L) and candidate/unlabelled set (D_U). A set of most informative (uncertain) samples (D_P) are selected from the unlabelled set using query strategy (Q) iteratively. Here, P represents the number of samples to have in D_P at each iteration. These samples are then appended to D_L after labelling by the supervisor (S) and removed from D_U. The key concept in AL is the type of Q which is used to search the most informative samples from D_U. The model (M) is initially trained on D_L. In every iteration, D_P is appended to D_L and then, M is retrained. These steps are reiterated until the pre-defined stopping criterion is not satisfied. Figure 1 depicts the block diagram of general AL technique. Further, the pseudocode of AL has been described in Algorithm 1.

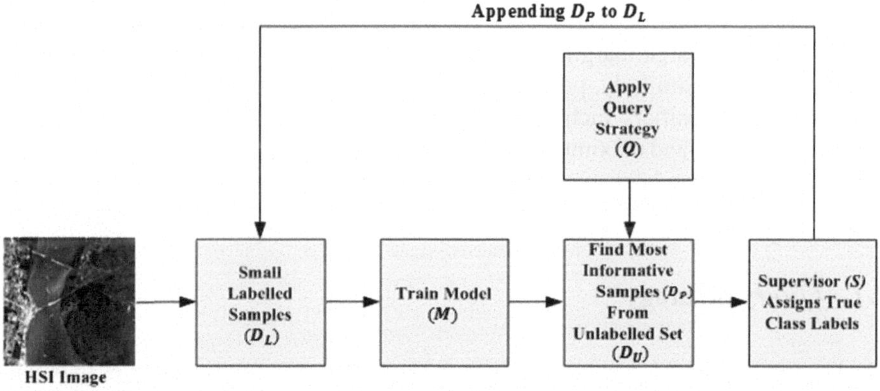

Fig. 1 Block diagram of general AL technique

Algorithm 1: Pseudocode of general AL framework.

Initialization:

1: Randomly divide the dataset into D_L and D_U.

2: Set P to be picked at every iteration.

3: Train Mon D_L.

4: Set iteration $(iter)$=1.

Repeat

5: Extract D_P from D_U.

6: Label D_P by S.

7: Append D_P to D_L and remove D_P from D_U.

8: Retrain M with updated D_L.

9: $iter = iter + 1$.

Until stopping criteria (number of *iter*) is not satisfied.

There are three major issues of concern in AL technique: (i) how to choose D_L to seed the AL process, (ii) a selection strategy to pick D_P from D_U which is known as query strategy/query function (Q) and (iii) a stopping criteria of AL process. Among the three, Q is the key component [8]. Based on Q, AL can be broadly grouped into: (i) committee based heuristic [9]; (ii) large margin based heuristic [10, 11] and (iii) probability based heuristic [12]. Any classifier can be utilized in committee based heuristic whereas large margin based heuristic is based on support vector machine (SVM) classifier. Similarly, probability based heuristic is based on posterior probabilities based classifiers such as discriminant classifier [13], multinomial logistic regression [12, 14] and maximum-likelihood classifier [15]. In the next section, three query strategies have been explained.

2.1 *Query Strategy*

The key issue in the AL technique is how to choose the informative samples to obtain the satisfactory classification performance by using a lesser number of samples as compared to the conventional passive learning [16–20]. This chapter discusses three query strategies: random sampling (RS), multiview with maximum disagreement (MV-Dis_{Max}) and entropy query by bagging (EQB) which have been briefly described in the following sub-sections.

2.1.1 Random Sampling Based AL

Random sampling based AL (RS-AL) is one of most common and simple Q. Here, D_P is selected randomly from D_U without applying any criterion. Then, D_P is appended in D_L and removed from D_U. The M is retrained with updated D_L iteratively. Algorithm 2 explains the RS-AL framework.

Algorithm 2: Pseudocode of RS-AL framework.

Initialization:

1: Randomly divide the dataset into D_L and D_U.

2: Set P to be picked at every *iter*.

3: Train M on D_L.

4: Set $iter = 1$.

Repeat

5: Extract $D_P \in D_U$ randomly.

6: Label D_P by S.

7: Discard D_P from D_U.

8: Append D_P to D_L.

9: Retrain M with updated D_L.

10: $iter = iter + 1$.

Until stopping criteria (number of *iter*) is not satisfied.

2.1.2 Multiview with Maximum Disagreement Based AL

The concept of multiview (MV) was given by [21]. Then, the MV based AL (MV-AL) approach is explored in HSI classification by [22] and [23]. Here, the way of constructing the multiple views is very important. One approach of MV generation in HSI is explained here. Let $X = \{x_1, x_2, \ldots, x_N\} \in R^{B \times N}$ be the HSI where, B be the number of bands and N be number of pixels. $Y = \{y_1, y_2, \ldots, y_N\} \in R^N$ be the label of the image where, each pixel $x_i \in X$ is defined by one of the classes (C) *i.e.* $y_i \in \{1, 2, \ldots, C\}$. In the supervised classification, the hypothesis $h : X \rightarrow Y$ must satisfy while learning and entire set of feature bands of the image are taken as one view. Whereas in MV, the features are segmented into multiple subsets as $(X_1 \times X_2 \times \ldots, X_{K_V})$ where K_V is number of views. On the basis of this principle, [17] has segmented the spectral bands into several continuous sub-band sets according to the band index.

It is assumed while constructing the views that each view is enough to classify HSI *i.e.* there should be diversity in the views. One method of measuring this diversity is "maximum disagreement" (Dis_{Max}) [18]. The classifier is trained over various constructed views. Each view is trained to classifier independently. The overall confusion of these trained classifiers on testing a sample ($x_i \in D_U$) is known as disagreement (Dis). Let the classification functions f_V^i is defined for view i, where $i = 1, 2, 3 \ldots, K_V$. Then, Dis_{Max} for each sample ($x_i \in D_U$) is expressed by Eq. (1).

$$Dis_{max} = \max_{x \epsilon D_U} Dis\left(x, f_V^1, f_V^2, \ldots, f_V^{K_V}\right) \tag{1}$$

Where $Dis(\cdot)$ is disagreement for each sample ($x_i \in D_U$) and it is expressed as below:

$$Dis\left(x, f_V^1, f_V^2, \ldots f_V^{K_V}\right) = \text{count}\left|f_V^i\right| \; for \; i = 1, 2, \ldots, K_V \tag{2}$$

Where, $count|.|$ calculates the number of unique elements in the set.

These set of views increases the performance of the learning model. Here, D_P is selected from D_U in every iteration by applying Dis_{max}. The flow diagram of MV-Dis_{max} has been shown in Fig. 2 and the Algorithm 3 explains its pseudocode.

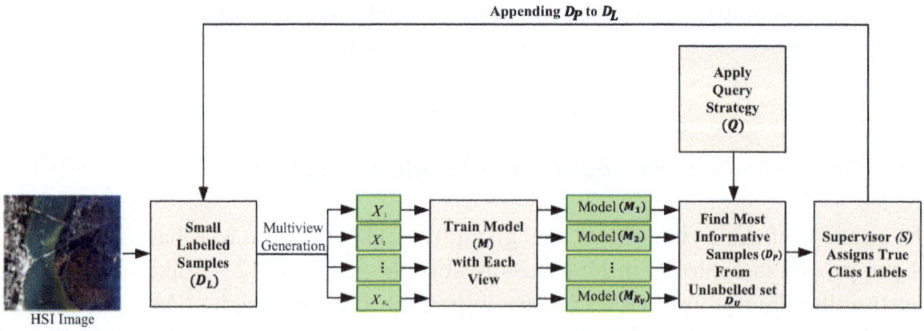

Fig. 2 Flow diagram of MV-Dis_{Max} based AL technique

Algorithm 3: Pseudocode of MV-$\boldsymbol{Dis_{max}}$ based AL framework.

Initialization:

1: Generate K_V views $(X_1, X_2, \ldots X_{K_V})$ from spectral features of X.

2: Randomly divide the dataset into D_L and D_U.

3: Set P to be picked at every *iter*.

4: Train M corresponding to each view *i.e.* generate models $M_1, M_2, \ldots, M_{K_V}$.

5: $iter = 1$.

Repeat

6: Test each sample $x_i \in D_U$ by models $M_1, M_2, \ldots, M_{K_V}$ respectively by applying Eq. (1).

7: Calculate Dis of $x_i \in D_U$ by applying Eq. (2).

8: Extract a set of D_P having Dis_{max} and remove them from D_U.

9: Label D_P by S and update D_L by appending D_P.

10: Retrain the models $M_1, M_2, \ldots, M_{K_V}$ with updated D_L.

11: $iter = iter + 1$.

Until stopping criteria (number of *iter*) is not satisfied.

2.1.3 Entropy Query by Bagging based AL

The idea in query by bagging is to build k training sets on bootstrap samples [24]. This conception has been extended into the multiclass classification using entropy heuristic; known as entropy query by bagging (EQB) [25]. A bootstrap (D'_{L_G}) is formed with the replacement of the original samples in every *iter*. Each set of bootstrap contains a defined subset of D_L. Each bootstrap is formed by picking samples randomly from D_L. These sets are used to train the corresponding labelling model and the trained models are reused to predict the class labels of unlabeled set D_U. Therefore, k possible class labels for each sample x_i was predicted from the unlabelled set. The entropy distribution $H(x_i)$ is calculated from the k class labels of x_i as follow:

$$H(x_i) = \sum_{C} -p_{i,c} \log(p_{i,c}) \tag{3}$$

Where $p_{i,c}$ is the probability to have the c class for $x_i \in D_U$. $H(x_i)$ is calculated for each $x_i \in D_U$. The maximum entropy is calculated using Eq. (4).

$$\hat{x} = \max_{x_i \in D_U} H(x_i) \tag{4}$$

Any classifier can be adopted in this method. The flow diagram of EQB based AL framework (EQB-AL) is shown in Fig. 3 and the Algorithm 4 describes its pseudocode.

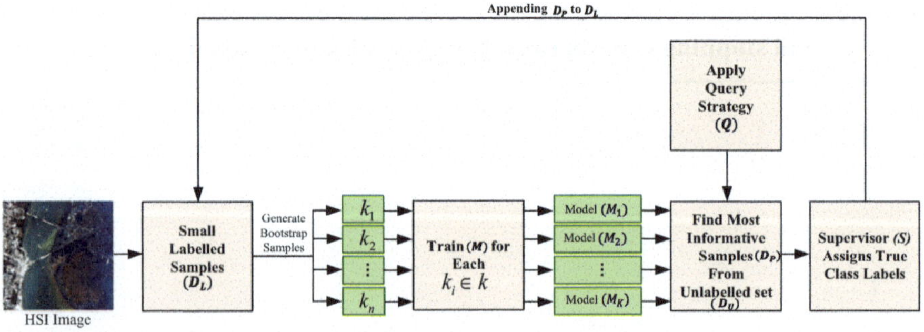

Fig. 3 Flow diagram of EQB-AL approach

Algorithm 4: Pseudocode of EQB-AL framework.

Initialization:

1: Randomly divide the dataset into D_L and D_U.

2: Set P to be picked at every *iter*.

3: Set *k*.

4: Set percentage of samples (pct) from D_L drawn into bootstrap samples.

5: Train M with the current set of D_L.

6: Set *iter =1.*

Repeat

For M =1 to k do

7: Obtain subset $D_{LG}^{'}$ by applying $pct * D_L$.

8: Train M^{th} model using $D_{LG}^{'}$.

End of for

9: Predict the class membership of $x_i \in D_U$ by each Mth model.

10: Compute the entropy for every $x_i \in D_U$ using Eq. (3).

11: Extract D_P form D_U with maximum entropy using Eq. (4) to further label by *S*.

12: Update the D_L by appending these extracted D_P.

13: Retrain M with updated D_L.

Until stopping criteria (number of *iter*) is not satisfied.

3 HSI Classification Using AL Approach with SVM Classifier

SVM is widely used in supervised machine learning [26]. Over the last few decades, kernel based SVM has been extensively used in the classification of HSI data [27, 28]. It is based on large margin separation which provides the decision boundary as hyperplane on the labelled training data. The samples present at this hyperplane are known as support vectors and therefore the classifier is known as SVM. In binary classification, the distance of the sample (x_i) from the hyperplane is given by:

$$f(x_i) = \sum_{j=1}^{n} \alpha_j y_j K(x_j, x_i) + b) \quad (6)$$

Where, $K(x_j, x_i)$ is a kernel function. α_j represents non-zero coefficients. y_j is the label of support vector viz. $+1$ for positive class and -1 for negative class. The higher the value of $f(x_i)$, higher is confidence. Various query heuristics have been developed so far on the basis of support vectors which are described in following sub-sections.

3.1 Margin Sampling (MS)

In SVM, the support vectors lie at the boundaries of the hyperplane [16]. The heuristic that uses this geometrical property is called MS [1, 29]. The concept of MS has been popularly implemented in HSI classification [27]. This query heuristic picks those samples closest to the hyperplane and considers them to be most informative (uncertain) samples. The sampling of the candidate in MS heuristic is performed by minimizing Eq. (7) as follows:

$$\hat{x}^{MS} = arg \min_{x_i \in D_U} \left\{ \min_{C} |f(x_i, c)| \right\} \quad (7)$$

Where $f(x_i, c)$ provides the distance of c class sample to the hyperplane. Readers may refer to [22, 25, 30–32] where MS has been used as a query heuristic in AL technique for HSI classification.

3.2 Multiclass Level Uncertainty (MCLU)

An extension of MS is the MCLU which uses the distances of the samples to the hyperplane for the two most probable classes [30] defined as below.

$$\hat{x}^{MCLU} = arg \min_{x_i \in D_U} \{ f(x_i)^{MC} \} \quad (8)$$

where,

$$f(x_i)^{MC} = \max_{c \in C} f(x_i, c) - \max_{c \in C \backslash c+} f(x_i, c) \quad (9)$$

where, c^{+} is the maximal confidence (MC) class. Its value provides certainty level of samples, *i.e.*, the samples which have higher value of this criterion provides high certainty whereas the samples having lower value of this criterion results into higher uncertainty. Readers may refer to [30–32] where MCLU has been used as a query heuristic in AL technique for HSI classification.

4 HSI Classification Using AL Approach with ELM Classifier

It is observed that AL approach leads to obtain better classification accuracy for HSI but the computation time is very high [33, 34]. The solution to this issue has been presented by [35] using ELM classifier in AL technique. Further, the ELM-AL approach has been applied first time in HSI classification by [36, 37] and demonstrated that ELM-AL approach has achieved comparable classification accuracy than SVM-AL approach with substantial decrement in computation time. The description of ELM classifier has been provided in following sub-section.

4.1 Extreme Learning Machine (ELM)

The ELM is a supervised learning classifier that was introduced by [35]. It is a single hidden layer feedforward network (SLFN). Here, weights and biases between input and hidden layer are generated randomly instead of tuning and only the weights and biases between hidden layer and output layer are updated during training. It has been popularly known for its three remarkable properties such as: (i) extremely fast training; (ii) good generalization ability and (iii) universal classification ability. ELM is very simple and can be easily solved using least square calculation. The ELM architecture is shown in Fig. 4.

The output of ELM for generalized SLFN is expressed as:

$$H\beta = Y \tag{14}$$

where $\beta = [\beta_1, \beta_2 \ldots .\beta_L] \in R^{L\times O}$ is the matrix of weights connecting nodes of hidden and output layer, $Y = [y_1, y_2 \ldots y_O]^T \in R^{N\times O}$ represents output weights, where N is the total features of a sample and H is the matrix of output weights of hidden layer and is expressed as below.

$$H(w_1, \ldots.., w_L, b_1, \ldots\ldots, b_{L,} x_1, \ldots\ldots x_N) = \begin{bmatrix} G(w_1, b_1, x_1) & \ldots & G(w_L, b_L, x_1) \\ \vdots & \vdots & \vdots \\ G(w_1, b_1, x_N) & \ldots & G(w_L, b_L, x_N) \end{bmatrix}_{N\times L} \tag{15}$$

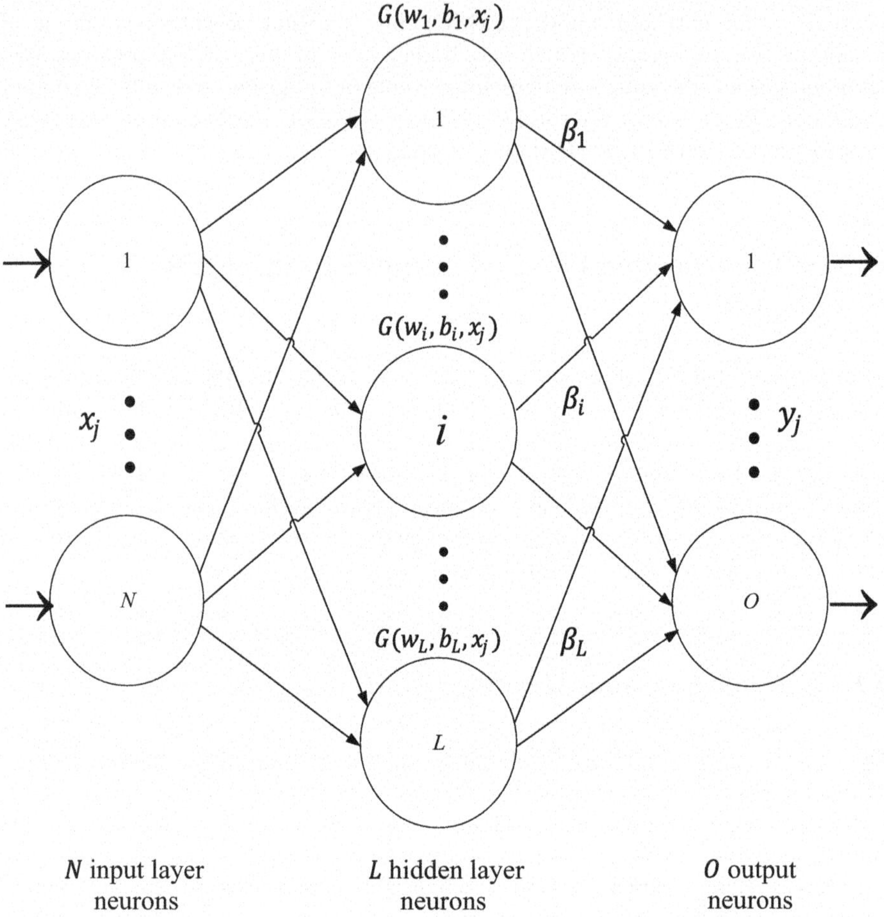

Fig. 4 Architecture of ELM

Where, w_i is the ith weight vector and b_i is the ith bias between the input and hidden layer. In ELM, the objective is to minimize the error ($\|H\beta - Y\|^2$). Thus, obtaining Eq. (16) from Eq. (15):

$$Minimize : \frac{1}{2}\|\beta\|^2 + R\frac{1}{2}\sum_{i=1}^{N}\xi_i^2 \tag{16}$$

$$\text{s.t.} \quad h(x_i)\beta = y_i^T - \xi_i^2 \qquad i = 1, 2 \ldots\ldots N$$

where, ξ is training error of sample x_i, R is a regularization parameter and $h(x_i)$ is the feature map with respect to x_i. There is one limitation of ELM that the classification accuracy may vary for different trials due to random weight assignment between input and hidden layer. Therefore, a kernel-based ELM (KELM) was suggested by [38]. The kernel matrix is expressed as:

$$K\left(x_i, x_j\right) = \left(h(x_i)h\left(x_j\right)\right) \tag{17}$$

where, $K\left(x_i, x_j\right)$ is a kernel function. It may be linear, polynomial and radial basis function etc. The output of ELM [39] can be obtained as follows:

$$f(x) = K_x\left(\frac{1}{C} + K\right)^{-1} Y \tag{18}$$

Where, $K = \left[K\left(x_i, x_j\right)\right]_{i,j=1}^{N}$ and $K_x = [K(x, x_1), \ldots\ldots, K(x, x_N)]$.

5 Dataset and Experimental Setup

5.1 Dataset Description

Various remote sensing HSI data are publically available [40]. In this chapter, the details of two HSI data has been provided: (i) Kennedy Space Centre (KSC) and Botswana (BOT). KSC dataset was captured by AVIRIS sensor. It contains spectral bands of 224 which reduced to 176 after eliminating noise. It consists of 13 classes with 5211 pixels. The image of the KSC dataset and its ground truth is shown in Fig. 5 and class details are provided in Table 1. BOT dataset was captured by Hyperion sensor. It contains spectral bands of 242 which reduced to 145 after eliminating noise. It consists of 14 classes with 3248 pixels. The image of BOT dataset and its ground truth is shown in Fig. 6 and class details of BOT are provided in Table 2.

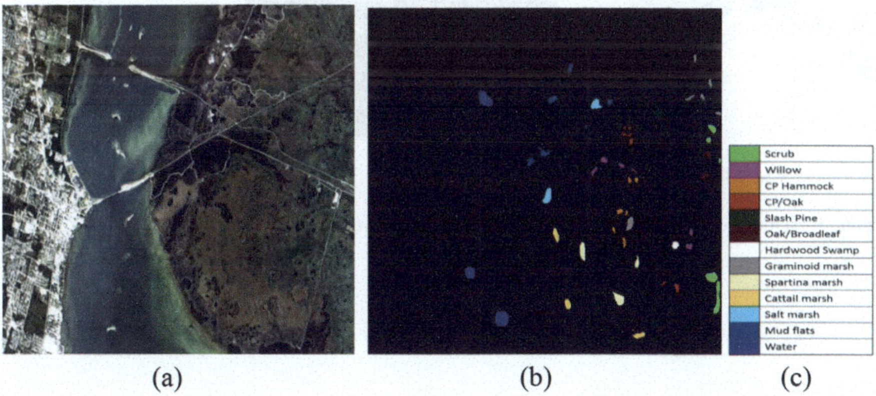

Fig. 5 KSC dataset: (**a**) original image, (**b**) ground truth image, (**c**) class labels

Table 1 Class-wise details of KSC Dataset

Class name	# Samples
Scrub	761
Willow swamp	243
Cabbage palm hammock	256
Cabbage palm/Oak hammock	252
Slash pine	161
Oak/Broad leaf hammock	229
Hardwood swamp	105
Graminoid marsh	431
Spartina marsh	520
Cattail marsh	404
Salt marsh	419
Mudd flats	503
Water	927
Total	5211

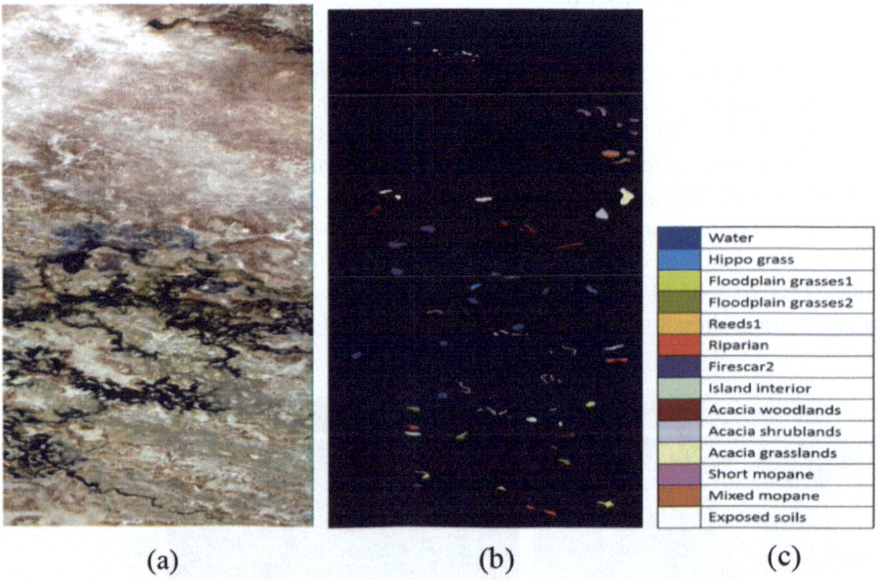

Fig. 6 BOT dataset: (**a**) original image, (**b**) ground truth image, (**c**) class labels

Table 2 Class-wise details of BOT Dataset

Class name	# Samples
Water	270
Hippo grass	101
Floodplain grasses 1	251
Floodplain grasses 1	215
Reeds 1	269
Riparian	269
Firescar2	259
Island interior	203
Acacia woodlands	314
Acacia shrublands	248
Acacia grasslands	305
Short mopane	181
Mixed mopane	268
Exposed soils	95
Total	3248

5.2 Experimental Setup

The performance comparison of six systems formed by crisscross combinations of two classifiers (SVM and ELM) and three query strategies (*i.e.* RS, MV and EQB) has been presented in this chapter. These six systems are: (i) SVM-RS; (v) SVM-MV; (iii) SVM-EQB; (iv) ELM-RS; (v) ELM-MV and (vi) ELM-EQB. The most informative samples were randomly selected in RS query strategy. While using MV query strategy, the views were generated by segmenting the available spectral features into continuous disjoint sub-bands according to band indices [22]. For example, the 5 views based on the band indices associated with KSC dataset are 1-11, 12-31, 32-96, 97-130 and 131-176 while in case of BOT dataset, it is 1-25, 26-61, 62-79, 80-110 and 111-145. In case of EQB query strategy, the parameter k for EQB has been set to four and every k^{th} bootstrap contains 60% of D_L. Then, the AL technique has been triggered by picking 5 pixels as initial D_L from every class for both datasets resulting $D_L = 65$ (5 pixels × 13 classes) and $D_L = 70$ (5 pixels × 14 classes) for KSC and BOT dataset respectively. We have fixed $P = 7$ and total number of iterations to be 100 for all six systems. Moreover, the experiments have been performed for 10 trials due to random partition of D_L, D_U and D_T. The overall accuracy and computation time were then calculated by averaging the accuracy and computation time of 10 trails respectively. The characteristics and experimental parameters of KSC and BOT dataset are presented in the Table 3. These initial experimental setups were

Table 3 Experimental setting for KSC and BOT dataset

Characteristics	KSC	BOT
Total number of samples (pixels)	5211	3248
Number of land covers (classes)	13	14
Initial samples from each class	5	5
Initial training set ($\boldsymbol{D_L}$)	65	70
No. of iterations ($\boldsymbol{iter}$)	100	100
Batch size ($\boldsymbol{P}$)	7	7
Initial unlabelled set ($\boldsymbol{D_U}$)	1840	854
Test set ($\boldsymbol{D_T}$)	2606	1624
Band Indices for MV	1-11, 12-31, 32-96, 97-130, 131-176	1-25, 26-61, 62-79, 80-110, 111-145

used to model the SVM and ELM classifier. The experiments have been performed in Matlab-2016 running in i5-2400, CPU@3.10 GHz. AL toolbox (ALTB) [41]. were used for implementation of AL technique and LIBSVM library has been utilized for executing SVM.

6 Results and Discussion

The performance of six systems: (i) ELM-RS; (ii) ELM-MV; (iii) ELM-EQB; (iv) SVM-RS; (v) SVM-MV; and (vi) SVM-EQB have been presented here on BOT and KSC datasets. The evaluation has been done based on two parameters: (a) classification accuracy and (b) computation time. Figure 7 presents the classification accuracy (learning curves) which are plotted between classification accuracy and number of iterations. Further, Table 4 shows computation time for all six systems. Apart from it, we have obtained the classification accuracy after completion of 100 iterations and depicted this result in Table 5 for two datasets and all six systems.

The observations from these results are as follows: (i) EQB shows better classification accuracy than MV and RS for both datasets for all six systems irrespective of the classifier; (ii) classification accuracy obtained using ELM based AL models are slightly less in comparison to SVM based AL models. However, the computation time in ELM based AL models is significantly less compared to SVM based AL models irrespective of the query strategy. The above two observations show that ELM-EQB technique can be a better choice in HSI classification where adequate classification accuracy can be achieved with significantly less computation time.

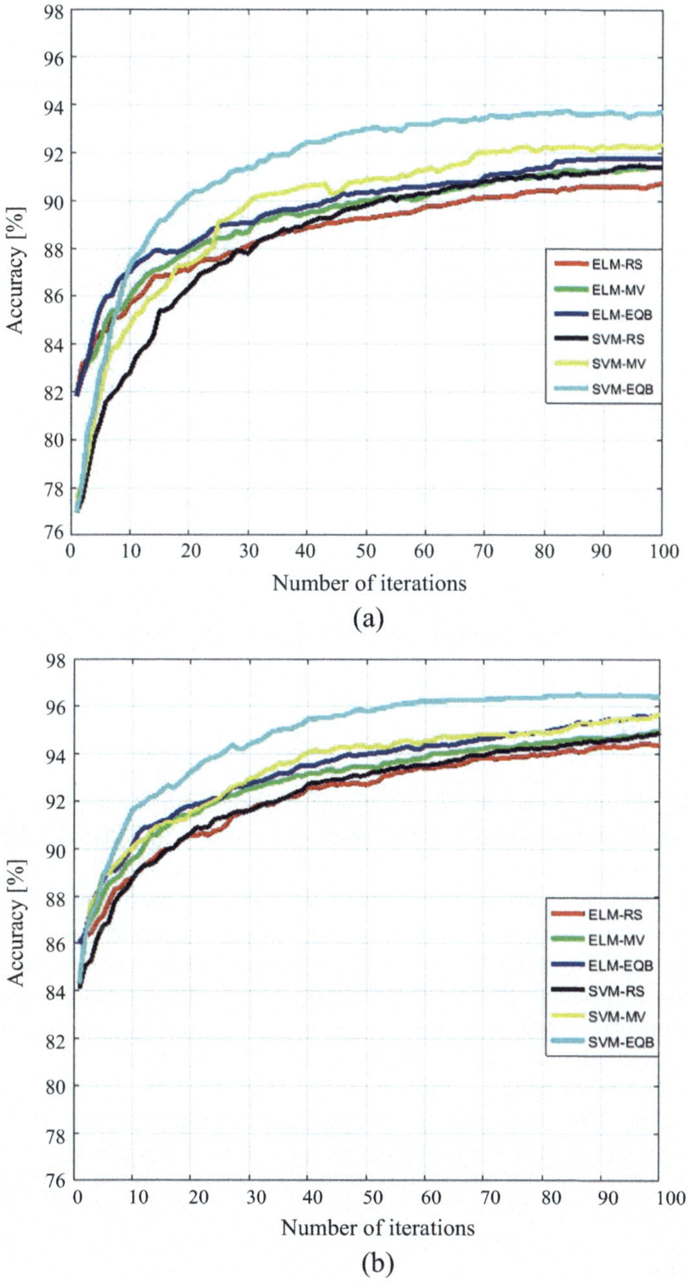

Fig. 7 Overall accuracy vs. number of iterations during training: (**a**) KSC and (**b**) BOT dataset

Table 4 Computation time of AL techniques

AL techniques	Computation time (sec.)	
	KSC	BOT
ELM-RS	0.101	0.070
ELM-MV	0.116	0.073
ELM-EQB	0.107	0.064
SVM-RS	190.371	139.037
SVM-MV	400.072	282.364
SVM-EQB	235.598	218.199

Table 5 Classification accuracy after completion of 100 iterations

AL techniques	Classification accuracy (%)	
	KSC	BOT
ELM-RS	90.77	94.19
ELM-MV	91.43	94.97
ELM-EQB	91.80	95.34
SVM-RS	92.03	95.41
SVM-MV	92.33	95.87
SVM-EQB	94.29	96.73

7 Conclusion

This chapter has presented an AL technique in detail with the application of HSI classification. With this objective, three query strategies (*i.e.* RS, MV, EQB) and how these query strategies can be integrated with two classifiers (ELM, SVM) has been discussed. Further, a comprehensive comparison has been done on the performance of six systems formed by crisscross combinations of three query strategies (*i.e.* RS, MV, EQB) and two classifiers (ELM, SVM). The results show that adequate classification accuracy has been achieved with significantly less computation time using ELM based AL models as compare to state-of-the-art SVM based AL models irrespective of the query strategies. In addition, the experiments demonstrated that EQB query strategy has produced better classification accuracy than RS and MV irrespective of the classifiers. Therefore, the encouraging outcome of this work suggests that ELM-EQB AL technique can be a better choice in HSI classification.

References

1. Schohn, G., & Cohn, D. (2000). Less is more: Active learning with support vector machines. In *ICML* (pp. 839–846).
2. Dasgupta, S., Hsu, D. J., & Monteleoni, C. (2008). A general agnostic active learning algorithm. In *Advances in Neural Information Processing Systems* (pp. 353–360).

3. Settles, B. (2009). *Active learning literature survey*. University of Wisconsin-Madison, Department of Computer Sciences.
4. Rajan, S., Ghosh, J., & Crawford, M. M. (2008). An active learning approach to hyperspectral data classification. *IEEE Transactions on Geoscience and Remote Sensing, 46*(4), 1231–1242.
5. Karaa, W. B. A., Ashour, A. S., Ben Sassi, D., Roy, P., Kausar, N. & Dey, N. (2016). Medline text mining: an enhancement genetic algorithm based approach for document clustering. In *Applications of Intelligent Optimization in Biology and Medicine* (pp. 267–287). Springer.
6. Chatterjee, S., Hore, S., Dey, N., Chakraborty, S., & Ashour, A. S. (2017). Dengue fever classification using gene expression data: A PSO based artificial neural network approach. In *Proceedings of the 5th International Conference on Frontiers in Intelligent Computing: Theory and Applications* (pp. 331–341).
7. Li, M., & Sethi, I. K. (2006). Confidence-based active learning. *IEEE Transactions on Pattern Analysis and Machine Intelligence, 28*(8), 1251–1261.
8. Camps-Valls, G., Tuia, D., Bruzzone, L., & Benediktsson, J. A. (2014). Advances in hyperspectral image classification: Earth monitoring with statistical learning methods. *IEEE Signal Processing Magazine, 31*(1), 45–54.
9. Bruzzone, L., & Bovolo, F. (2013). A novel framework for the design of change-detection systems for very-high-resolution remote sensing images. *Proceedings of the IEEE, 101*(3), 609–630.
10. Gómez-Chova, L., Camps-Valls, G., Munoz-Mari, J., & Calpe, J. (2008). Semisupervised image classification with Laplacian support vector machines. *IEEE Geoscience and Remote Sensing Letters, 5*(3), 336–340.
11. Tuia, D., Persello, C., & Bruzzone, L. (2016). Domain adaptation for the classification of remote sensing data: An overview of recent advances. *IEEE Geoscience and Remote Sensing Magazine, 4*(2), 41–57.
12. Gamba, P., Dell'Acqua, F., & Trianni, G. (2007). Rapid damage detection in the Bam area using multitemporal SAR and exploiting ancillary data. *IEEE Transactions on Geoscience and Remote Sensing, 45*(6), 1582–1589.
13. Benediktsson, J. A., Pesaresi, M., & Amason, K. (2003). Classification and feature extraction for remote sensing images from urban areas based on morphological transformations. *IEEE Transactions on Geoscience and Remote Sensing, 41*(9), 1940–1949.
14. Inglada, J., & Mercier, G. (2007). A new statistical similarity measure for change detection in multitemporal SAR images and its extension to multiscale change analysis. *IEEE Transactions on Geoscience and Remote Sensing, 45*(5), 1432–1445.
15. Fauvel, M., Tarabalka, Y., Benediktsson, J. A., Chanussot, J., & Tilton, J. C. (2013). Advances in spectral-spatial classification of hyperspectral images. *Proceedings of the IEEE, 101*(3), 652–675.
16. Tong, S., & Koller, D. (2001). Support vector machine active learning with applications to text classification. *Journal of Machine Learning Research, 2,* 45–66.
17. de Sa, V. R. (1994). Learning classification with unlabeled data. In *Advances in Neural Information Processing Systems* (pp. 112–119).
18. Jagatheesan, K., Anand, B., Samanta, S., Dey, N., Ashour, A. S., & Balas, V. E. (2017). Particle swarm optimisation-based parameters optimisation of PID controller for load frequency control of multi-area reheat thermal power systems. *International Journal of Advanced Intelligence Paradigms, 9*(5–6), 464–489.
19. Das, S. K., & Tripathi, S. (2018). Adaptive and intelligent energy efficient routing for transparent heterogeneous ad-hoc network by fusion of game theory and linear programming. *Applied Intelligence, 48*(7), 1825–1845.
20. Das, S. K., & Tripathi, S. (2019). Energy efficient routing formation algorithm for hybrid ad-hoc network: A geometric programming approach. *Peer-to-Peer Networking and Applications, 12*(1), 102–128.
21. Muslea, I., Minton, S., & Knoblock, C. A. (2006). Active learning with multiple views. *Journal of Artificial Intelligence Research, 27,* 203–233.

22. Di, W., & Crawford, M. M. (2012). View generation for multiview maximum disagreement based active learning for hyperspectral image classification. *IEEE Transactions on Geoscience and Remote Sensing, 50*(5), 1942–1954.
23. Zhou, X., Prasad, S., & Crawford, M. (2014). Wavelet domain multi-view active learning for hyperspectral image analysis. In *2014 6th Workshop on Hyperspectral Image and Signal Processing: Evolution in Remote Sensing (WHISPERS)* (pp. 1–4).
24. Efron, B. (1992). Bootstrap methods: another look at the jackknife. In *Breakthroughs in statistics* (pp. 569–593).
25. Tuia, D., Ratle, F., Pacifici, F., Kanevski, M. F., & Emery, W. J. (2009). Active learning methods for remote sensing image classification. *IEEE Transactions on Geoscience and Remote Sensing, 47*(7), 2218–2232.
26. Vapnik, V. N. (1999). An overview of statistical learning theory. *IEEE Transactions on Neural Networks, 10*(5), 988–999.
27. Mitra, P., Shankar, B. U., & Pal, S. K. (2004). Segmentation of multispectral remote sensing images using active support vector machines. *Pattern Recognition Letters, 25*(9), 1067–1074.
28. Dey, N., et al. (2015). Parameter optimization for local polynomial approximation based intersection confidence interval filter using genetic algorithm: An application for brain MRI image de-noising. *Journal of Imaging, 1*(1), 60–84.
29. Campbell, C., Cristianini, N., & Smola, A. (2000). Query learning with large margin classifiers. In *ICML* (Vol. 20, pp. 0).
30. Demir, B., Persello, C., & Bruzzone, L. (2011). Batch-mode active-learning methods for the interactive classification of remote sensing images. *IEEE Transactions on Geoscience and Remote Sensing, 49*(3), 1014–1031.
31. Tuia, D., Volpi, M., Copa, L., Kanevski, M., & Munoz-Mari, J. (2011). A survey of active learning algorithms for supervised remote sensing image classification. *IEEE Journal of Selected Topics in Signal Processing, 5*(3), 606–617.
32. Xu, J., Hang, R., & Liu, Q. (2014). Patch-based active learning (PTAL) for spectral-spatial classification on hyperspectral data. *International Journal of Remote Sensing, 35*(5), 1846–1875.
33. Schroder, M., Rehrauer, H., Seidel, K., & Datcu, M. (1998). Spatial information retrieval from remote-sensing images. II. Gibbs-Markov random fields. *IEEE Transactions on Geoscience and Remote Sensing, 36*(5), 1446–1455.
34. Nasrabadi, N. M. (2014). Hyperspectral target detection: An overview of current and future challenges. *IEEE Signal Processing Magazine, 31*(1), 34–44.
35. Huang, G.-B., Zhu, Q.-Y., & Siew, C.-K. (2006). Extreme learning machine: Theory and applications. *Neurocomputing, 70*(1–3), 489–501.
36. Pradhan, M. K., Minz, S., & Shrivastava, V. K. (2018). Fast active learning for hyperspectral image classification using extreme learning machine. *IET Image Proccessing, 13,* 549–555.
37. Pradhan, M. K., Minz, S., & Shrivastava, V. K. (2019). A kernel-based extreme learning machine framework for classification of hyperspectral images using active learning. *Journal of the Indian Society of Remote Sensing, 47,* 1693–1705.
38. Huang, G.-B., Zhou, H., Ding, X., & Zhang, R. (2012). Extreme learning machine for regression and multiclass classification. *IEEE Transactions on Systems, Man, and Cybernetics, Part B (Cybernetics), 42*(2), 513–529.
39. Zhou, Y., Peng, J., & Philip Chen, C. L. (2014). Extreme learning machine with composite kernels for hyperspectral image classification. *IEEE Journal of Selected Topics in Applied Earth Observations and Remote Sensing, 8*(6), 2351–2360.
40. http://www.ehu.eus/ccwintco/index.php/Hyperspectral_Remote_sensing_Scenes (2017). Accessed 22 Sept 2017.
41. https://github.com/IPL-UV/altoolboox (2017). Accessed 15 Jan 2017.

A Smart Document Converter: Conversion of Handwritten Text Document to Computerized Text Document

Ranjit Kumar Behera and Biswajeet Padhi

Abstract With the advancement of technologies many time-consuming tasks are automated using different software without direct involvement of human beings. Particularly in Educational Service industry, teachers and students are making so many good handwritten notes for future references. Even though they spent a good amount of time to prepare their notes, but they find it very difficult to update if required. At the same time if they want to share their notes among friends or students then the only option is either to make photo copies (Xerox) or by taking pictures and share as images. In both the cases, the note cannot be updated. Although many solutions are exists for converting text image to computerized text, but all are not efficient in terms user feedback. Hence, in this research article, a novel approach is presented to convert handwritten images to computerized text document. In this approach, first the handwritten characters are extracted from the input image using digital image processing techniques and then these characters are recognized by using machine learning techniques. In this paper convolution recurrent neural network (CRNN) technique is used.

Keywords Smart Document · Machine Learning · Convolution Neural Network · Recurrent Neural Network · Convolution Recurrent Neural Network

1 Introduction

Optical character recognition is one of oldest research area in the field of computer vision and handwriting recognition is a well-known problem. Broadly, two methodologies were used for training and recognition of hand-written words. One is holistic

R. K. Behera (✉) · B. Padhi
School of Computer Science and Engineering, National Institute of Science and Technology (Autonomous), Institute Park, Pallur Hills, Berhampur 761008, India
e-mail: ranjit.behera@gmail.com

B. Padhi
e-mail: biswajeetpadhi1999@gmail.com

S. K. Das et al. (eds.), *Machine Learning Algorithms for Industrial Applications*, Studies in Computational Intelligence 907,
https://doi.org/10.1007/978-3-030-50641-4_9

and another is analytic. Holistic schemes employ top-down approaches for recognizing the complete word, which eliminates the segmentation problem [1]. On the other hand, the analytic-strategies applied bottom-up approaches normally starts at character or stroke level moving towards constructing a meaningful text. For this strategy either explicit [2] or implicit [3] segmentation of word is employed so that character or strokes can be generated from a word. With the help of segmentation the recognition problem is greatly reduced to a mere identification of isolated characters or strokes. In the early stages of computer vision the feature extraction task is done by human. They take the help of image processing to extract features from images. In last decade, the applications of soft computing increases rapidly due to its soft and flexible nature in several areas in terms of bio-inspired, nature-inspired and also some intelligence services [4–9]. Deep Neural Network (DNN) is a part of machine learning which is subset of soft computing used for analyse and prediction methodologies. In this paper, two key elements are used such as DNN and Convolution Neural Network (CNN) where feature extraction task is done by CNN [10]. CNN features are more powerful than traditional human recognized features because in CNN the model will learn itself from the images and it improves its performance [11].

If any model uses traditional approach then it need to extract lines from document, and there after each line converted into multiple words then characters are extracted from words then the model is trained based on these features. [12]. But this is inefficient and also requires lots of human interaction. So instead of following the aforementioned approach, a better approach can be by training a single CNN model, which extracts the features from documents and train itself using those features through back propagation method. This will save lots of time and requires less human interaction and at the same time it will perform better than traditional approach. There are so many researches going on CNN and the most advanced CNN models which generate more accurate result then the traditional approach [13] is incorporated in our proposed work. There are some websites are available which takes an image and convert that into document [14, 15]. But reviews from the user are not satisfactory as these converters are unable to give the desired outcome. Hence, through this research article, we presented a model which can successfully convert a scanned text image into digital text with free of cost. In this application two types of Neural Network models are used. One is Convolution Neural Network which is shortly known as CNN and other one is bidirectional Long Short Term Memory (LSTM) which is shortly known as bidirectional LSTM. CNN is used to extract features from image data and bidirectional recurrent neural network (RNN) is responsible for generating digitized output text [16]. The analysis is done on a pre-existing model and here researches are done to get better performance than the existing model. In this work, more hyper parameter tuning are done and based on some experiment the dimension of kernel-size, stride-dimension, and feature-map-size are decided. Instead of taking GRU unit [17] bidirectional LSTM units are used which give better result than simple GRU unit.

The remainder of this paper is organized as follows: Sect. 2 describes the overview of the proposed model by giving a detail explanation of the each step followed to convert the handwritten-text to computerized-text. Model architecture and bidirectional LSTM is given in Sect. 3. In Sect. 4 the Simulation set up and result discussion has been presented. Finally the conclusion and future work is given in Sect. 5.

2 Model Overview

Optical character recognition (OCR) technique comprises of several of several steps. The process starts with scanning the handwritten document and then after pre-processing, segmentation, feature-extraction, and character recognition is done in the middle stages. Finally, the output is produced in the form of digital document. All the steps are clearly shown in the Fig. 1 and the detail explanation pertaining to each section is given in the following sub-sections.

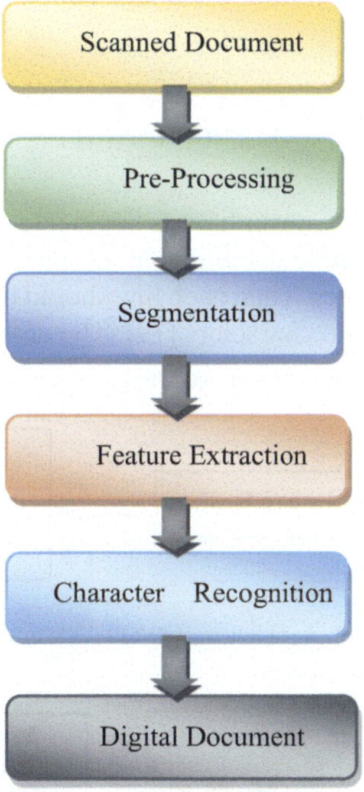

Fig. 1 Overview of model

2.1 Scanned Document

The first step is the Scanned Document phase. In this phase we take the photo copy of the document through camera devices or somebody can take an input already stored from a computer by browsing the entire path.

2.2 Pre-processing

The next phase is the Pre-Processing phase. This is one of the most important phases and the overall operation is shown in Fig. 2. In this phase, we resize the image so that it should fit to our model. Then we convert the image into gray scale image because here we don't need colour information [18]. Then noise should be removed from the image. We also adjust contrast and brightness of the image so that our model can learn from the image better.

2.3 Segmentation

The next phase is Segmentation phase. A machine learning model can't process the whole document at a time. So we need to break the whole document into multiple lines. Then each line should be break into multiple words. Again we break each word into characters [19]. The entire steps of segmentation are shown in Fig. 3.

Fig. 2 Document pre-processing

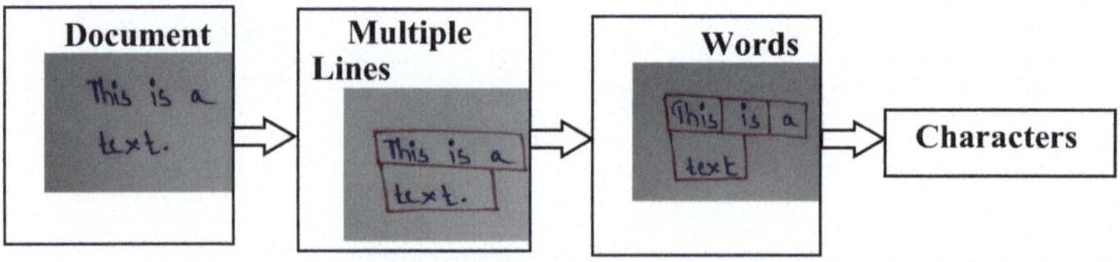

Fig. 3 Document segmentation

2.4 Feature Extraction

Feature Extraction is the most important phase in this process. In this phase our CNN model will take the image and convert them into features. Using these features the model will learn to recognize characters from the document [20]. Initially feature extraction tasks are done using image processing. But due to advancement of deep Neural Network more advanced model like CNN, RNN are developed. Hence, CNN is used to extract features from image data. It internally uses image processing kernels but these kernels are learned as part of training. CNN model is biologically inspired from human visual cortex. When the image passes through several convolution layer feature maps are generated. Initially these feature maps are trying to detect edges present in the image then when the image passes through deep CNN try to figure out what are the different shapes are present in the image then colour then try to find the regions in the image.

2.5 Character Recognition and Digital Document Generation

In the Character Recognition phase our model will recognize characters from the learned features. It will identify each character from the image and store it in a document and will generate a digital document.

In our application Segmentation and Feature extraction task is done by 7 layer CNN model. A convolution layer is used to extract features from image. In convolution layer the first step is applying convolution operation which identifies edges from the image and these edges are responsible for identifying characters in the image. It is shown in Fig. 4 [21].

After this we apply activation function to the feature map which is the output of convolution layer. Then we apply max pooling layer which take the maximum value in a grid. The convolution layer with max pooling is shown in Fig. 5 [21].

This layer is also responsible for avoiding over fitting the model to the image data. Here Batch Normalization is added to normalize the data before going into the next layer. If it is not done, then there may be chance of getting data with different distribution then the input data which is applied to the model. Here, dropout may not be needed because no of parameters is less. The output of 7 layer convolution layer is inserted into bidirectional LSTM layer [22]. There are 2 bidirectional LSTM layers present, which is shown in Fig. 6. The main task of bidirectional LSTM is to predict sequence of characters from the features generated from 7 layer CNN model.

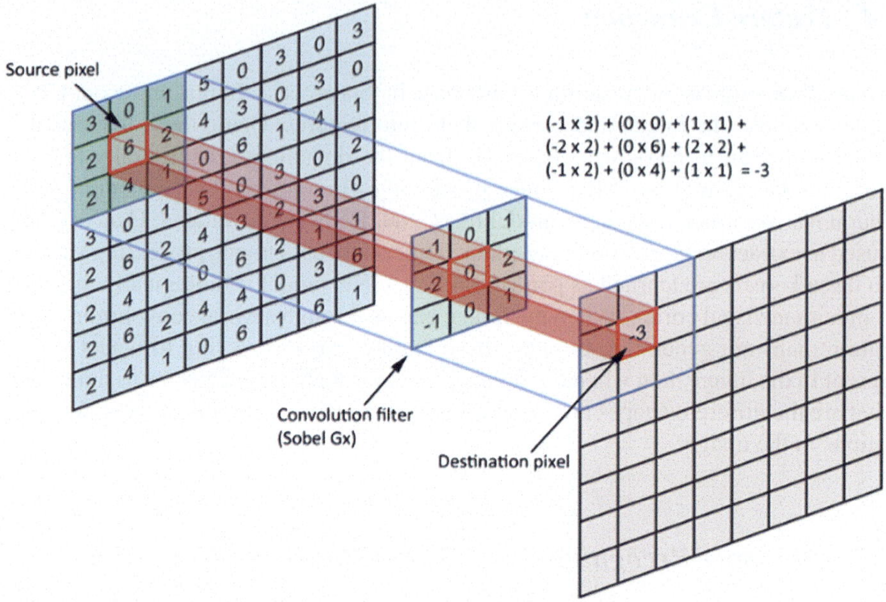

Fig. 4 Convolution layer

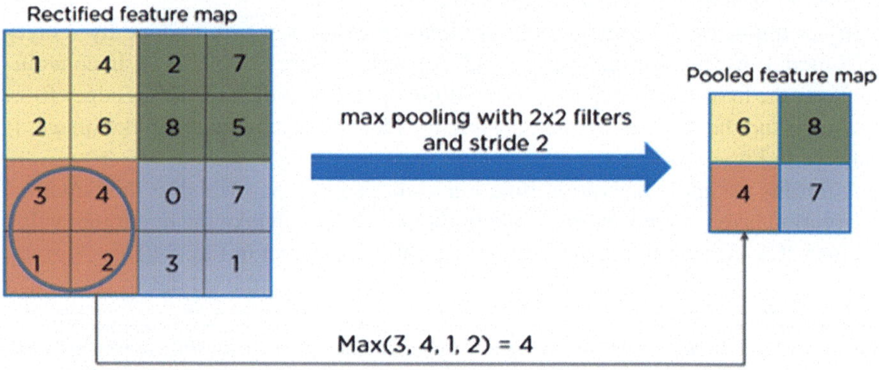

Fig. 5 Max pooling

3 Model Architecture

In the proposed model, first convolution layer is added which take the image of size $64 \times 128 \times 1$. Thereafter, convolution operation is applied with kernel size 3×3. As there are 64 such kernels which in turn will produce a tensor of dimension $128 \times 64 \times 64$. After that, max pooling layer is applied which will give $64 \times 32 \times 64$ dimension tensor. Then second convolution layer is applied having 128 kernels each

Fig. 6 Bidirectional LSTM

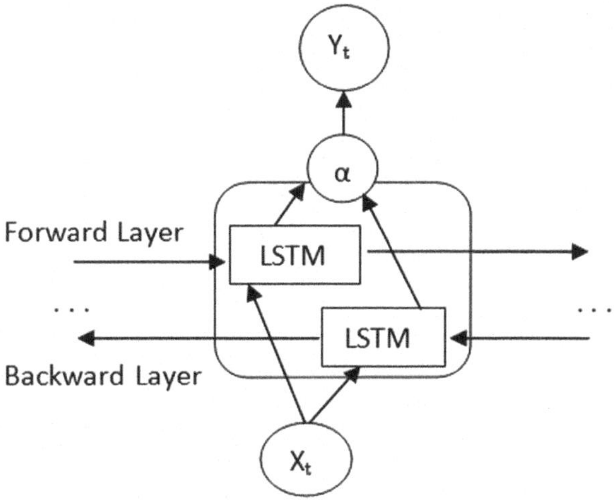

of size 3 × 3 which will generate 64 × 32 × 128 dimension tensor. Then once again max pooling is applied of pool size 1 × 2 which will generate output tensor of size 64 × 16 × 128. Then convolution layer of 256 kernels each of size 3 × 3 is applied which will give 64 × 16 × 256 dimension tensor. Similarly another 4 convolution layers are added which will produce final output tensor of size 64 × 4 × 512. Now this output will go to bidirectional LSTM. In this juncture the output generated by CNN model is reshaped and it is done from 64 × 8 × 512 dimension to 64 × 2048 dimensions. The detailed block diagram is shown in Fig. 7. After that dense layer is added which give 64 × 64 dimensions. We convert the output to this dimension because we have assumed that each line contains at most 64 characters and to identify each character 64 dimension features are needed.

This dimension is feed into bidirectional LSTM layer which consists of 256 LSTM units which is also known as memory cell or memory unit. Here the activation function is ReLU. So in order to initialize the weights we use "he_normal" initialize [23]. Previously people were used sigmoid, tanh activation functions. But due to this they ran into problem of vanishing gradient problem. To avoid this problem researchers find an most popular and powerful activation function called Rectified Linear Unit (ReLU). The rectified linear activation function is a piecewise linear function that will output the input directly if is positive, otherwise, it will output zero [24] shown in Eq. (1).

$$F(x) = x^{+} = \max(0, x) \tag{1}$$

This will generate output of dimension 64 × 256 dimensions. This will go to second bidirectional LSTM layer which will produce output of dimension 64 × 512. Now we again add 79 neuron dense layers which will generate output of size 64 × 79. Here we assume that each line consists of maximum of 64 characters and each character is one of the 79 characters which

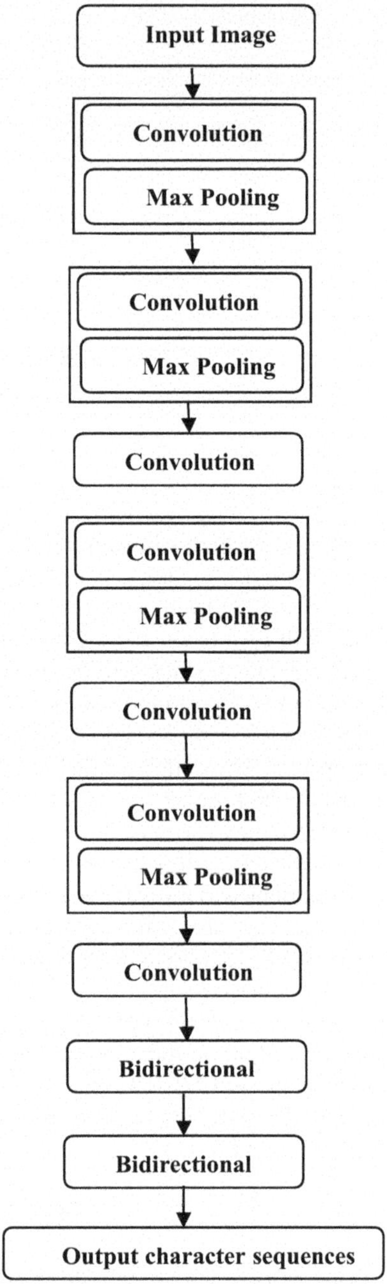

Fig. 7 Detailed model block diagram

are "!"#&'()*+,−./0123456789:;?ABCDEFGHIJKLMNOPQRSTUVWXYZabcdefghijklmnopqrstuvwxyz". So here we use multi class classification which is softmax and the loss function is categorical cross entropy and the optimizer is adam with accuracy matrix. The entire model is presented by splitting it into two algorithms. Algorithm 1 mainly defines the CRNN model, whereas Algorithm 2 defines the steps required to convert the image to text.

Algorithm 1: CRNN model.

```
Model( Image ):
    model = add convolution layer and take image as input
    model = add batch normalization layer
    model = add ReLU activation function to the current layer
    model = add max pooling layer
    model = add convolution layer
    model = add batch normalization layer
    model = add ReLU activation function to the current layer
    model = add max pooling layer
    model = add convolution layer
    model = add batch normalization layer
    model = add ReLU activation function to the current layer
    model = add convolution layer
    model = add batch normalization layer
    model = add ReLU activation function to the current layer
    model = add max pooling layer
    model = add convolution layer
    model = add batch normalization layer
    model = add ReLU activation function to the current layer
    model = add convolution layer
    model = add batch normalization layer
    model = add ReLU activation function to the current layer
    model = add max pooling layer
    model = add convolution layer
    model = add batch normalization layer
    model = add ReLU activation function to the current layer
    model = add bidirectional LSTM layer
    model = add bidirectional LSTM layer
    return model
```

Algorithm 2: Steps to concert image-to-text

```
Convert(Image):
    Lines = Break document into multiple lines
    document = open file
    model = Model(image)   // trained model
    For each line in lines:
        Digital_text_line = model(line)
        For each character in Digital_text_line:
            Append character into document
    Return document
```

4 Results and Discussions

The input, output and comparison with other methods are discussed in the following sub-sections.

4.1 Input

It is a gray-value image of size 128 × 32. Usually, the images from the dataset need to be resize it without distortion as all the images may not be of size 128 × 32. Then the resized image is copied into a target image of size 128 × 32. Finally, the gray values of the image are normalized, which simplifies the task for the Neural Network. Data augmentation can easily be integrated by copying the image to random position instead of aligning it to the left or by randomly resizing the image.

4.2 Output

Once the capture image passes into this model, it goes through the seven convolution layer where the feature extraction is done. Then the extracted features pass through two bidirectional LSTM which predict the sequence of characters. Then the sequence of characters can be stored in a document with different extension such as .txt, .docs, .pdf etc. But in the Fig. 8, we have shown the text format only as output.

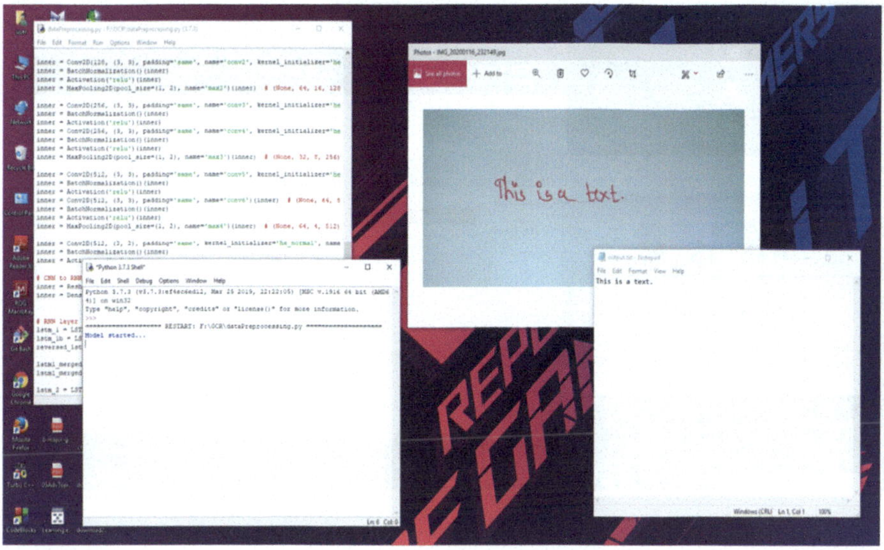

Fig. 8 Result after running the program with an input file

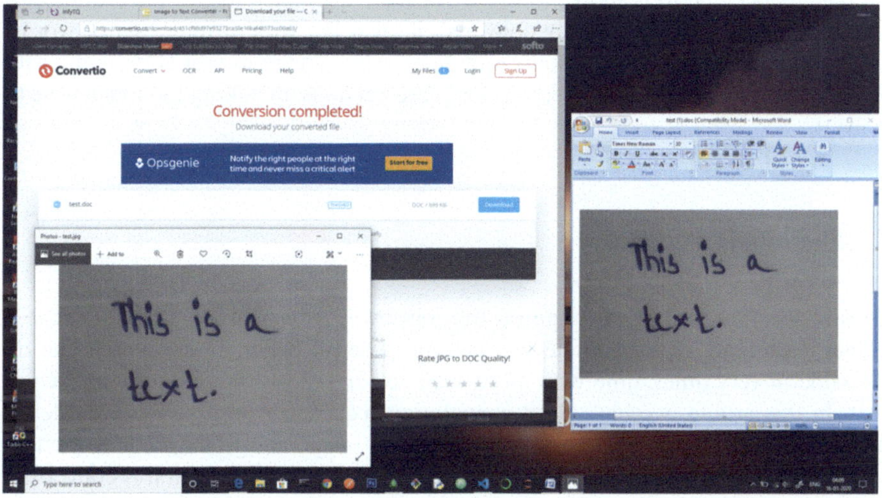

Fig. 9 Result of convertio [25] with an input file

4.3 Result Comparison

Some websites and mobile apps are available, which can also convert a scanned image into document. But most of them are not giving satisfactory result. For example, a website convertio.co [25] with an application name convertio, which takes the

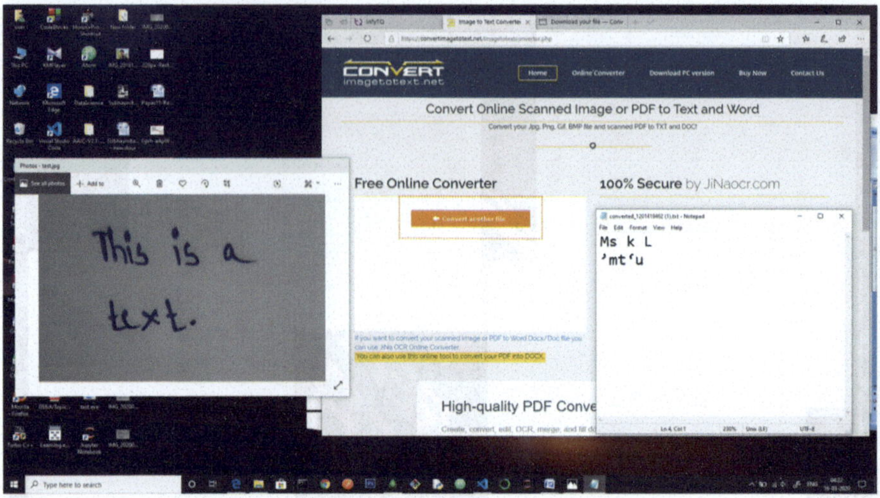

Fig. 10 Result of convertimagetotext.net [26] with an input file

scanned text image as input and then simply save the image into Microsoft word document. The output is shown in Fig. 9. There is another website convertimagetotext.net [26], which successfully converts into text but with different set of words. The result is shown in Fig. 10. Some mobile apps like image-to-word [15] come with options like paid version and free version. Free versions are taking too much time to convert.

5 Conclusions

To save the time and effort required for converting a hand-written document to a digital document, in this article we have proposed a better model which can do the same in very quick time with more accuracy and most importantly with free of cost. In this article, we have used convolution recurrent neural network model in which CNN extracts the feature from the digital image and RNN predicts the text sequence using the extracted features. From the result, we found that it is converting the handwritten document to its corresponding computerized document with more than 79% accuracy. In future, if we are planning to train our model with large data-set and tune the model accordingly, so that the accuracy can improve. For more accuracy one can also add more CNN layers and can use Connectionist temporal classification (CTC) model.

References

1. Guillevic, D., & Suen, C. Y. (1998). Recognition of legal amounts on bank cheques. *Pattern Analysis and Applications, 1*(1), 28–41.
2. Wang, J., & Jean, J. (1994). Segmentation of merged characters by neural networks and shortest path. *Pattern Recognition, 27*(5), 649–658.
3. Mohamed, M., & Gader, P. (1996). Handwritten word recognition using segmentation-free hidden Markov modeling and segmentation-based dynamic programming techniques. *IEEE Transactions on Pattern Analysis and Machine Intelligence, 18*(5), 548–554.
4. Behera, R. K., Reddy, K. H. K., & Roy, D. S. (2020). A novel context migration model for fog-enabled cross-vertical IoT applications. In *International Conference on Innovative Computing and Communications* (pp. 287–295). Singapore: Springer.
5. Roy, D. S., Behera, R. K., Reddy, K. H. K., & Buyya, R. (2018). A context-aware fog enabled scheme for real-time cross-vertical IoT applications.". *IEEE Internet of Things Journal, 6*(2), 2400–2412.
6. Dey, N., Ashour, A. S., Beagum, S., Pistola, D. S., Gospodinov, M., Gospodinova, E. P., et al. (2015). Parameter optimization for local polynomial approximation based intersection confidence interval filter using genetic algorithm: an application for brain MRI image de-noising. *Journal of Imaging, 1*(1), 60–84.
7. Chatterjee, S., Hore, S., Dey, N., Chakraborty, S., & Ashour, A. S. (2017). Dengue fever classification using gene expression data: a PSO based artificial neural network approach. In *Proceedings of the 5th International Conference on Frontiers in Intelligent Computing: Theory and Applications* (pp. 331–341). Singapore: Springer.
8. Jagatheesan, K., Anand, B., Samanta, S., Dey, N., Ashour, A. S., & Balas, V. E. (2017). Particle swarm optimisation-based parameters optimisation of PID controller for load frequency control of multi-area reheat thermal power systems. *International Journal of Advanced Intelligence Paradigms, 9*(5–6), 464–489.
9. Karaa, W. B. A., Ashour, A. S., Sassi, D. B., Roy, P., Kausar, N., & Dey, N. (2016). Medline text mining: an enhancement genetic algorithm based approach for document clustering. In *Applications of Intelligent Optimization in Biology and Medicine* (pp. 267–287). Cham: Springer.
10. LeCun, Y., Boser, B., Denker, J. S., Henderson, D., Howard, R. E., Hubbard, W., et al. (1989). Backpropagation applied to handwritten zip code recognition. *Neural Computation, 1*(4), 541–551.
11. Bojarski, M., Del Testa, D., Dworakowski, D., Firner, B., Flepp, B., Goyal, P., Jackel, L.D., & et al. (2016). End to end learning for self-driving cars. arXiv preprint arXiv:1604.07316.
12. Mori, S., Nishida, H., & Yamada, H. 1999. *Optical character recognition.* New York: Wiley.
13. Nanni, L., Ghidoni, S., & Brahnam, S. (2017). Handcrafted vs. non-handcrafted features for computer vision classification. *Pattern Recognition, 71,* 158–172.
14. https://www.img2go.com/convert-to-document.
15. https://play.google.com/store/apps/details?id=com.cometdocs.imagetoword&hl=en_IN.
16. Wigington, C., Stewart, S., Davis, B., Barrett, B., Price, B., & Cohen, S. (2017). Data augmentation for recognition of handwritten words and lines using a CNN-LSTM network. In *14th IAPR International Conference on Document Analysis and Recognition (ICDAR)* (Vol. 1, pp. 639–645). IEEE.
17. https://github.com/sushant097/Handwritten-Line-Text-Recognition-using-Deep-Learning-with-Tensorflow.
18. Shi, M., Fujisawa, Y., Wakabayashi, T., & Kimura, F. (2002). Handwritten numeral recognition using gradient and curvature of gray scale image. *Pattern Recognition, 35*(10), 2051–2059.
19. Lienhart, R., & Effelsberg, W. (2000). Automatic text segmentation and text recognition for video indexing. *Multimedia Systems, 8*(1), 69–81.
20. Wang, T., Wu, D. J., Coates, A., & Ng, A. Y. (2012). End-to-end text recognition with convolutional neural networks. In *Proceedings of the 21st International Conference on Pattern Recognition (ICPR2012)* (pp. 3304–3308). IEEE.

21. https://www.freecodecamp.org/news/an-intuitive-guide-to-convolutional-neural-networks-260c2de0a050/.
22. Graves, A., Jaitly, N., Mohamed, A. -R. (2013). Hybrid speech recognition with deep bidirectional LSTM. In *2013 IEEE Workshop on Automatic Speech Recognition and Understanding* (pp. 273–278). IEEE.
23. Li, Y., Yuan, Y. (2017) Convergence analysis of two-layer neural networks with ReLu activation. In *Advances in Neural Information Processing Systems* (pp. 597–607).
24. Agarap, A. F. (2018). Deep learning using rectified linear units (ReLU). arXiv preprint arXiv: 1803.08375.
25. https://convertio.co/jpg-doc/. Accessed on 12 Mar 2020 at 10:30AM.
26. https://convertimagetotext.net/imagetotextconverter.php. Accessed on 12 Mar 2020 at 10:30AM.

GRNN Based an Intelligent Technique for Image Inpainting

Debanand Kanhar and Raunak Chandak

Abstract Image inpainting is usually framed as a constrained image generation problem. It is a method that helps to reconstruct the lost or deteriorated parts of images as well as video. The main focus in image inpainting techniques is how precisely to can generate the corrupted pixels in an image. In this paper, we tried using a single pass learning algorithm which greatly reduce time to train the model. The objective of the proposed model is to reconstruct large continuous regions of missing or deteriorated parts of an image. In this paper, GRNN based model along with some image inpainting techniques is being used. Each image is divided into two sections: the missing part that is to be reconstructed, and the context. The network would work identically for arbitrary removals not just for regions having particular shapes such as square or rectangles. Final evaluation is based on the Mean Squared Error (MSE) and Peak Signal to Noise Ratio (PSNR) between the corrupted image and the original image for the regions which is to be regenerate.

Keywords Image Inpainting · Deep Learning · General Regression Neural Network · Image Processing

1 Introduction

Image inpainting is a technique of modifying image in such a way that it is unidentifiable to human eyes [1, 2]. Various sophisticated algorithms are used to regenerate the corrupted part of the image or remove some defects in such a way that it looks natural. Inpainting can be done by using the surrounding regions of the corrupted area and texture of the image. There are various applications of inpainting such as in

D. Kanhar · R. Chandak (✉)
School of Computer Science and Engineering, National Institute of Science and Technology (Autonomous), Institute Park, Pallur Hills, Berhampur 761008, Odisha, India
e-mail: raunak.chk@gmail.com

D. Kanhar
e-mail: devanand@nist.edu

S. K. Das et al. (eds.), *Machine Learning Algorithms for Industrial Applications*, Studies in Computational Intelligence 907,
https://doi.org/10.1007/978-3-030-50641-4_10

photographs and cinemas, it can be used to remove defects in photographs or remove object from the film to create some special effect or for de-censoring images or to remove some objects from videos. Various methods to perform inpainting are given as:

(a) **Structural method of inpainting:** Images that are smooth and have clear distinct and strong borders can be restored using this method. It is based on idea that similar shape and domain can be used to regenerate geometry of lost or corrupted regions. One such way is to start from the border of corrupted area and keep passing the information inwards.
(b) **Textural method of inpainting:** While the former method works well with smooth images, but if the images are highly textured then this method of inpainting works better. The obvious reason is because texture can be defined as repetition of some pattern and cannot be regenerated by filling the gap based on information from boundaries. For a textured image one might need to combine frequency and spatial domain information to get desired texture. One way is to segment the image and selecting areas from the image which are corresponding to the region which is to be inpainted. Another could be to assign priorities to boundary pixels of inpainting domain. Regions which higher probability ensures that linear structure would be extended first.
(c) **Combination of structural and textural:** Combine the previous two methods and to get an algorithm which works even better than them. It uses the boundary information to generate structural data while scanning the image for texture and applying it to the corrupted region. Here information is first gathered from corrupted domain and then image smoothness is determined by Laplacian operator and is propagated by anistrophic diffusion. The reason it works so well is that it combines the positive of both the algorithm into one. One excellent example would be exemplar based method of inpainting. Here, one can search a suitable patch from surrounding region to fill in the corrupted regions. The Partial Differential Equations (PDE's) can also be used for this purpose.

The proposed is based on soft computing [3] approach where General Regression Neural Network (GRNN) is used as a soft computing method for image inpainting for irregular holes. There are several soft computing methods are available, among them fuzzy logic, genetic algorithm, particle sward optimization, etc. play an important role [4–10]. These are rapidly used in several application.

The remaining of the paper is organize as follows. Section 2 describe some existing works as related works. Section 3 describe as general regression neural network. The proposed method describe in the Sect. 4. Section 5 describe result and analysis part. Finally, conclusion described in Sect. 6.

2 Related Works

In image processing encoding of image is required to represent the image in compressed form, to provide the security by changing the visual information of the image content or hiding the content. Dey et al. [11] proposed a hybrid method of image encoding and hiding. This method embeds the spirally encode secret image into HH component of cover image using alpha blending technique. Dey et al. [12] introduced multi-level encoding scheme of gray scale image by applying Walsh table and spiral encoding respectively. Palodhi et al. [13] 2020 proposed a hybrid image processing technique based on convolution technique. The basic aim of this proposal is to analyze etch-pit image in detector system i.e. nuclear track. It contains several shapes and sizes for analyzing the same things. In 2016, Pradhan et al. [14] have introduced a novel content-based image retrieval (CBIR) system which automatically discards the irrelevant background information to improve the visual feature quality. Here, the authors have utilized the graph-based visual saliency approach to locate the object region from which the final features have been extracted to perform the image retrieval task. In 2017, Varish et al. [15] have proposed a new texture and color feature extraction scheme for CBIR. Initially, they have computed the probability color histogram of the image and applied a cumulative probability-based scheme to divide the histogram into some non-uniform chunks. Next, they have computed the dual-tree complex wavelet transforms (DT-CWT) based texture features from the intensity image. Finally, they have combined the texture features along with the computed statistical parameters from chunks of the image histogram to create a final feature vector for CBIR. The advent of multimedia storage, capturing, and communication accumulated a wide spectrum of heterogeneous video data available to enterprise and consumer applications. In the meanwhile, storage and retrieval of multimedia data, and in particular image and video data in particular, needs more than connecting with data repositories and delivering data. In this context, the literature possess limited tools and techniques to organize, and retrieve image video data. Manual annotation of video content in the repositories is time consuming and subjected to inaccuracy and user's domain knowledge dependency. Content based image and video retrieval is one of the most recent and highly potential technique to index and retrieve the most relevant image or videos based on the visual characteristics from the repositories. Feature extraction is one of the most important aspect in retrieval and efforts have been made to reduce the curse of dimensionality reduction. Authors of [16] provided a dimensionality reduction approach using reduced Scale Invariant Feature Transform (SIFT) feature for effective image retrieval. Image and video retrieval is a highly challenging domain and video retrieval is the more challenging than image. In this context, a video retrieval method is proposed [17] which can retrieve the relevant videos from the repositories with reduced frames and minimal computational cost. Video summarization based on clip segmentation for key frame extraction is employed and color correlation for matching the corresponding keyframes between the query video and videos in the repositories. But, this work lack spatial and temporal consistency which are essential and well defined structures to define the content. Samanta et al. [18] introduced the idea of application of genetic operator to encode the gray scale image. Samanta et al. [19] extended their previous idea for color image. Samanta et al. [20] proposed cuckoo search based image multi-level image segmentation technique. This paper used correlation as fitness function, where high correlation indicates good quality of segmentation. Result of this paper supports that the proposed method achieved the good quality of segmentation. A new kind of algorithm which is based on food collecting mechanism of ant introduced in Samanta et al. [21] this paper also shown the effectiveness of their algorithm for multilevel image segmentation. The method is very effective to select the optimum threshold values with high correlation. Singh et al. [22] 2020 proposed a machine learning method for predict the weight and size of the image. It basically uses two basic parameters such as mass and size of the rice kernels. This is a recursive method for identifying the rice kernels and estimation the size. Kumar et al. [23] have used a block-level feature extraction scheme to further improve the quality of the extracted visual features of the image. They have decomposed the input image into non-overlapping blocks and employed the discrete cosine transforms (DCT) on each image block. Next, from all the significant DCT components they have computed the GLCM features to perform the final image retrieval. Next, in 2018, Kumar et al. [24] have introduced another CBIR system in which they have utilized the color and shape feature together to improve retrieval accuracy. Here, they have used color autocorrelogram and gray-level co-occurrence matrix in the YCbCr color domain to extract the high-quality visual features. Later, in 2018, Pradhan et al. [25] have further improved the image retrieval efficiency by using the combination of color and texture features together. Here, the authors have employed color histograms and DT-CWT to extract color and texture features for image retrieval. Next, in 2018, Raj et al. [26] have introduced an image fusion approach that can be used for medical imaging and image retrieval tasks. In this method, the author suggested a four neighborhoods Shannon entropy and non-sub sampled contour let transform based image fusion scheme. To maintain such structure and retrieve the relevant video, authors of [27] demonstrated a hierarchical spatiotemporal technique for effective and efficient video retrieval technique. High retrieval accuracy in this method is achieved by pattern generation and angular distribution density approach of encoded frames with queue pool provide the temporal consistency and matching. Similarly, a computational mechanism based on motion vector is illustrated in [28] which extract the key frames using outlier concept. The spatial and temporal consistency is maintained by employing spatial pyramid matching which divide the key frames into sub regions and compute the feature. The objective evaluation and subjective visual perception of this method make it an effective and highly accurate approach for video retrieval. The rapid growth of multimedia system and its effective storage and retrieval bring new challenge which create copyright protection issue of images and video in the repositories. Copyright protection is basically the authentication of image and video content and ownership which can identify illegal copies. One solution is to embed an image by adding an invisible structure or image known as a digital watermark to the image or videos while the other is privacy preserving in CBIR. In order to identify illegal copies in an image authors of [29] proposed a watermarking scheme to balance the trade-off between imperceptibility and security by employing singular value decomposition and visual cryptography. This highly secure method is highly secure and is evident from the fact that even when the watermark is in intruder's possession, it is impossible even for the most sophisticated cryptanalysts or intruder to decrypt the embedded watermark without the key share owned by the authorized image owner. In [30], the authors described an image steganographic approach that can hide a large amount of secret bits into the cover colour image by maintaining good perceptibility. The algorithm focuses on solving the pixel overflow problem in the stego image as exists in (Pixel Value Differencing) PVD method. The proposed method also increases security so that intruder cannot identify the presence of secret data in stego-image. In [31], the authors have merged the (Pixel Value Differencing) PVD method with (Least Significant Bit) LSB method to improve the message hiding capacity. The specific area of the cover image are selected for hiding data using both the methods whereas the rest are embedded with only PVD method. It enhances the security measure as well. In [32], the proposed method had been designed for contrast enhancement of any raw image by employing directional morphological filters. This algorithm efficiently enhances the contrast of an image containing oriented features specially.

3 General Regression Neural Network

General Regression Neural Network (GRNN) is a variation of radial basis neural network which is specifically designed for regression and function approximation. It does not require any error based back propagation learning. Therefore it is a type of one pass learning algorithm as all the parameters are determined in one complete iteration of dataset. High accuracy is generally obtained as it uses Gaussian estimation. Another advantage is that it can handle noise in inputs. However, as the size of model is huge, it can be computationally expensive. One use case of this algorithm is that it can be applied in any regression problem where assumption of linearity is not justified. GRNN model consists of four layers which are given as:

(a) **Input layer:** As the name suggests this layer is the first layer of GRNN model. Number of nodes in this layer equals to the number of independent features in the data.
(b) **Pattern or hidden layer:** The training pattern which is going to be used is represented in this layer. Number of neural nodes used defines pattern on which model will be trained on. Therefore, has to select an optimal number of nodes in this layer as it will affect our training.
(c) **Summation layer:** There are only 2 nodes in this layer which are fully connected to nodes in pattern layer. First node acts as a numerator where summation of all the information from previous layer is done. Let weights for this layer connecting ith node of pattern layer and first node of this layer be represented as w_i. Second node acts as denominator but the weights of all nodes connected to this unit

is 1. Therefore, A and B define value of first node and second nodes given as Eqs. (1) and (2):

$$\sum_{i=1}^{n} w_i \cdot \exp\left(\frac{-D_i^2}{2\sigma^2}\right) \tag{1}$$

$$\sum_{i=1}^{n} \exp\left(\frac{-D_i^2}{2\sigma^2}\right) \tag{2}$$

where D_i is an Euclidean distance between X (for which output is to be found out) and inputs; n is the number of inputs and σ is a spread parameter which determines smoothness of approximation. More the value of this parameter, smoother the image formed will be. Optimal value can be decided using various value of the parameter and choosing the one which has least error.

(d) **Output layer:** This layer gives the output of the model. As this is a regression model, the output is a real value. Suppose y represents the output of the model, then value of the y shown in Eq. (3).

$$y = \frac{A}{B} \tag{3}$$

where A and B are obtained from summation layer.

3.1 Training Procedure

It involves finding out the value of σ. This can be determined using Mean Square Error (MSE). Evaluate MSE for different values of σ and then use that value which has minimum MSE by using Table 1.

Table 1 Example input and their corresponding values

Input	Output
1	4
3	7
5	9
7	10

Let input is 4, then, first calculate D's based on Eqs. (4) and (7).

$$D1 = (4 - 1)^\wedge 2 = 9 \tag{4}$$

$$D2 = (4 - 3)^\wedge 2 = 1 \tag{5}$$

$$D3 = (4 - 5)^\wedge 2 = 1 \tag{6}$$

$$D4 = (4 - 7)^\wedge 2 = 9 \tag{7}$$

Let us take value of smoothing factor as 1 for calculating weights using formula given in Eq. (8).

$$\exp\left(\frac{-D_i^2}{2\sigma^2}\right) \tag{8}$$

where values of w1 to w4 are 0.01, 0.6, 0.6, and 0.01, hence, the value of B and A are given in Eqs. (9) and (10).

$$B = w1 + w2 + w3 + w4 = 1.22 \tag{9}$$

$$A = 4^* w1 + 7^* w2 + 9^* w3 + 10^* w4 = 9.74 \tag{10}$$

Therefore, for input x = 4, y = 7.98 = 8
Advantages of GRNN algorithm are as follows:

(a) Because of its one pass learning nature, it takes very short amount of time to learn.
(b) It does not get struck in local optima.

3.2 GRNN for Image Inpainting

Mathematically image can be defined as function g(x,y) in 2-dimensional plane where each pixel takes on some value from g(x,y). Hence if [x,y] belongs to known region than z = g(x,y) and if it belongs to unknown/corrupted region than z = g′(x,y) where g′(x,y) is approximated from g(x,y). Now train the model such that g(x,y) can predict z for known regions accurately. Also one can expect that it would expect it produces result which visually pleasant.

4 The Proposed Method

The main problem is now to determine how to train the model, suppose if every known pixel is used as an input to predict unknown regions than it would be computationally expensive. GRNN network stores all information from input layer in its pattern layer and use that to form a model which identifies the unknown regions first and sorts them based on areas in descending manner. To fill up a region start from the outermost pixel of the unknown region and start moving inwards progressively. One can use it's nearest outermost known layer to predict the values.

One can use the property of dilution and erosion to catch the previous outermost known regions and use them as inputs. The outermost unknown region in that region is filled pixel by pixel using the known layer and then move onto next inner layer of unknown regions. Let mask be the same size as image which is being used for inpainting. This mask is binary and represents 1 for known region and 0 for unknown regions. The proposed algorithm shown in Algorithm 1.

Algorithm 1: The proposed algorithm.

Step 1: Identify the corrupted region in the image.
Step 2: Create a binary mask with same dimensions as the image and where 0 represents the corrupted region and 1 represents known regions.
Step 3: For each value of σ
Step 4: Find all region which needs to be filled and try to fill them simultaneously.
Step 5: For each region apply dilution and erosion properties in the binary mask to get the outermost layer of the region and store as area_to_be_filled.
Step 6: Similarly use dilution and erosion to get nearest layer with known values surrounding the region.
Step 7: Create training set with for all as input and output y=g(x,y) which is the pixel value at that location.
Step 8: Use canny edge detection on the surrounding area.
Step 9: Incorporate those edges (if found) along with GRNN model giving priorities as required.
Step 10: Build a GRNN model using the training set.
Step 11: Using this model areas_to_be_filled, calculate the predicted value g'(x,y).
Step 12: Update the corresponding region in the mask also.
Step 13: Goto Step 4.

In GRNN based inpainting, one has to find value of one unknown parameter σ, which is the spread parameter and controls the smoothing done by approximate function. In the proposed method, we have used two loss functions to evaluate our method given as:

4.1 Mean Squared Error (MSE)

MSE averages the squares of difference between predicted result (i.e. estimated values) and actual result (i.e. what is estimated?). The fact that MSE is almost always strictly positive (and not zero) is because the estimator does not account for information that could produce a more accurate estimate. The equation of MSE shown in Eq. (11).

$$\text{MSE} = \frac{1}{n}\sum_{i=1}^{n}\left(y_i - \hat{y}_i\right)^2 \tag{11}$$

where n represents the number of data for training set.

4.2 Peak Signal to Noise Ratio (PSNR)

PSNR is used to measure the ratio between the maximum possible power of a signal (in our case brightness of pixel) and the power of corrupting noise that affects the fidelity of its representation. It measure the peak error. The equation of PSNR shown in Eq. (12).

$$\text{PSNR} = 10.\log_{10}\left(\frac{255 * 255}{MSE}\right) \tag{12}$$

5 Results and Discussion

The proposed method simulated in the Python by given system configuration shown in Table 2.

Table 2 Simulation parameters

Software/Hardware	Specification
Windows OS	8.1
MS Office	Office 16.0
Python	3.6
Processor	Intel i5
Speed	2.4 GHz–3.8 GHz
RAM	8 GB
Hard Disk	I TB

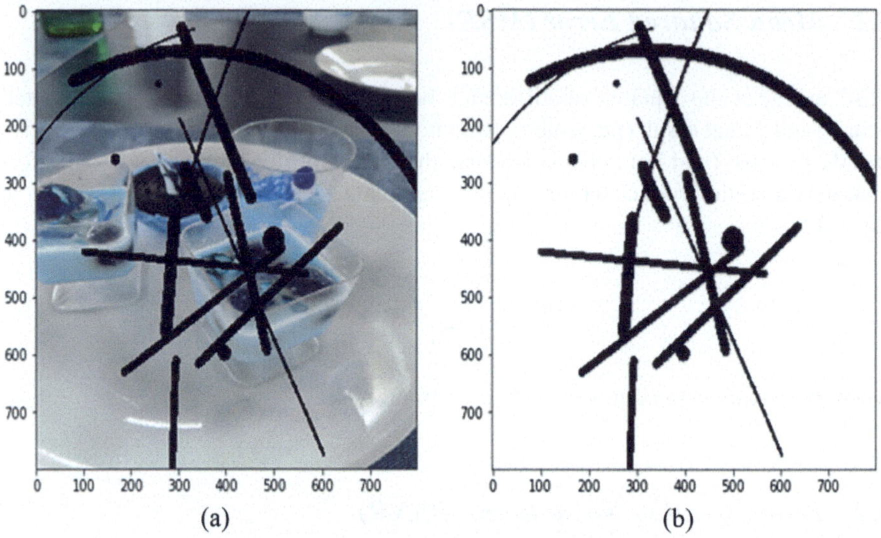

Fig. 1 The corrupted image and its corresponding mask

The simulation performed in 30 iteration which is depicted by few iteration such as iteration 0, iteration 20, and iteration 30. Figures 1(a) and (b) shows corrupted image and its corresponding mask. Figure 1(a) is the original image which is corrupted. Corrupted regions are represented by black patch. This is the region which needs to be remade by inpainting. Figure 1(b) is the corrupted region in the image is represented in binary format. Here, 0 indicates all the corrupted region and 1 denotes undamaged region.

In iteration 10, the corrupted region in the mask starts shrinking gradually shown in Fig. 2. Figure 2(a) shows gradual decrease in area occupied by corrupted regions as the model slowly starts to fill required regions. Figure 2(b) corresponding mask showing the areas which are required to be filled and which are already filled up. Comparing with last mask can see that the corrupted regions are slowly decreasing.

In iteration 20, shows the shrinking continues shown in Fig. 3. Figure 3(a) the status of image after current iteration. More regions have been filled up. Slowly moving towards completion. Figure 3(b) shows the status of mask after current iteration. Here, the changes made in masks are quite visible now.

In iteration 30, corrupted regions have almost been identified and fixed which is shown in Fig. 4. In Fig. 4(a), the task is almost complete. Nearly all the regions have been filled up. Very few corrupted region is left to be computed. In Fig. 4(b), very thin line or spots of black region can be seen which suggests that very few corrupted regions are now left.

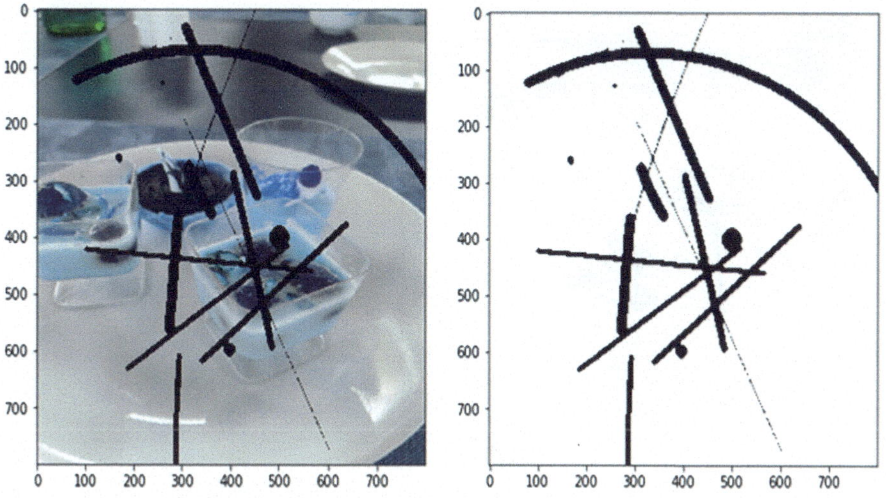

(a) (b)

Fig. 2 The corrupted region in the mask starts shrinking gradually

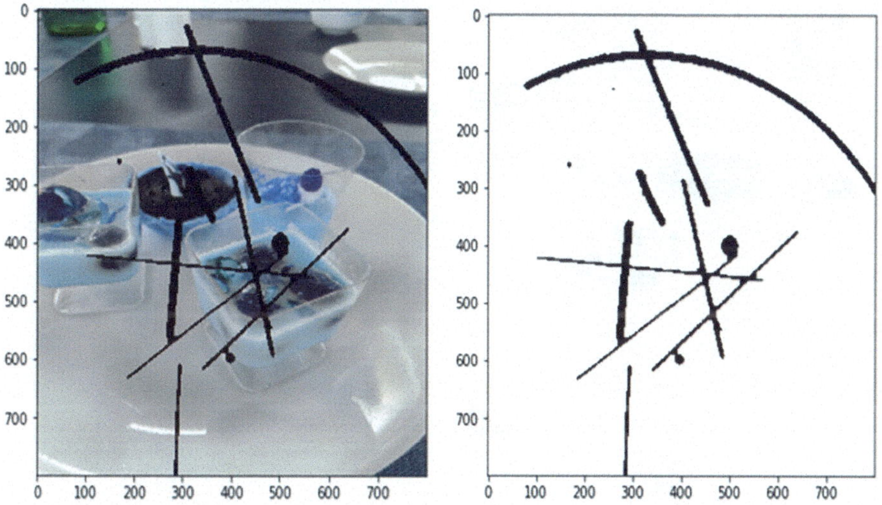

(a) (b)

Fig. 3 The shrinking continues

In final iteration, one can see that the regions are completely filled now and the generated image does not look like it has been modified shown in Fig. 5. Figure 5(a) shows result after final iteration. The model gives an inpainted image after it tries to fill up all the corrupted regions. Figure 5(b) shows final mask after completion of the algorithm. As compared with mask in other iteration(s), one can see that there are no corrupted regions (denoted by 0) left which has to be redeemed.

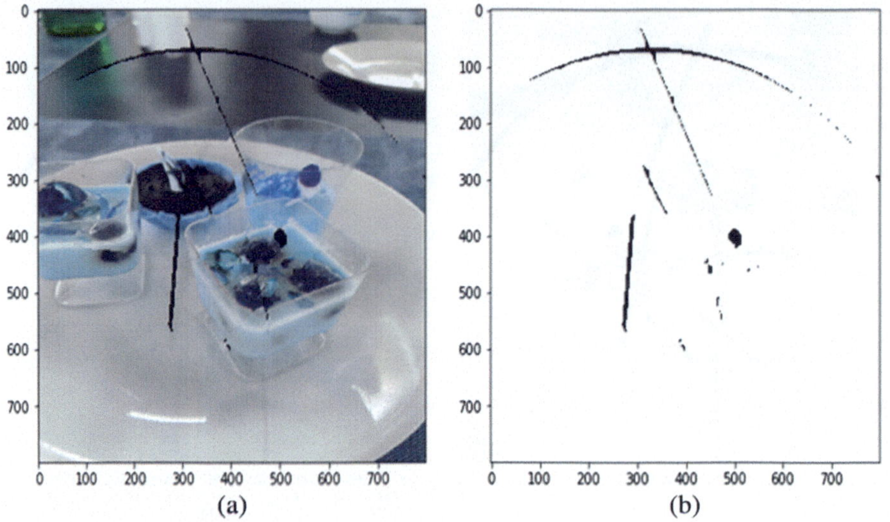

Fig. 4 The corrupted regions have almost been identified and fixed

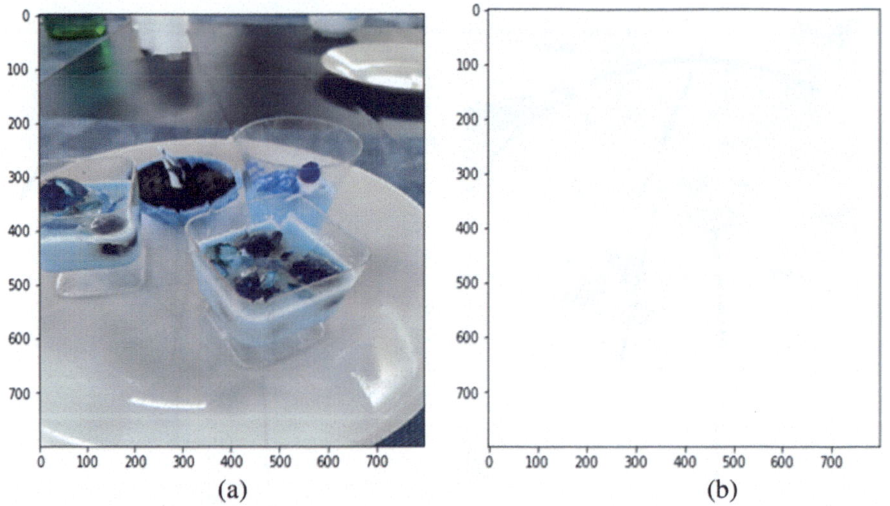

Fig. 5 Regions are completely filled and the generated image does not look like it has been modified

6 Conclusion

We presented a method of image inpainting using GRNN model combining it with Edge detection to produce fine results. It used variety of defective input images and produced their corresponding inpainted output along with step by step of traversal how the algorithm is working and the final output is reached. Assuming the one needs a model which can regenerate an image to such extend to get majority of lost information but have shortage of time to go through training of model and the images are very few in number than this technique could provide an alternative. As research is ongoing in this particular field, better algorithm could be invented which would generate results which provide much better results while taking considerably less time.

References

1. Shi, K., & Guo, Z. (2020). On the existence of weak solutions for a curvature driven elliptic system applied to image inpainting. *Applied Mathematics Letters, 99,* 106003.
2. Hu, W., Ye, Y., Zeng, F., & Meng, J. (2019). A new method of Thangka image inpainting quality assessment. *Journal of Visual Communication and Image Representation, 59,* 292–299.
3. Das, S. K., Kumar, A., Das, B., & Burnwal, A. P. (2013). On soft computing techniques in various areas. *Computer Science and Information Technology, 3,* 59.
4. Das, S. K., & Tripathi, S. (2019). Energy efficient routing formation algorithm for hybrid ad-hoc network: A geometric programming approach. *Peer-to-Peer Networking and Applications, 12*(1), 102–128.
5. Das, S. K., & Tripathi, S. (2018). Adaptive and intelligent energy efficient routing for transparent heterogeneous ad-hoc network by fusion of game theory and linear programming. *Applied Intelligence, 48*(7), 1825–1845.
6. Das, S. K., & Tripathi, S. (2017). Energy efficient routing formation technique for hybrid ad hoc network using fusion of artificial intelligence techniques. *International Journal of Communication Systems, 30*(16), 1–16. e3340.
7. Chatterjee, S., Hore, S., Dey, N., Chakraborty, S., & Ashour, A. S. (2017). Dengue fever classification using gene expression data: a PSO based artificial neural network approach. In *Proceedings of the 5th International Conference on Frontiers in Intelligent Computing: Theory and Applications* (pp. 331–341). Singapore: Springer.
8. Jagatheesan, K., Anand, B., Samanta, S., Dey, N., Ashour, A. S., & Balas, V. E. (2017). Particle swarm optimisation-based parameters optimisation of PID controller for load frequency control of multi-area reheat thermal power systems. *International Journal of Advanced Intelligence Paradigms, 9*(5–6), 464–489.
9. Dey, N., Ashour, A. S., Beagum, S., Pistola, D. S., Gospodinov, M., Gospodinova, E. P., et al. (2015). Parameter optimization for local polynomial approximation based intersection confidence interval filter using genetic algorithm: an application for brain MRI image de-noising. *Journal of Imaging, 1*(1), 60–84.
10. Karaa, W. B. A., Ashour, A. S., Sassi, D. B., Roy, P., Kausar, N., & Dey, N. (2016). Medline text mining: an enhancement genetic algorithm based approach for document clustering. *Applications of Intelligent Optimization in Biology and Medicine* (pp. 267–287). Cham: Springer.
11. Dey, N., Samanta, S., & Roy, A. B. (2011). A novel approach of image encoding and hiding using spiral scanning and wavelet based alpha-blending technique. *International Journal of Computer Technology and Applications, 2*(6), 1970–1974.

12. Dey, N., Samanta, S., & Roy, A. B. (2011). A novel approach of multilevel binary image encoding using walsh table and spiral scanning. *International Journal of Engineering Trends and Technology, 2*(3), 12–14.
13. Palodhi, K., Chatterjee, J., Bhattacharyya, R., Dey, S., Ghosh, S. K., Maulik, A., et al. (2020). Convolution based hybrid image processing technique for microscopic images of etch-pits in Nuclear Track Detectors. *Radiation Measurements, 130,* 106219.
14. Pradhan, J., Pal, A. K., & Banka, H. (2016, December). A prominent object region detection based approach for CBIR application. In *2016 Fourth International Conference on Parallel, Distributed and Grid Computing (PDGC)* (pp. 447–452). IEEE.
15. Varish, N., Pradhan, J., & Pal, A. K. (2017). Image retrieval based on non-uniform bins of color histogram and dual tree complex wavelet transform. *Multimedia Tools and Applications, 76*(14), 15885–15921.
16. Verma, M. K., Dwivedi, R., Mallick, A. K., & Jangam, E. (2018). Dimensionality reduction technique on SIFT feature vector for content based image retrival. *International Conference on Recent Trends in Image Processing and Pattern Recognition* (pp. 383–394). Singapore: Springer.
17. Mallick, A. K., & Maheshkar, S. (2016, December). Video retrieval based on color correlation histogram scheme of clip segmented key frames. In *2016 Fourth International Conference on Parallel, Distributed and Grid Computing (PDGC)* (pp. 213–218). IEEE.
18. Samanta, S., Dey, N., De, A. (November, 2011) A case study of image encoding using genetic operators. *International Journal of Emerging trend is Engineering and Development 3*(1), 278–285.
19. Samanta, S., Roy, A. B., Dey, N. (2012) Color image encoding using genetic operators. *International Conference on Innovation in Management and Information Technology, (ICIMIT 2012), on 07th April, 2012.*
20. Samanta, S., Dey, N., Das, P., Acharjee, S., & Chaudhuri, S. S. (2012) Multilevel threshold based gray scale image segmentation using cuckoo search. *International Conference on Emerging Trends in Electrical, Communication and Information Technologies-ICECIT, 12–23 Dec 2012, ELSEVIER Proceedings.*
21. Samanta, S., Acharjee, S., Mukherjee, A., Das, D., & Dey, N. (2013, December). Ant weight lifting algorithm for image segmentation. In *2013 IEEE International Conference on Computational Intelligence and Computing Research* (pp. 1–5). IEEE.
22. Singh, S. K., Vidyarthi, S. K., & Tiwari, R. (2020). Machine learnt image processing to predict weight and size of rice kernels. *Journal of Food Engineering, 274,* 109828.
23. Kumar, S., Pradhan, J., & Pal, A. K. (2017, December). A CBIR scheme using GLCM features in DCT domain. In *2017 IEEE International Conference on Computational Intelligence and Computing Research (ICCIC)* (pp. 1–7). IEEE.
24. Kumar, S., Pradhan, J., & Pal, A. K. (2018). A CBIR technique based on the combination of shape and color features. *Advanced Computational and Communication Paradigms* (pp. 737–744). Singapore: Springer.
25. Pradhan, J., Kumar, S., Pal, A. K., & Banka, H. (2018, January). Texture and color visual features based CBIR using 2D DT-CWT and histograms. In *International Conference on Mathematics and Computing* (pp. 84–96). Singapore: Springer.
26. Raj, A., Pradhan, J., Pal, A. K., & Banka, H. (2018, March). Multi-scale image fusion scheme based on non-sub sampled contourlet transform and four neighborhood Shannon entropy scheme. In *2018 4th International Conference on Recent Advances in Information Technology (RAIT)* (pp. 1–6). IEEE.
27. Mallick, A. K., & Maheshkar, S. (2018). Near-Duplicate Video Retrieval Based on Spatiotemporal Pattern Tree. In Proceedings of 2nd International Conference on Computer Vision & Image Processing, pp. 173–186, Springer, Singapore.
28. Mallick, A. K., & Mukhopadhyay, S. (2019, March). Video retrieval based on motion vector key frame extraction and spatial pyramid matching. In *2019 6th International Conference on Signal Processing and Integrated Networks (SPIN)*, (pp. 687–692). IEEE.

29. Mallick, A. K., & Maheshkar, S. (2016). Digital image watermarking scheme based on visual cryptography and SVD. In *Proceedings of the 4th International Conference on Frontiers in Intelligent Computing: Theory and Applications (FICTA) 2015* (pp. 589–598). New Delhi: Springer.
30. Mandal, J. K., & Das, D. (2012). Colour image steganography based on pixel value differencing in spatial domain. *International Journal of Information Sciences and Techniques*, *2*(4), 83–93.
31. Mandal, J. K., & Das, D. (2012, August). A novel invisible watermarking based on cascaded PVD integrated LSB technique. In *International Conference on Eco-friendly Computing and Communication Systems* (pp. 262–268). Heidelberg: Springer.
32. Das, D., Mukhopadhyay, S., & Praveen, S. S. (2017). Multi-scale contrast enhancement of oriented features in 2D images using directional morphology. *Optics Laser Technology, 87,* 51–63.

Data Analysis and Prediction

Content-Based Airline Recommendation Prediction Using Machine Learning Techniques

Praphula Kumar Jain and Rajendra Pamula

Abstract Presently, We are living in the 21st Century, with a population of around 7.7 Billion people in the world. There is a rapid increase in the number of customers traveling by airplane day by day. The number of airplane companies has also increased considerably, in order to meet the requirement of the customers. When there is a number of companies serving the same purpose, it creates confusion among people, which one is to choose. So it has become very important for the travelers to know, which airline could be the best for them as per their demand and budget. It can be known by the honest reviews submitted by the travelers by sharing their previous experience. The reviews submitted not only help the customers to choose the appropriate airline but also help the airline companies to know their shortcomings and improve their quality of service. Since airline reviews data is online and its huge amount of data, so it becomes difficult to manage things manually. Due to this fact, there is a need for a model that can help in the recommendation. This chapter aims at distinguishing reviews as positive or negative from the content of the online customer reviews submitted by the previous customer and providing a recommendation. In this paper, we have compared three machine-learning algorithms namely Logistic Regression, Stochastic Gradient Descent (SGD) and Random Forest Classifier. It also predicts the accuracy of recommendation done by the ML Techniques.

Keywords Logistic Regression · Stochastic Gradient Descent · Random Forest · Machine Learning · Sentiment Analysis

P. K. Jain (✉) · R. Pamula
Department of Computer Science and Engineering, Indian Institute of Technology (ISM), Dhanbad, Dhanbad 826004, JH, India
e-mail: praphulajn1@gmail.com

R. Pamula
e-mail: rajendrapamula@gmail.com

S. K. Das et al. (eds.), *Machine Learning Algorithms for Industrial Applications*, Studies in Computational Intelligence 907,
https://doi.org/10.1007/978-3-030-50641-4_11

1 Introduction

One of the most important aspects of every business organisation is to understand consumer intention to a product or service. The organizations need to deal with the feedback both positive and negative along with their consequences. They should also focus on techniques as well as the solutions to handle it. The need for measuring consumer promoter score (NPS) and consumer loyalty (CL) has been extensively documented in extant literature. Both NPS and CL have a great impact not only on profitability but also for the success of an organization [2]. The organisations measure metrics like Customer Satisfaction (CSAT) and NPS where Consumer reviews and their recommendations acts as one of the major sources of information [1].

Presently, Human Beings are more comfortable online and they prefer to share their opinions through online platforms, whereby they are expressing their views, opinions, and experiences, giving ratings or textual reviews on various parameters to different services, and openly recommending or discouraging the purchase of products and services. While some websites have standardized fields where the consumers can share their opinion based on which ratings have to be given, others have textual reviews or both [3]. Such mixed reviews make the operation of NPS difficult for organizations. There are numerous ways to asses and address customer satisfaction, and behavioral intentions. Managers generally rely on customer feedback to identify future managerial goals and to monitor the performance of a firm through NPS and CL [4]. The International Air Transport Association (IATA) provides a passenger satisfaction benchmarking study called Airs@t.

The online reviews are an important source of information, which plays an important role to enhance the decision-making capabilities of stakeholders in this context i.e. presumptive consumers and service providers. Firstly, to examine whether the textual reviews can successfully and completely determine the reviewer's assessment for recommending that product or not. Secondly, the ability to disentangle the recommendation decision to examine how specific service aspects are being evaluated by prior consumers and how they drive the overall recommendation decision can further inform the stakeholders. This chapter helps in analyzing distinct service aspects and inferring recommendations expressed in online reviews is thus needed to address these relevant issues.

The vital role of online reviews as a key data source in tourism has also been shown in recent studies that focus on the use of online reviews and their role in influencing consumer behavior and decision making. Travel blogs are a form of digital storytelling or word-of-mouth, which are self-published to disseminate travel narratives and adventures [5].

This chapter is organized as follows "literature review" in this section different theoretical and methodological points related to the prediction of airline sentiment have been presented. In "methodology" section, a details description of the related dataset and machine learning techniques are given. "Results and Discussion" section, the result analysis presented with detail description. The chapter concluded in the last "conclusion" section.

2 Literature Review

Travelers usually prefer to have proper research about their preferred destination before the actual travel [6]. Not just travelers, every class of Business more than half of its consumers do online research before investing in any product or service [7]. But Tourism has the highest percentage of pre-online searching before purchase where 73% of the travelers did online research before making travel decisions. Going through previous studies, it has very well explained that the stance of experienced consumers in online reviews plays an important role in influencing the purchase decision of consumers [8]. Author [9] investigated the ascendancy of experienced consumer feedback on a product or service and they too concluded the same. Since the growth of travel service is directly proportional to experience, opinions in online reviews have a very strong impact on travelers [10].

These online reviews are helpful, not only for business promotion and growth but also to bring about innovation in the product or service and improve customer experience. The author suggested a word of mouth convention about the product or services providers in the field of consumer recommendations. The author [5] also suggested NPS is the most realistic parameter to predict organizational growth. NPS is defined as the promoters and detractors percentage difference. This proposition has received much scrutiny by academics and practitioners alike over the past decade. Previous research has found that growth in NPS is correlated with growth in business.

There are a large number of travel bloggers all around the world and an infinite count of online reviews floating all over the internet. This count is beyond any human being could visualize. In order to make proper use of these reviews, there is a need to find an innovative technique that can handle this large amount of data with minimal human influence. This technique must be able to analyze the visual attitude automatically as in sentiment classification can do the task of understanding the online reviews [11, 12, 19]. Mining opinions is a complex task wherein, first of all, we need to extract these reviews from the websites and secondly, we have to separate out reviews from no reviews. Author [13, 18] found that traditional text mining algorithms do not perform well on sentiment classification as it does in the case of Topic Based Categorization. Keywords can be used to identify topics but sentiments are expressed in an adroit manner.

Sentiment Classification aims to extract the required from textual reviews and classify the reviews as positive or negative based on the polarity of the review [14, 15]. With the results of sentiment classification, consumers would know the necessary information to determine which products to purchase and sellers would know the response from their customers and the performances of their competitors. With the wide adoption of computing technology, sentiment classification of reviews has become one of the foci of recent research endeavors. The method has been attempted in different domains such as movie reviews, product reviews, customer feedback reviews, and legal blogs [13, 17].

In relation to opinion mining applications, the extant literature indicates two types of techniques have been utilized, including machine learning and semantic orientation [13, 20, 21]. The machine learning approach that is applicable to this problem mostly belongs to supervised classification in general, and text classification techniques in particular, for opinion mining [16]. In supervised learning, prior training of the dataset is required for the machine to actually do sentiment classification automatically.

3 Methodology

3.1 Dataset Description

The online airline reviews dataset collected form the airlinequality.com website. This is a freely online platform to collect airline reviews dataset. Our collected dataset contains 2000 rows of different customer feedback all around the world. In this dataset customer feedback is in the form of content, along with the content positive or negative recommendation is also mentioned in the form of 0 or 1. Where 0 indicate negative recommendation and 1 shows positive recommendation.

3.2 Data Preprocessing

While implementation we require Pandas, an inbuilt library in Python plays an important role. Pandas is an inbuilt package in Python providing fast, flexible, and expressive data structures designed to make working with "relational" or "labeled" data both easy and intuitive. As described above, the recommended score defines whether a review is positive or negative. Therefore, count the total number of positive and negative reviews. As per the dataset, there are 1186 positive reviews and 814 negative reviews.

We drop the unnecessary columns like "airline name", "author country" and "cabin flown". Now we have only two columns "content" and "recommended". Then we count the content length of all the reviews. Below given table shows some instances of the content length i.e. length of the reviews.

We need to separate out the positive and negative reviews in order to do sentiment analysis. After this, we create separate lists to store negative and positive words. We then represent them in a word cloud for a better understanding of the frequent and non-frequent terms.

3.3 Word Cloud

Word Cloud (also known as text cloud or tag cloud) is a way of visual representation of frequently used words in a collection of text files. The size of each word in this picture is an indication of the frequency of occurrence of the word in the entire text. Such diagrams are very useful when doing text analytics. It provides a general idea of what kind of words are frequent in the corpus, in a sort of quick and dirty way. In the Figs. 1 and 2, the words bigger in size represent the words that are common to almost all the reviews like flight, seat etc. while the words which are small in size represent the negative words and positive words respectively.

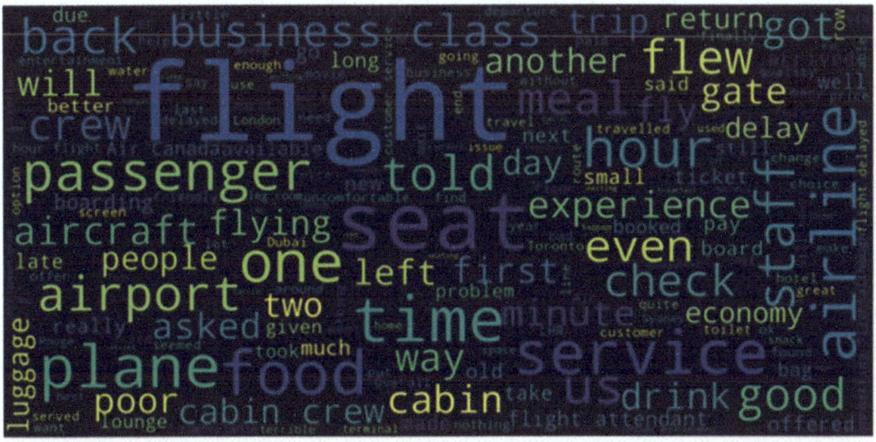

Fig. 1 Word cloud representing the negative text

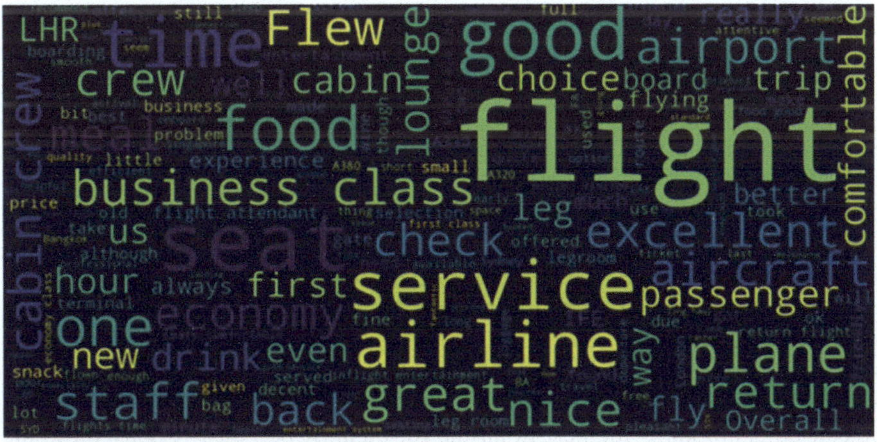

Fig. 2 Word cloud representing the positive text

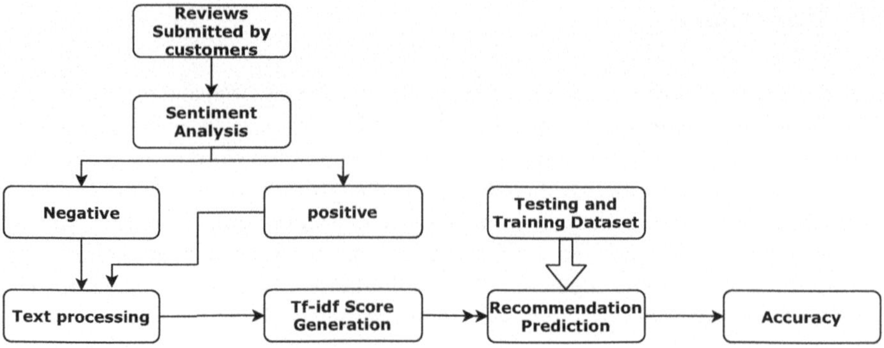

Fig. 3 Flow diagram of proposed approach

3.4 Approach

As described above, we have two types of reviews, positive and negative. The machine distinguishes them from the recommendation value. If the value is 0 then that is a negative review and if the value is 1 then that review is positive. The machine uses this feature to do sentiment analysis. The machine is trained on a set of data to do sentiment analysis. When the machine is trained, it does sentiment analysis on another set of data that just have the reviews and give recommendation scores to it. In the end, it predicts its accuracy, precision, f1-score and support. There is also a matrix showing the number of reviews in the testing dataset being predicted correctly with respect to actual values. We use the Logistic regression, SGD Classifier, Random Forest Classifier algorithms to reach our goal. Flow diagram of proposed approach is shown in Fig. 3.

4 Result and Discussion

With the increasing number of airline companies, reviews play a very important role. The present generation has serious trust issues, to them, experiences matter more. They do not believe in the services promised by the airline companies, They believe in authentic reviews i.e. they trust the point of view of the experienced people or the people who have traveled by that airline. With the advancement in science and Technology, Online platforms are the best way to share individual views and experiences. Even the airline companies themselves support it, because there is no better critic than the audience, in this case, the passengers. These reviews help the companies to understand and rectify their flaws. These reviews also help the consumers to know the best-fit airline for them. It becomes very useful to the consumers if they can already get a list of recommended airline companies when they search for an airline. For that, the machine first needs to be able to distinguish between positive and negative reviews and give a recommendation score to it, so that it becomes easier to filter the features for customers based on their choice. In this chapter, we have used 3 algorithms namely Logistic Regression, Stochastic Gradient Descent Classification and Random Forest Classification as discussed below.

4.1 Logistic Regression

Logistic Regression is a statistical method of predictive analysis that was used in the biological sciences in the early twentieth century. It was then used in many social science applications. Logistic Regression is used when the dependent variable(target) is categorical. This is used to describe data and to explain the relationship between one dependent binary variable and one or more nominal, ordinal, interval or ratio-level independent variables. When we apply Logistic Regression classification on our dataset, we get 86% accuracy with an 87.7% Precision score. We have very few contents that have been wrongly classified, i.e., only 8 positive reviews have been classified as negative ones and 48 negative reviews are classified as positive. The machine displays "Model Performance Metrics" are shown in Table 1 we have obtained obtain on applying the Logistic Regression Algorithm.

4.2 Stochastic Gradient Descent (SGD) Classification

Stochastic Gradient Descent (SGD) is a very simple and efficient approach to discriminative learning of linear classifiers under convex loss functions such as (linear) Support Vector Machines and Logistic Regression. SGD has been known in the machine learning community for a long time, and it has recently received a considerable amount of attention in the context of large-scale learning. SGD has been successfully applied to large-scale and sparse machine learning problems often encountered in text classification and natural language processing. Using SGD Classifier we get 88% accuracy and 87.9% precision. Like the Logistic Regression this also displays the Model Performance Metrics in Table 1.

4.3 Random Forest Classifier

Random forest, like its name implies, consists of a large number of individual decision trees that operate as an ensemble. Each individual tree in the random forest spits out a class prediction, and the class with the most votes becomes our model's prediction. The fundamental concept behind random forest is a simple but powerful one the wisdom of crowds. In data science-speak, the reason that the random forest model works so well is: A large number of relatively uncorrelated models (trees) operating as

Table 1 Comparative result analysis for implemented model

Classifier	Accuracy	Precision	Recall	F1 score
Logistic Regression	0.86	0.87	0.86	0.85
SGD Classification	0.88	0.87	0.88	0.87
Random Forest Classifier	0.80	0.80	0.80	0.80

a committee will outperform any of the individual constituent models. The machine displays "Model Performance Metrics" are shown in Table 1 we have obtained on applying the Random forest classifier.

5 Conclusion

With the increasing number of airline companies, online reviews play a vital role in sentiment analysis. The present generation has serious trust issues, to them, experiences matter more. They do not believe in the services promised by the airline companies, they believe in authentic online customers reviews i.e. they trust the point of view of the experienced people or the people who have traveled by that airline. With the advancement in science and Technology, Online platforms are the best way to share individual views and experiences. Even the airline companies themselves support it, because there is no better critic than the audience, in this case, the passengers. These online customer reviews also help the consumers to know the best-fit airline for them. Earlier, everything was done manually but with the increasing number of companies and with the increasing population, it is very difficult to maintain websites that are completely handled manually. It becomes very useful to the consumers if they can already get a list of recommended airline companies when they search for an airline. For that, the machine first needs to be able to distinguish between positive and negative reviews and give a recommendation score to it, so that it becomes easier to filter the features for customers based on their choice.

In this chapter our aims to do sentiment analysis and distinguish between reviews and by allotting recommendation scores classify them as positive or negative. In the end, the accuracy and prediction scores of the used algorithms presented hered. In this chapter, I have used three machine learning algorithms namely Logistic Regression, Stochastic Gradient Descent (SGD) Classification and Random Forest Classification as discussed above. Out of the three, SGD Classification gives the most accuracy i.e 88%, Logistic Regression comes in second with 86% accuracy and Random forest Classification comes last with 80.25% accuracy. On comparing these three algorithms we can understand that SGD Classification gives us the most accurate results.

In future, this work can be extended to be collaborated to an airline search Engine where if a review is positive it can suggest airline to the consumers as per the filter chosen by them. This implementation can also be implemented for different service provider industries like hotel, tourist guides, and security.

References

1. Ho-Dac, N. N., Carson, S. J., & Moore, W. L. (2013). The effects of positive and negative online customer reviews: Do brand strength and category maturity matter? *Journal of Marketing, 77*(6), 37–53.
2. Reichheld, F. F. (2003). The one number you need to grow. *Harvrd Business Review, 81*(12), 46–55.
3. Siering, M., Deokar, A. V., & Janze, C. (2018). Disentangling consumer recommendations: Explaining and predicting airline recommendations based on online reviews. *Decision Support Systems, 107*, 52–63.
4. Morgan, N. A., & Rego, L. L. (2006). The value of different customer satisfaction and loyalty metrics in predicting business performance. *Marketing Science, 25*(5), 426–439.
5. Xiang, Z., & Gretzel, U. (2010). Role of social media in online travel information search. *Tourism Management, 31*(2), 179–188.
6. Lo, A., Cheung, C., & Law, R. (2002). Information search behavior of Hong Kong's inbound travelers-a comparison of business and leisure travelers. *Journal of Travel & Tourism Marketing, 13*(3), 61–81.
7. Ye, Q., Zhang, Z., & Law, R. (2009). Sentiment classification of online reviews to travel destinations by supervised machine learning approaches. *Expert Systems with Applications, 36*(3), 6527–6535.
8. Godes, D., Mayzlin, D., Chen, Y., Das, S., Dellarocas, C., Pfeiffer, B., et al. (2005). The firm's management of social interactions. *Marketing Letters, 16*(3–4), 415–428.
9. Zhu, F., & Zhang, X. (2006). The influence of online consumer reviews on the demand for experience goods: The case of video games. In *ICIS 2006 Proceedings* (p. 25).
10. Moon, J., Chadee, D., & Tikoo, S. (2008). Culture, product type, and price influences on consumer purchase intention to buy personalized products online. *Journal of Business Research, 61*(1), 31–39.
11. Barbosa, L., & Feng, J. (2010). Robust sentiment detection on twitter from biased and noisy data. In *Proceedings of the 23rd International Conference on Computational Linguistics: Posters*. Association for Computational Linguistics.
12. Turney, P. D. (2002). Thumbs up or thumbs down? Semantic orientation applied to unsupervised classification of reviews. In P. Isabelle (Ed.), *Proceeding of Association for Computational Linguistics 40th Anniversary Meeting, Philadelphia, PA, USA* (pp. 417–424) ACL.
13. Pang, B., Lee, L., & Vaithyanathan, S. (2002). Thumbs up? Sentiment classification using machine learning techniques. In P. Isabelle (Ed.), *Proceeding of 2002 Conference on Empirical Methods in Natural Language, Philadelphia, USA* (pp. 79–86). Association for Computational Linguistics.
14. Dave, K., Lawrence, S., & Pennock, D. M. (2003). Mining the peanut gallery: Opinion extraction and semantic classification of product reviews. In G. Hencsey & B. White (Eds.) *Proceeding of 12th International Conference on World Wide Web* (pp. 519–528). Budapest, Hungary: ACM Press.
15. Okanohara, D., & Tsujii, J. (2005). Assigning polarity scores to reviews using machine learning techniques. In R. Dale, K. F. Wong, J. Su, & O.Y. Kwong (Eds.), *Natural Language Processing IJCNLP 2005. Lecture Notes in Computer Science (LNAI)* (vol. 3651, pp. 314–325).

16. Conrad, J. G., & Schilder, F. (2007). Opinion mining in legal blogs. In A. Gardner (Ed.), *Proceedings of the 11th International Conference on Artificial intelligence and Law, Stanford, California* (pp. 231–236). ACM Press.
17. Beineke, P., Hastie, T., & Vaithyanathan, S. (2004). The sentimental factor: Improving review classification via human-provided information. In *Proceedings of the 42nd Annual Meeting of the Association for Computational Linguistics (ACL) 2004* (pp. 263–270). Association for Computational Linguistics.
18. Turney, P. D., & Littman, M. L. (2003). Measuring praise and criticism: Inference of semantic orientation from association. *ACM Transactions on Information Systems*, *21*(4), 315–346.
19. Chaovalit, P., & Zhou, L. (2005). Movie review mining: A comparison between supervised and unsupervised classification approaches. In R. R. Sprague (Ed.), *Proceedings of the 38th Hawaii International Conference on System Sciences, Big Island Hawaii* (pp. 1–9). IEEE.
20. Ye, Q., Zhang, Z., & Law, R. (2008). Sentiment classification of online reviews to travel destinations by supervised machine learning approaches.
21. Jain, P. K., et al. (2019). Airline recommendation prediction using customer generated feedback data. In *4th International Conference on Information Systems and Computer Networks (ISCON)*. IEEE.

A Meta-heuristic Based Approach for Slope Stability Analysis to Design an Optimal Soil Slope

Jayraj Singh and Haider Banka

Abstract Slope stability analysis is routinely performed due to geological instability found in earth slopes. It has been an important role in geotechnical engineering for all the times. Although the efficiency of modern technology has significantly improved, the slope stability analysis is becoming more difficult due to the existence of imprecise, uncertainties and discontinuous function in the actual scenario. Moreover, given the presence of local minima points, the position of the critical failure surface in slope stability analysis is made inaccurate and cumbersome. This work proposes a meta-heuristic based approach to locate critical failure surfaces under prevailing conditions and constraints. The reliability and efficiency of the approach are examined by investigating the benchmark case studies. The outcome results indicate that, the proposed approach could acquire acceptable performance over existing methods and attain a better solution quality in terms of accuracy and efficiency.

Keywords Meta-Heuristic Algorithms · Slope Stability Analysis · Critical Failure Surface · Firefly Algorithm

1 Introduction

Finding the critical surface is very important in the assessment of slope stability to determine the most sliding surface and to observe the condition under which any structured shape of the slide mass can fail. The Limit Equilibrium Analysis (LEM) approach is among the most widely adopted Stability Factor Evaluation Procedure [1]. This procedure gain popular because of its less number of required parameters like; slope geometry, geological, topography, dynamic, static, loads, geotechnical parameters and hydrological condition and its simplicity. In this way, some LEM's methods, such as Fellenius [2], Bishop [3], Janbu methods [4] and others successfully computes factor of safety (FoS) value with some assumption conditions.

J. Singh (✉) · H. Banka
IIT(ISM) Dhanbad, Dhanbad, India
e-mail: jayrajsinghit@gmail.com

H. Banka
e-mail: haider.banka@gmail.com

S. K. Das et al. (eds.), *Machine Learning Algorithms for Industrial Applications*, Studies in Computational Intelligence 907,
https://doi.org/10.1007/978-3-030-50641-4_12

The demanding of these methods is to divide the possible slide mass into a fixed number of vertical slices to resolve the various forces acting on the slice face [5]. Various researchers and engineers have successfully used the application of these methods of slices and explained their pros and cons [6]. In some methods a good initial guess or trial and error process have been used to find critical failure surface. However, in absence of engineer's experience, the critical slip surface evaluation is still a challenging task. In addition to this, the factor of safety function associated with a slip surface is often highly multi-modal and non-smooth due to its discontinuous function [7]. Hence, the stochastic approaches could produce efficiently the optimal result to such problems. Various stochastic approaches have been focused due to their elegant efficiency and performance. Many researchers in the following literatures [8–13], have applied different stochastic search techniques based on Monte Carlo technique, grid search, GA, ACO and ICA techniques. In the present study, a firefly algorithm is considered to analyze the critical slip surface associated with minimum factor of safety (FoS) value. Janbu method from the LEM procedure has described as fitness function for the algorithm using the Firefly algorithm. With this analysis, a better technical solution can be found, which will reduce an reasonable level of risk towards slope damage. The robustness and efficacy of the present algorithm is tested by solving a benchmark problem from the literature.

1.1 Motivation and Contribution

The problem of slope stability analysis by locating critical surface is usually very complex and become a challenging task due to the following reasons; (i) Given n-surfaces in a potential mass, there are m-possible solutions in the form of failure surfaces. Thus finding an optimal solution to a critical slip surface requires high computational time and storage. (ii) It is an unconstrained global optimization problem, which is also referred as NP-complete problem [7]. Therefore, in these circumstance, the metaheuristic approach could effectively obtain better solution [14–17]. This chapter is mainly contribute to the application of FA, a well-known meta-heuristic based approach. The techniques suggested provides an empirical analyses to obtain critical failure surfaces for slope stability analysis. By this study, the engineers and scientists are able to label the susceptible slopes to include the preventive remedy for saving from severe life loss and high economic impact. To minimize the enormous disruption from the slope failures or the risk of landslides, this study may provide a vital role in geotechnical industry to evaluate slope stability.

2 Preliminaries Notation

In this Chapter, Some preliminary notations which are used in the study have been described in Table 1.

Table 1 Description of notations

Notation	Description
FoS	Factor of Safety
α	Angle between the tengent to center of the base of each slice and horizontal
W	Weight of each slice
b	Width of each slice
γ	Unit weight (Density) of the material
T	Mobilized shear force
N	Normal force
C	Cohesion of the material
μ	Pore pressure at slice base
h	Mean height of the slice
(X_l, Y_l)	Lower intersection point of the slip surface
(X_u, Y_u)	Upper intersection point of the slip surface

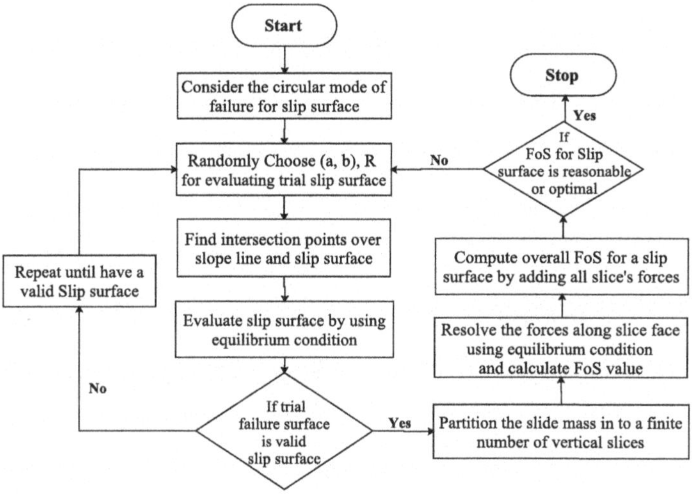

Fig. 1 Process diagram for slope stability evaluation

3 Slope Stability Analysis

The critical surface is analysed to perceive the most falling surface in a moving mass. This can be achieved by generating n- trial slip surfaces for which the corresponding FoS value are successively computed by comparing the available resisting and driving moments along the surface [18]. The analysis for slip surface is formulated in the below procedure and can be illustrated using Fig. 1.

3.1 Phrasing of Slip Surface in Terms of (a, b) and R

In order to evaluate the factor of safety corresponding to a failure surface, a circular arc surface on slide mass (ABCD) is generated by choosing random values of center point (a, b) and radius (R) under its all kinematic and geometrical constraint. The valid intersection points $(x_l, y_l)(x_u, y_u)$ are then calculated. If these intersection points are not found as per the geometric constraint or not placed within its boundary, then it would be either regenerated or corrected for being correct points. Further, the whole slide mass is partitioned into n- vertical slices.

Let $(x_1, y_1)(x_2, y_2), \ldots, (x_n, y_n)$ are the middle points of n- slices and slope hight is reflected by $y = h$.

$$x_{l,u} = \frac{bm + a - dm \pm \sqrt{\partial}}{m^2 + 1} \tag{1}$$

$$y_{l,u} = \sqrt{(R^2 - (x_{l,u} - a)^2)} + b \tag{2}$$

where , $\partial = r^2(1 + m^2) - (b - ma - d)^2$. Now, based on the geometry, the tangential force and normal force on each slice are calculated as;

$$\text{Tengential force on the slice } (T) = \gamma * h * b * Sin(\alpha) \tag{3}$$

$$\text{Normal force on the slice } (N) = \gamma * h * b * Cos(\alpha) \tag{4}$$

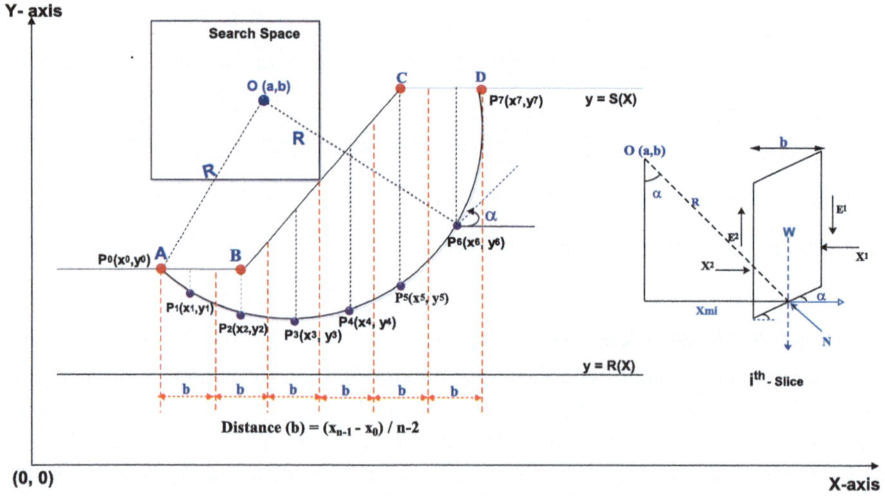

Fig. 2 Free body diagram for a slip surface evaluation

here, the base angle (α) at points (X_{mi}, Y_{mi}) between tangents and center line of every slice (shown in Fig. 2 of the ith-slice FBD) will be computed as;

$$\text{Base angle } (\alpha) = sin^{-1}\left(\frac{X_{mi}}{R}\right) \quad where, \; x_{mi} = x_i + b/2 \tag{5}$$

Now, by substituting all the above respective values in the Eq. 7, the FoS value associated with the slip surface is calculated. This evaluation by generating n- trial slip surfaces are successively repeated until it does no longer attains the minimum criteria.

3.2 Modeling the Fitness Function

The modeling of the fitness function has been accomplished here for employing proposed algorithm. The factor of safety (FoS) function computed using method of slice (Janbu method) from LEM's procedure is described as the fitness function. The factor of safety equation by janbu method calculates on the basis of Mohr's coulomb criterion ($\tau = c + \sigma \tan \phi'$). Therefore, different assumptions have been made regarding to equilibrium conditions [19]. Table 2 illustrates the assumption criteria of Janbu method for the assessment of slope stability.

3.2.1 Janbu Method

Janbu method allows any shape of the failure surface from either of circular or non-circular. Here, it is considered that the horizontal forces are equal on each slices and are only responsible to derive factor of safety. Where as, the inter slice forces (shear force 'T') are neglected and therefore, the obtained factor of safety equation are expressed as:

$$FoS = f_o * FS_{JanbuSimplified} \tag{6}$$

$$FS = \frac{1}{\sum tan\alpha} \cdot \sum \left[(w - ub) \left[\frac{1}{\cos\alpha + \frac{\tan\phi' . \sin\alpha}{F}} \right] \tan\phi' + c'.b \right] cos\alpha \tag{7}$$

here, janbu introduced a correction factor (f_o) to compensate for the fact that the method satisfies only force equilibrium while shear interstice forces are assumed

Table 2 Assumption criteria of considered Janbu methods

LEM's method	Equilibrium conditions		
	Horizontal force	Vertical force	Moment force
Janbu	Yes	Yes	No

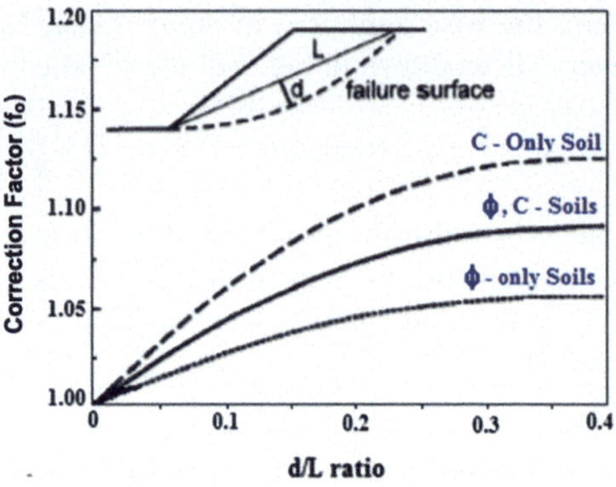

Fig. 3 Variation of correction factor value as a function

to zero. This correction factor accommodate the effects of inter-slice shear forces. The plotting curve shown in Fig. 3 demonstrated the variation of this factor (f_o) as a function between strength parameters of soil materials and slope geometry (i.e., d and L).

4 Firefly Algorithm

The firefly algorithm (FA) is a stochastic based approach, which is inspired by the rhythmic flashing behavior of fireflies [20]. Although Firefly algorithm works similar as PSO, ABC and ACO algorithms. But due to its ease of implementation and elegant efficiency, the FA algorithm receiving a lot of attention and are widely adopted to competent in outperforming highly nonlinear, unconstrained optimization type problems. The algorithm works on the phenomenon of the bioluminescent communication, where a firefly is attracted by another fireflies regarding to their sex. Their attractiveness is directly depends to the brightness and it decreases as the distance among them if increase. Therefore, less brighter firefly will be attracted towards more brighter firefly. However, with increased distance of two fireflies reduces their attractiveness. The algorithm initiates with the random selection of the fireflies. The brightness of the fireflies is associated with type of fitness function which is to be optimized for considered problem. This objective function is evaluated throughout the generation until to meet optimal solution. The basic steps of the firefly algorithm is demonstrated using flow diagram in Fig. 4.

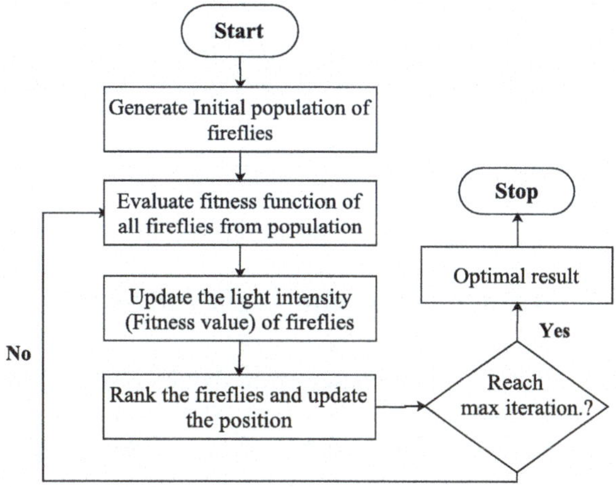

Fig. 4 Work flow diagram of Firefly algorithm

5 Implementation of Firefly Algorithm

In this study, the FA is implemented to analyze the slope stability by locating optimal slip surface associated with minimum FoS. The firefly algorithm is introduced here to minimize the FoS function which describes the objective function of the approach. The proposed Algorithm is proceeded through following steps:

5.1 Initialization of the Firefly Population

In Firefly algorithm, potential solutions of the problem represented by fireflies or agents are initialized randomly in the population. The number of fireflies are corresponding to the number of solutions (N_f) $i.e.,$ $F_i\ \forall_i\ 1 \le i \le N_f$. The pictorial view of an agent (firefly) is demonstrated shown in Fig. 5.

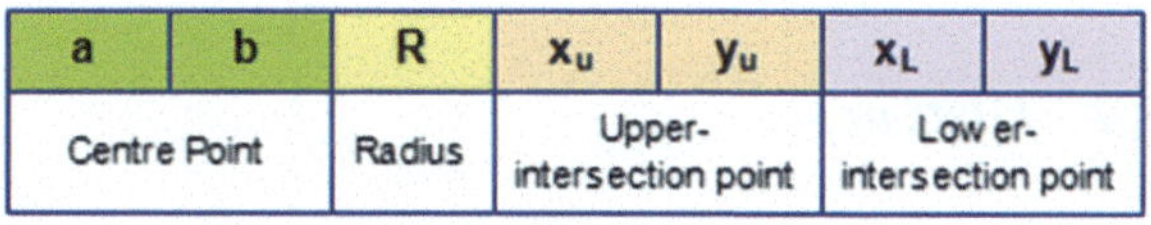

Fig. 5 Pictorial representation of a firefly agent

5.2 Position

Each solution in the firefly population has one of the dimension as position $((x_{id}))$, which is shown by Eq. 8.

$$X_i = x_i^1, \ldots x_i^d, \ldots, x_i^n \tag{8}$$

$Where,\ i = 1, 2, \ldots, n$ and x_i^d represent an ith agent's position in d^{th} dimension.

5.3 Attractiveness and Light Intensity

Two most important phases of firefly algorithm are described as: Variation of light intensity and attractiveness (i.e., Movement towards attractive). Every firefly is attracted towards more brighter glow neighbor fireflies. This attractiveness between fireflies decreases as their distance increases. Let the intensity of light at distance 'r' is I_r and intensity of source firefly is I_s. As we know, this light intensity varies as per the law of inverse which is shown in Eq. 9.

$$I_r = \frac{I_s}{r^2} \tag{9}$$

The light intensity in a medium is determined as follow:

$$I_r = I_0 e^{-\gamma . r} \tag{10}$$

The singularity at r=0 in Eq. 9 is avoided by combining the effects of the inverse square law. The gaussian approximate the equation as follow:

$$I_r = I_0 e^{-\gamma . r^2} \tag{11}$$

Finally after this all calculation, The attractiveness β is formulated using Eq. 12. This attractiveness is directly proportional to intensity of light of the adjacent fireflies.

$$\beta = \beta_0 e^{-\gamma . r^m} \quad (m \geq 1) \tag{12}$$

5.4 Distance

The Cartesian distance $(r_{i,j})$ is calculated between the i and j- fireflies at x_i and x_j respectively as follow:

$$X_{i,j} = \sqrt{\sum_{k=1}^{d} (x_{i,k} - x_{j,k})^2} \tag{13}$$

Here, $x_{i,k}$ represents k^{th}- component of the spatial coordinate in d^{th} dimension.

5.5 *Movement*

In order to enhance the solution quality, less brighter firefly is attracted towards different more brighter firefly. This movement is formulated using Eq. (14) as follow:

$$x_i = x_i + \beta_0 e^{-\gamma . r_{i,j}^2}(x_j - x_i) + \alpha\epsilon \tag{14}$$

where, 'α' is a random parameter and 'ϵ' is a vector of random numbers drawn from a Gaussian distribution. If $\beta_0 = 0$ then it is simple random walk of the firefly.

6 Validation Using Numerical Study

A benchmark study was reviewed from the literature to demonstrate the efficacy of the firefly algorithm. The benchmark study was adopted from the study of Yamagami and Ueta [14, 21], which was appropriate to analyze the safety factor with homogeneous material. The designed model for this benchmark case study depicted in Fig. 6. The geo-technical parameters of the soil material are given as pore water pressures (μ) is 0, internal friction (ϕ) is taken to 10 degree, unit weight (γ) is set to 17.64 kN/m3 and cohesion (c') is 9.8 KPa. Here, the firefly algorithm is implemented to minimize the safety factor in stability analysis. In this study Janbu method from LEMs procedure has been used to analyze the slope stability and performance of the firefly algorithm. Initially, the FoS values is examined using Genetic algorithm (GA), whereas a chromosome in the population is encoded in the binary form using 24 bits (01110100|11011010|00010010), which are associated to dimension variables (a, b) and R of the individual. Every individual chromosome represents a valid solution in the solution space. In the similar way, PSO algorithm is also implemented for comparing the superiority of the algorithms. In PSO algorithm, the particles in the population are associated with a encoded solution, which are initially generated

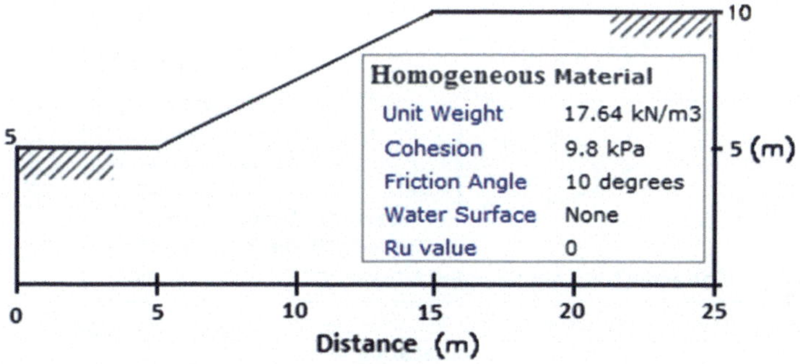

Fig. 6 Geometry of homogeneous soil slope model

random as taken similar as in Firefly algorithm [15, 16]. The chromosome and particle representation of the encoded solution for slip surface evaluation are depicted in Fig. 7. In their further steps, these algorithms (GA and PSO) are followed by its specified operations to continuously improve the result. In the next step, FA algorithm is introduced to observe the optimal surface which is known as critical surface. The findings of the result indicated that, the FA produces more efficient result with high convergence rate. The parameters of firefly algorithm, which are observed for fine tune the results are illustrated in Table 3.

The critical slip surface located by FA- algorithm over the considered LEM method is depicted using Fig. 8.

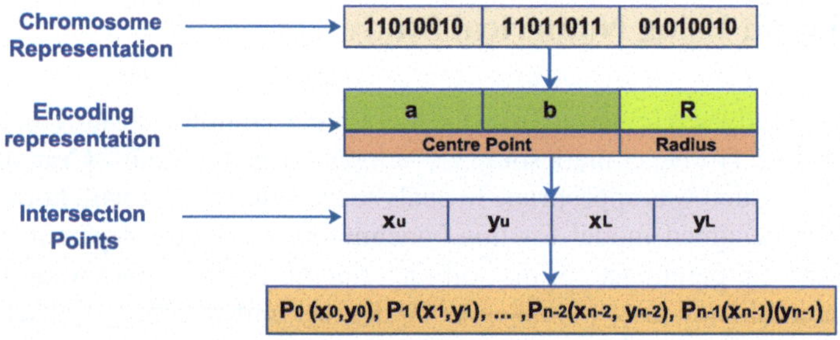

Fig. 7 Pictorial view of an encoded solution in the population

Table 3 Tuned parameters for Firefly algorithm

Tuning parameters	Janbu method
Absorption coefficient (γ)	1
population size	100
β	0.2
β_0	1.4
Number of generation	100

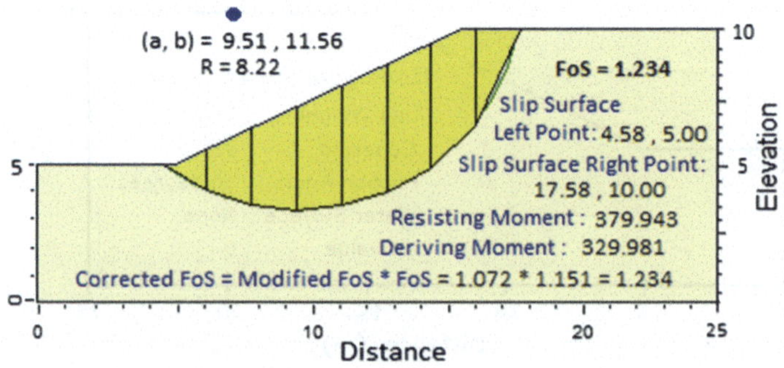

Fig. 8 Interpreted view for critical failure surface on janbu method

7 Result Comparison and Discussion

In order to evaluate and compare the performance of the above algorithms, the algorithms must follow a successive number of iterations until better solution is no longer produced. The minimum FoS obtained using FA method is stabilized at nearly 1.234 on Janbu method over different generations. the comparative results with previous studies are shown in Table 4.

To demonstrate the algorithm's superiority, an error is evaluated between the FoS values obtained through these stochastic methods and the slide tool. The comparative analysis of this study is tabulated in Table 5. The results indicate that approximately 9.9% of errors occur in the FA approach, which are the lowest error values in respect to GA and PSO algorithm as measured 18.24% and 16.36%. This all claim to be a higher stability analysis by the algorithm. Additionally, it uses good convergence rate with local minima avoidance. The proposed algorithm is relatively appropriate to have better solution quality on the basis of these advantages.

Table 4 A summarized FoS values of the benchmark problem with previous study

Reference	Optimization method	LEM's	FoS
Yamagami and Ueta [21]	BFGS, DFP	Spencer	1.338
Greco et al. [6]	MC, PS	Fellenius	1.326–1.333
Malkawi et al. [9]	MC	Fellenius	1.238
Solati and Habibgahi [22]	GA	Janbu	1.380
Jianping et al. [10]	GA	spencer	1.324–1.327
Kahatadeniya et al. [12]	ACO	Line	1.311
Current study	Firefly algorithm	Janbu	1.234

Table 5 The statistical analysis for maximum error (%) with Janbu method

LEM's methods	No. of Gen.	Stochastic methods			Rocscience slide tool			Error (in %)		
		GA	PSO	FA	GA	PSO	FA	GA	PSO	FA
Janbu method	10	1.314	1.334	1.341	1.454	1.507	1.461	10.654	12.968	8.948
	40	1.294	1.296	1.293	1.443	1.583	1.422	11.514	22.145	9.976
	70	1.261	1.245	1.245	1.409	1.46	1.38	11.736	17.269	10.843
	100	1.253	1.239	1.234	1.458	1.465	1.357	16.360	18.240	9.967

8 Conclusion

The thorough review of this chapter discusses the feasibility of the firefly algorithm and its efficacy. The FA algorithm acquire factor of safety more consistent with higher precision value to observe the critical slip surface. Moreover, the procedure excludes all invalid surfaces in order to evaluate a reasonable surface within its kinematic and geometric constraints. The algorithm's outcomes have proved to be superior in terms of local minima avoidance and convergence rate. The FA approach could achieve better quality solutions and stable convergence characteristics over other approaches, such as GA and PSO. Nevertheless, refinement in the procedures is still required to continuous improve the efficiency. In addition, non-homogeneous material with different layer strengths and water-saturated slopes may be attempted to explore further in the future.

References

1. Fredlund, D. G., & Krahn, J. (1977). Comparison of slope stability methods of analysis. *Canadian Geotechnical Journal*, *14*(3), 429–439.
2. Fellenius, W. (1936). Calculation of stability of earth dam. In *Transactions. 2nd Congress Large Dams, Washington, DC* (Vol. 4, no. 7-8, pp. 445–462).
3. Bishop, A. W. (1955). The use of the slip circle in the stability analysis of slopes. *Geotechnique*, *5*(12), 7–17.
4. Janbu, N. (1973). Slope stability computations. In R. C. Hirschfeld & S. J. Poulos (Eds.), *Embankment-dam engineering textbook* (Vol. 12, no. 4, pp. 67). John Wiley and Sons Inc., Thomas Telford Ltd.
5. Chen, Z. Y., & Shao, C. M. (1988). Evaluation of minimum factor of safety in slope stability analysis. *Canadian Geotechnical Journal*, *25*(4), 735–748.
6. Greco, V. R. (1996). Efficient Monte Carlo technique for locating critical slip surface. *Journal of Geotechnical Engineering, American Society of Civil Engineers*, *122*(7), 517–525.
7. Cheng, Y. M., Li, L., & Chi, S. C. (2007). Performance studies on six heuristic global optimization methods in the location of critical slip surface. *Computers and Geotechnics*, *34*(6), 462–484.
8. Zolfaghari, A. R., Heath, A. C., & McCombie, P. F. (2005). Simple genetic algorithm search for critical non-circular failure surface in slope stability analysis. *Computers and Geotechnics*, *32*(3), 139–152.
9. Malkawi, A. I. H., Hassan, W. F., & Sarma, S. K. (2001). Global search method for locating general slip surface using Monte Carlo techniques. *Journal of Geotechnical and Geoenvironmental Engineering*, *127*(8), 688–698.
10. Sun, J., Li, J., & Liu, Q. (2008). Search for critical slip surface in slope stability analysis by spline-based GA method. *Journal of Geotechnical and Geoenvironmental Engineering*, *134*(2), 252–256.
11. Sengupta, A., & Upadhyay, A. (2009). Locating the critical failure surface in a slope stability analysis by genetic algorithm. *Applied Soft Computing*, *9*(1), 387–392.
12. Kahatadeniya, K. S., Nanakorn, P., & Neaupane, K. M. (2009). Determination of the critical failure surface for slope stability analysis using ant colony optimization. *Engineering Geology*, *108*(1), 133–141.
13. Kashani, A. R., Gandomi, A. H., & Mousavi, M. (2016). Imperialistic competitive algorithm: A metaheuristic algorithm for locating the critical slip surface in 2-dimensional soil slopes. *Geoscience Frontiers*, *7*(1), 83–89.

14. Singh, J., Banka, H., & Verma, A. K. (2019). A BBO-based algorithm for slope stability analysis by locating critical failure surface. *Neural Computing and Applications*, *31*(10), 6401–6418.
15. Cheng, Y. M., Li, L., Chi, S., & Wei, W. B. (2007). Particle swarm optimization algorithm for the location of the critical non-circular failure surface in two-dimensional slope stability analysis. *Computers and Geotechnics*, *34*(2), 92–103.
16. Singh, J., Banka, H.,& Verma, A. K. (2018). Analysis of slope stability and detection of critical failure surface using gravitational search algorithm. In *Fourth International Conference on Recent Advances in Information Technology (RAIT)* (pp. 1-6). IEEE.
17. Singh, J., Banka, H., & Verma, A. K. (2019). Locating critical failure surface using meta-heuristic approaches: A comparative assessment. *Arabian Journal of Geosciences*, *12*(9), 307.
18. Singh, J., Verma, A. K., & Banka, H. (2018). Application of biogeography based optimization to locate critical slip surface in slope stability evaluation. In *Fourth International Conference on Recent Advances in Information Technology (RAIT)* (pp. 1–5) IEEE.
19. Aryal, K. P. (2006). *Slope stability evaluations by limit equilibrium and finite element methods*. Ph.D. thesis, Norwegian University of Science and Technology.
20. Fister, I., Fister, I, Jr., Yang, X., & Brest, J. (2013). A comprehensive review of firefly algorithms. *Swarm and Evolutionary Computation*, *13*, 34–46.
21. Yamagami, T. (1988). Search for noncircular slip surfaces by the Morgenstern-Price method. In *Proceedings of the 6th International Conference Numerical Methods in Geomechanics* (pp. 1335–1340).
22. Solati, S., & Habibagahi, G. (2006). A genetic approach for determining the generalized inter-slice forces and the critical non-circular slip surface. *Iranian Journal of Science and Technology, Transaction B, Engineering, Shiraz University*, *30*(1), 1–20.

An Application of Operational Analytics: For Predicting Sales Revenue of Restaurant

Samiran Bera

Abstract Operational analytics improves existing operations of a firm by focusing on process improvement. It acts a business tool for resource management, data streamlining which improves productivity, employee engagement, customer satisfaction and provide investment opportunities. Crucial insights into the problem can be obtained which aids to determine key business strategy through various stages of data analysis and modeling, such as exploratory data analysis, predictive modeling, documentation and reporting. In this work, a real world dataset is considered for the study, where the sales revenue of restaurant is predicted. A second stage regression model built upon base regression models which are linear regression, ridge regression, decision tree regressor. Based on the results obtained, the following findings are reported: (i) annual sales revenue trend, (ii) food preference in cities, (iii) demand variability i.e. effect of first week and weekend, and (iv) comparison against ensemble methods in terms of prediction accuracy. This work also suggest avenues for future research in this direction.

Keywords Operational Analytics · Exploratory Data Analysis · Predictive Modeling · Feature Engineering · Ensemble

1 Introduction

Operational Analytics (OA) is a process of improving existing operations of a firm, i.e. by focusing on process improvement and cost management. It's a stem grown from business analytics, which depends on real-time data to gain actionable insight through various data mining and aggregation tools. Thus, it is also known as Information Technology Operations Analytics (ITOA).

S. Bera (✉)
Indian Institute of Technology (Indian School of Mines) Dhanbad, Police Line Rd, Sardar Patel Nagar, Dhanbad 826004, Jharkhand, India
e-mail: samironbera@gmail.com

S. K. Das et al. (eds.), *Machine Learning Algorithms for Industrial Applications*, Studies in Computational Intelligence 907,
https://doi.org/10.1007/978-3-030-50641-4_13

1.1 Motivation

Operational Analytics allows decision makers to get more transparency over the business planning and manage daily operations to meet customer expectations. Thus, it has become a strategic priority to gain competitive advantage. Several reports suggest that operational analytics improves firms performance significantly, which can be summarized as follows,

1. **Improvement in productivity**: Operational analytics can identify performance bottlenecks, prepare streamlining and suggest alternatives to reduce costs and increase efficiency. Thus, operational analytics is employed to maximize profit, improve productivity or minimize cost.
2. **New investment opportunities**: With an increase in efficiency, firms can scale up its business, opening new avenues of opportunities making it more resilient to market conditions.
3. **Effective business decision tool**: Executive decisions are data driven that prevents business from suffering inefficiencies. This allows firms to make up for employees having any lack of experience.
4. **Efficient resource management**: It allows business to understand and interpret results from data, i.e. get actionable intelligence for efficient resource utilization. Further, it helps to identify the key focus areas and consider investment possibilities.
5. **Improved customer satisfaction**: Most performance issues can be quickly identified through root-cause analysis. An improved response time to (singular or group) performance failure increases production throughput and customer satisfaction.
6. **Smooth data streamlining**: It increases the ability to share information with much less complexity among employees. This implies individuals can receive relevant insights and reporting to stakeholders much easily.
7. **Enhance employee engagement**: Operational analytics promotes collaboration to employees and encapsulate entire organization as a team. This empower employees with knowledge which improves overall productivity.

With advance in business technology and its adoption in supply chain, data generated through different channels are vital to the entire process. It is because the strength of data-driven decisions relies heavily on the quality of data. Therefore, organization nowadays invests heavily in data and various analysis.

Further, to maintain performance, it has become imperative to stay updated with recent data which can be collected from day-to-day basis. Thus, Operational Analytics is the tool that aids firms from different sectors (manufacturing and service) to improve its performance.

1.2 Contribution of This Study

The key contribution of this study can be enumerated as follows:

1. In this study, a systematic data analysis process which comprises of data collection and wrangling, predictive modeling and optimization through hyperparameter tuning is performed on a real-world dataset to study strength, weakness, opportunities and threats to business.
2. To optimize model performance, a second stage model is developed over base regression models and tuning key problem parameters, and compared against ensemble methods using R-Squared Error to select the best model.

The structure of this work is organized as follows. Section 2 contains relevant literatures. Section 3 provides an overall concept on data analysis process with techniques adopted and performance metrics used for this study. Section 4 describes the problem in this study, and provides results and discussion based on exploratory data analysis and predictive modeling. Section 5 concludes by summarizing the work suggesting possible avenues for future work.

2 Literature Review

Even though a vast area of research is devoted to data analytics, only few attempts were made towards operational analytics. Therefore, operational analytics remains unfairly scarce in terms of exploration and study. In this section, relevant literature on the components of data/operational analytics is discussed, which establishes the foundation of this study.

The distribution of relevant literature is organized as (i) exploratory data analysis, (ii) base regression models and (iii) ensemble models, as individual segment comprises of vast body of knowledge which can be pursued for study. Each segment is vital for data analysis work, and therefore numerous branches have grown over the last decade. However, this section restricts the discussion pertaining to the problem in this study.

2.1 Exploratory Data Analysis

Exploratory Data Analysis (EDA) is a process to obtain insights from data using statistical approach and graphical representation [1]. It was developed by [2] and [3]. Since then, several visualization techniques have been explored by academician and industrial engineers. Today EDA is critical to business process as

1. It identifies hidden patterns and relationship among variables/features which determines hypothesis to be validated, and

2. It aids in strategic business decision which are guided by insights obtained from EDA [4].

Visualization tools [5–8] have application in wide range of domain, some of which are in genome-sequence data analysis [5], meteorological data [6], predictive analysis [8], exploration of tabular data [9–11]. This shows the spectrum EDA serves and its rate of adoption, which implies that EDA is critical for any business to succeed.

As any area of research, requirement of EDA have evolved over the years. With increase in volume, variety and velocity of data, the primary concern of researchers is perform EDA in an optimized fashion [12], This implies, scalability, flexibility and robustness of EDA process. Further, EDA should allow developer identify and analyze complex relationships among variables (through bivariate analysis & multivariate analysis), accurately with limited analytical knowledge [1]. This could assist in several domains including medicare [13, 14], telecom [15–17], manufacturing [18], news industry [19], food industry [20], and more. This urges to acknowledge the importance of EDA and need for further development.

2.2 *Feature Engineering*

Several studies use only the raw features in carrying their analysis [21, 22]. However, raw data is subject to incompleteness and inconsistency. This may yield inferior models which fails to produce optimal results. To this end, feature engineering allows to improve data quality [23, 24]. Feature engineering, therefore, is crucial to exploratory data analysis and for training machine learning model. However, it is quite labor-intensive as entire process heavily relies on data processing and designing pipelines [25]. Existing features can be extended through transformation [26].

Feature engineering finds its application in technologies such as deep learning [27] which leverages heavily on feature engineering, image and signal processing to find interesting pattern [25], dimensionality reduction [28, 29], and as auto-encoders [30]. Feature engineering have also attracted winning contributions in several competitions such as Kaggle, ACM's KDD cup [31].

2.2.1 Feature Selection

Feature selection is the process of filtering relevant subset of features from an original feature set as per certain criterion. Doing so, redundant and irrelevant features are removed compressing data processing scale. The merits of implementing feature selection through pre-process learning algorithms is improved learning accuracy, low learning time, and simplified output [32, 33].

The application of feature selection can be found in a wide variety of areas such as, image recognition [34, 35], image retrieval [36], text mining [37, 38], bioinformatic data analysis [39], statistics [40, 41], information theory [42, 43], manifold [43, 44], and rough set [45, 46].

2.3 *Predictive Models*

2.3.1 Base Regression Models

The most popular base regression models are linear regression, ridge regression, decision tree regressor and lasso regression. However, lasso regression is not considered in this study. The following regression considered are,

1. *Linear regression*: a statistical tool used in predictive modeling [47]. Linear regression is used for finding linear relationship between dependent/target and one or more independent variable/predictors. There are two types of linear regression,
    a. Simple regression
    b. Multiple regression.
    Regression model finds the best fit line that reduces prediction error (i.e. difference between actual and predicted value) as much as possible.
2. *Ridge Regression*: it analyzes multiple regression data which suffer from multicollinearity. It implies that it corrects the bias and variance of the model by imposing penalty on the objective function [48].
3. *Decision tree algorithm*: a supervised learning which solves both regression and classification problems. Decision tree is represented by a tree to solve the problem where each leaf node belongs to a group [49, 50].

These algorithms widely used in literature [51, 52]. Besides these, several other algorithms such as KNN algorithm [53], KMeans algorithm [54], COVR-AHC algorithm [55] and several other methods can be used for forming groups as clusters.

2.3.2 Ensemble Models

It is the combination of more than one model to leverage of merits of base models [56]. Ensemble models usually produces better performance [57–59], but at the cost of computation. In contrast to solo models, ensemble models are less susceptible to variance in output. Further, the advantage of ensembles with respect to single models has been reported in terms of increased robustness and accuracy [60]. Few popular ensembling techniques [61], which have receive most attention are, (i) Random forest [62], (ii) Bagging and Boosting [63], and (iii) Stacking [64].

3 Research Methodology

Operational analytics provides data-driven solution, therefore, relies heavily on the integration of data processing and modeling. This implies that the performance of predictive model also depends on the quality of data. To achieve an integration of processes, the following steps are followed which is shown in Fig. 1.

Fig. 1 Steps in data analysis process

3.1 Data Collection

Data collection is the process of gathering data from one or multiple sources. This process can be classified based on (i) source of data, (ii) type of data, and (iii) structure of data. An overview is provided in Table 1 followed by discussion.

Table 1 Data collection

Source	*Primary*	*Secondary*	*Synthetic*
	1. Questionnaires & Surveys 2. Observations	1. Documents & Records	1. Generated
Type	1. Quantitative	2. Qualitative	
Structure	1. Structured	2. Unstructured	3. Semi-structured

3.1.1 Based on Data Source

Depending on the source, data can be categorized into *three* groups, primary, secondary and synthetic.

Primary data is collected via either physically (in-person) or electronically (over telephone/internet). These data are obtained through questionnaire & surveys which consists of relevant questions, i.e. in a formal setting. However, data can be collected in an informal setting as well, based on observation of events. Data collection can be regular (daily, weekly, quarterly, season based) or conducted in phases (based on design of experiment), both relying heavily on the business objective and human resource. The time required to collect data varies from few days to years depending on the study. For example, time series required data to be collected over time, thus is longer compared to when data collection is obtained from a target group.

Secondary data is obtained from previously collected data, which can be found in form of documents and records as journals, books, reports. These data are either available in printed form or published online. The time required to collect data depends on data availability i.e. permission to access and format of information required. For example, time required to retrieve information from an online database is much faster than print documents.

Synthetic data is generated based on problem knowledge when either data collected is unavailable (i.e. absent, irrelevant) or insufficient (i.e. incomplete, inconsistent, unreliable). These data are mostly used to compensate unavailability of primary or secondary data. However, it is particular useful to study specific aspects of a problem which existing data fails to capture. The time required to generate data depends on data generation complexity and computational power.

3.1.2 Based on Data Type

Depending on the type of data, it can be categorized into two groups, quantitative and qualitative.

Quantitative data implies data that is expressed as numbers. Various aspects of a business problem are easily captured in these dimensions—time, cost, variety, volume using numerical values i.e. continuous/categorical. Besides these, several aspects are captured as strings (such as city, country, etc) which can be encoded as numbers i.e categorical. Further, quantitative data can either be nominal or ordinal depending on the context.

Qualitative data implies data that is expressed as words. Several aspects of a business problem that cannot be expressed or encoded as numbers directly, are captured in non-numeric descriptive form. Obtained data is later transformed using different scales i.e categorical. Further, qualitative data can be ordinal.

3.1.3 Based on Data Structure

Depending on the structure of the data, it can be categorized into three groups, structured, unstructured, and semi-structured.

Structured data are organized in a database format, i.e. table having rows and columns. Relational keys are used to map pre-designed fields. As structured data is schema dependent, it is less flexible but however very robust. Example of structured data can be found in relational data.

Unstructured data do not follow any pre-defined format, as seen for structured data. Due to lack of schema, unstructured data is very flexible. However, it requires data format specific knowledge to store, retrieve and process data, thus isn't as robust compared to structured data. Example of unstructured data can be found in word document, portable document format (PDF), text, media (images, videos) and logs.

Semi-structured data does not follow tabular format as structured data, however exhibit some organization properties. Flexibility and robustness depends on the complexity of data structure and query. Example of semi-structured data can be found in XML data, spreadsheet.

However, today firms leverages on more than one data structure to capture wide variety of data. With advanced technologies, unstructured and semi-structured data is converted based on domain knowledge into structured format for further analysis.

3.2 *Data Wrangling*

Data wrangling is the process of cleaning, transforming and enriching data into more usable format. Data wrangling is also known as data munging. This process can be classified broadly into four steps. An overview is provided in Table 2 followed by discussion.

Table 2 Data wrangling

Data Exploration			
1. Shape of dataset	2. Type of features	3. Missing & Unique values	
4. Measure of central tendency	5. Measure of dispersion	6. Measure of association	
Feature Engineering			
1. Standardization	2. Normalization	3. Binarization	
4. Encoding	5. Imputation	6. Customized	
Feature Selection			
Based on trend	*Based on relation*	*Based on relevance*	*Based on score*
1. Univariate analysis	1. Correlation 2. Duplicacy 3. Derived 4. Multivariate analysis	1. All in 2. Backward elimination 3. Forward selection 4. Stepwise elimination	1. Score Comparison
Target variable			
1. Continuous	2. Categorical		

3.2.1 Data Exploration

Data exploration and analysis deals with assessment of the dataset from a statistical point of view i.e. basic information and potential relationship. The data composition is reported in form of summary statistics and graphical representation(s). The following assessments are performed while exploring data,

1. *Primary assessment*: It deals with

   (i) *Shape of dataset*, i.e sample size and number of features
   (ii) *Type of features* i.e. continuous or categorical

   (iii) *Data availability* i.e. missing values
   (iv) *Unique values* for categorical variable to identify data groups (if any)

2. *Secondary assessment*: It deals with

   (i) *Measure of central tendency* i.e. mean, median and mode
   (ii) *Measure of dispersion*, i.e. range, variance, standard deviation, skewness and kurtosis
   (iii) *Measure of association*, i.e. correlation

3.2.2 Feature Engineering

Feature Engineering is also known as applied machine learning. Feature engineering uses domain knowledge to extract and discover information which is vital to improve model performance. To engineer features, either one or more than one techniques can be followed,

1. *Feature scaling* through normalization (i.e. scaling data values between [0, 1]) and standardization (i.e. scaling data values to $\{\mu = 0, \delta = 1\}$)
2. *0–1 Binarization* by classifying values into two classes
3. *Label encoding* to form groups of information
4. *Imputing missing values*, outliers and ambiguous values
5. *Transformation functions* designed for specific transformation

It should be noted that feature engineering strictly depends on feature composition and model in use.

3.2.3 Feature Selection

Relevant features are used to build models, and by eliminating irrelevant features before and during training the model. This reduces memory usage and improves performance. Features can be selected or removed,

1. *Based on trend*: Through univariate analysis, each feature is represented to evaluate any trend or pattern, i.e. increasing/decreasing/concave/convex/inflated/random
2. *Based on relation*: It implies that a feature is a function of another feature(s). It is expresses as,

   (i) *Correlation*:
      - Positive correlation: when two feature follow same trend,
      - Negative correlation: when two feature follow opposite trend,
      - No correlation: when two feature do not influence each other,

(ii) *Derivation*:
  - Unit conversion
  - Date extraction
  - Difference (increment, decrement) or combination (sum, product)
  - Label encoding

(iii) *Identity*:
  - Duplicate column
  - Inverse column (True/False, 0/1, Yes/No, etc)

(iv) *Transformation*:
  - Power transform (Box-Cox)
  - Logit transformation
  - Data formats, long format to wide format and vice versa

3. *Based on relevance* i.e. by
   - Including selected feature(s) to an empty feature set
   - Eliminating irrelevant feature(s) from universal feature set, or
   - Through step-wise sequential fashion
4. *Based on score*, i.e. by executing each model with distinct combination of feature(s) to identify the set of feature(s) that provides the best score

3.2.4 Target Variable

The identification of the dependent variable (also known as the target variable) from the problem statement guides the model selection. It implies that, based on the type of target variable,

1. *Regression model* is developed to predict dependent variable/ target variable when it is continuous
2. *Classification/clustering model* is developed to predict dependent variable/ target variable when it is categorical

3.3 Modeling

Data modeling consists of key process: model selection based on descriptive statistics, model validation using training and testing dataset, and hyper-parameter tuning using cross-validation. An overview is provided in Table 3 followed by discussion.

3.3.1 Model Selection

Model selection depends upon multiple factors, which can be grouped into (i) type of dependent/target variable, (ii) feature composition i.e. type and relationship among features, and (iii) performance. Selection of model is, thus,

Table 3 Modeling

Selection			
Regression	*Classification*	*Clustering*	*Mixed/Hybrid*
Linear regression	Logistic regression	K-means	Combination of more than one modeling approach
LASSO regression	Support vector machines	Hierarchical	
Ridge regression	Naive Bayes classifier	Mean-shift	
------------ Decision tree -----------		DBSCAN	
------------ Random forest ----------		EM using GMM	
---------------- Bagging & Boosting -------------			
Validation			
1. Re-substitution	2. Hold-out	3. K-fold cross	
4. Leave one out	5. Random sampling	6. Bootstrapping	
Tuning			
1. Grid search CV	2. Randomized search CV	3. Customized	

1. *Based on dependent/target variable*: A regression model is developed when the dependent/target variable is continuous. And, classification/clustering model is developed when the dependent/target variable is categorical. However, classification model is chosen when target classes are known (i.e. used in supervised learning) which in contrary to clustering is unknown (i.e. used in unsupervised learning).
2. *Based on feature composition*: One or more model features can become insignificant and thus eliminated in a model. For example, LASSO regression eliminates features which are insignificant by reducing their coefficient value by imposing penalty. Appropriate selection of model allows to leverage such benefits, and thus improves model performance.
3. *Based on performance:* Ensemble methods such as random forest, bagging and boosting combines multiple models (in sequential/parallel fashion) into one prediction model (homogeneous/heterogeneous). It can decrease variance (as in bagging), bias (as in boosting), and improve prediction (as in stacking). The performance of ensemble, however, depends on the accuracy and diversity of the base models.

3.3.2 Model Validation

Model validation ensures that the model performs as expected, i.e. according to business objective defined considering the assumption(s) undertaken and limitation(s) imposed. To validate the model, the dataset is split into three groups,

(i) *Training set*—Portion of dataset used to train the model
(ii) *Testing set*—Portion of dataset used to test performance of trained model
(iii) *Validation set*—Portion of dataset used to validate the result post-testing

where all sets are mutually exclusive and collectively complete.

To generate testing set, this operation can be performed via by splitting the dataset in a random fashion (as in random sampling) or selecting a set of records (as in KFold Validation, Leave-One-Out Validation) or re-selecting previously chosen record i.e. selection with replacement (as in re-substitution & bootstrap sampling) as testing set. Similarly, validation set can be formed (as hold-out sample).

3.3.3 Model Tuning

To optimize model performance and accuracy, model parameters can be trained i.e. defining hyper-parameters. This can be achieved in the following ways:

1. *Grid Search CV*: All possible combination of parameters are considered to find the optimal parameters. Thus, Grid Search CV is computationally expensive and time consuming
2. *Random Search CV*: Addresses the demerits of Grid Search CV by considered only few of the parameter setting(s). Thus, it is computationally less expensive
3. *Customized*: Select parameter setting(s) are chosen based on tool knowledge. Thus, it least computationally expensive.

3.4 Performance Measure

Performance of the model can be determined based on accuracy of predicted values. Different performance metric can be used based on the type of problem (regression, classification, clustering) and data definition (such as data imbalance, percentage missing data, etc). An overview is provided in Table 4 followed by discussion.

Table 4 Performance measure

Regression	*Classification*	*Clustering*
1. Mean absolute error	1. Accuracy score	1. Adjusted rand score
2. Mean squared error	2. Confusion matrix	2. Homogeniuty score
3. R^2 score	3. Classification report	3. V score measure
4. Adjusted R^2 score		

3.4.1 For Regression Model

The measure of performance for regression model is the difference between actual (y_i^a) and predicted values (y_i^p) for samples $i = 1 \ldots n$. The metrices which can be used are as follows,

1. Mean square error (MSE): $\frac{1}{n}\sum_{i=1}^{n}(y_i^a - y_i^p)^2$
2. Mean absolute error (MAE): $\frac{1}{n}\sum_{i=1}^{n}(|y_i^a - y_i^p|)$
3. Mean absolute percentage error (MAPE): $\frac{1}{n}\sum_{i=1}^{n}|\frac{y_i^a - y_i^p}{y_i^a}| \times 100$
4. R^2 Score: $R^2 = 1 - \frac{S_R}{S_T}$, where S_R, S_T are sum of squared residual (SSR), sum of squared total (SST) respectively. SSR and SST is computed as follows,

   $S_R = \sum_{i=1}^{n}(y_i^a - y_i^p)^2$
   $S_T = \sum_{i=1}^{n}(y_i^a - \bar{y^a})^2$

5. Adjusted R^2 score: $1 - (1 - R^2)(\frac{n-1}{n-p-1})$, where p is penalty implied.

3.4.2 For Classification Model

The measure of performance for classification model is the mismatch between actual (y_i^a) and predicted values (y_i^p) for samples $i = 1 \ldots n$. The metrices which can be used are as follows,

1. *Confusion matrix*: It represents the cross-tabulation of actual and predicted values by aggregating (i) number of true positives (TP), (ii) number of true negatives (TN), (iii) number of false positives (FP), (iv) number of false negatives (FN), which is shown in Fig. 2.
2. *Accuracy score*: It is the percentage of correct predictions compared to all prediction, as $\frac{TP+TN}{TP+TN+FN+FP}$
3. *Classification report*: It is used to for problem detection and support easier interpretation using some of the following metrics,

   (i) *Sensitivity/Recall*: $\frac{TP}{TP+FN}$
   (ii) *Specificity*: $\frac{TN}{TN+FP}$
   (iii) *Precision*: $\frac{TP}{TP+FP}$

Fig. 2 Formation of Confusion Matrix

		Predicted Value	
		True	*False*
Actual Value	*True*	TP	FN
	False	FP	TN

3.4.3 For Clustering Model

The measure of performance for clustering model is the intra and inter group mismatch among all samples. The metrices which can be used are, Adjusted rand score, Homogenuity score, and V score measure.

3.5 Documentation and Reporting

Analysis and findings are summarized in form of documents and reports when model deployment. The type of documentation and reporting depends on the target audience, an overview of which is briefly discussed in next section.

3.5.1 Dashboard

A single slide dashboard is prepared when reporting to senior management/top executive, mentioning key-points that are crucial for business decision without consuming much time.

3.5.2 Story Telling

A detailed report in form of story telling is prepared when reporting to middle management, mentioning all information required to

1. Assess fulfillment of business objective
2. Understand cause and effect of features (individual or through interaction)
3. Identify and construct focus groups
4. Provide actionable intelligence to senior executives and middle management

The flowchart representing the entire data analysis procedure which is described above is shown in Fig. 3. This will allow readers to understand the sequence of actions required to solve an operational analytics problem.

3.6 Positioning of This Study

In this study, structured data is obtained from secondary sources having quantitative values. Exploratory data analysis is performed to obtain crucial information such as shape of dataset, types of features, amount of missing and unique values, measures of central tendency, dispersion and association. Prior to modeling, several feature values are encoded and binarized, and missing values are imputed. Further, select features were chosen based on trend and relation. Regression model(s) is chosen for

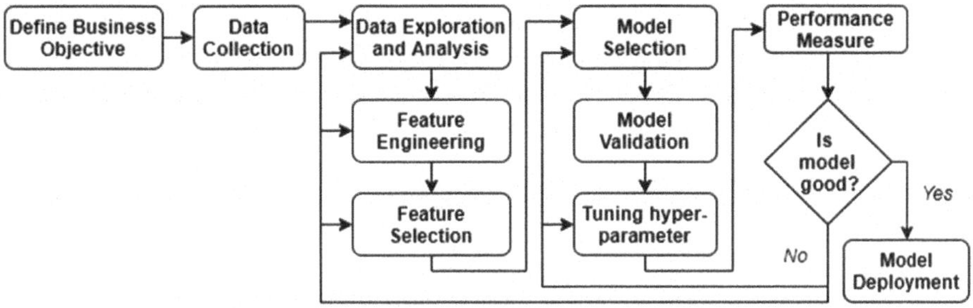

Fig. 3 Flowchart representing data analysis process

analysis as the target variable is identified to be continuous, and performance Model is trained using test-train split into 30:70 ratio, and the performance is improved through hyper-parameter tuning.

In the next section, a real world case study is selected to explain the application of above discussed process in detail.

4 Case Study: Restaurant Revenue Prediction Dataset

Kaggle is a subsidiary of Google LLC comprising of the world's largest online community of data scientists. It offers machine learning practitioners with powerful tools and resources, and real-world dataset(s) which are used in various competitions. This attracts people across the industry and academia to participate, contribute and expand the existing knowledge pool.

To understand the data analysis process, it is beneficial to have a dataset which represents a real-world problem. Therefore, the 'Restaurant Revenue Prediction' dataset, which is popular among data science enthusiast is considered for this purpose. Next, follows a brief description of the dataset.

4.1 Problem Description

Globally, TFI have more than 1200 quick service restaurants. Some of its popular brands are Burger King, Sbarro, Popeyes, Usta Donerci, and Arby's. Therefore, TFI have a massive workforce which is around 20000 across Europe and Asia. Further, TFI is keen on expanding the chain of restaurants to new territories that requires significant investments.

Developing new restaurant sites not only requires high investment of capital, but time as well. Thus, selection of site is an executive decision. As, it is an extremely difficult task to determine the optimal site location due to geographical and cultural differences, decision are based on subjective process i.e. personal judgement and experience of the development team. The team considers not only on the site selection, but focuses on brand selection which have the potential to penetrate consumer market and succeed. However, if either wrong location or wrong brand is chosen, the project will not succeed. This may result in shutting down the restaurant site incurring massive operating losses which is within 18 months.

4.1.1 Preference of Analytics over Mathematical Optimization

Similar problem can be solved using operations (quantitative techniques) such as facility-location/facility-allocation problem. This approach requires formulating a mathematical model to solve a facility-location problem and evaluate the effectiveness of investment in restaurant site(s). This will provide TFI to focus on other key business areas such as innovation, sustainability and human resource management. Although, operations can provide optimal restaurant location site, it can be computationally expensive, labor intensive and time consuming, especially for dataset which consists 100,000 regional location.

4.2 Research Methodology

The restaurant revenue prediction dataset is studied using simple regression techniques to predict sales revenue. Necessary data processing have been performed prior to modeling i.e. training and testing. The entire work mechanism is encapsulated in algorithm 1 and represented adequately by Fig. 4 as shown below.

It can be noted that, more complex models can be built to achieve better performance, however, it is not in the scope of this section. The purpose of this section is to provide an overall overview of predictive modeling, and thus is limited to data exploration and analysis, feature engineering and selection, and building base data models to make predictions.

4.3 Results and Discussion

4.3.1 Insights Through Exploratory Data Analysis

The dataset consists of two sets: training and testing. The training (testing) set have 137 (100000) samples and 43 (42) features. Compared to testing set, training set have one addition feature which is the dependent/target variable, and it is continuous. Therefore, regression models are considered for prediction.

It is interesting to observe, that the testing set have much more samples than training dataset. Thus, the level of accuracy obtained during model training is likely

Algorithm 1: Operational Analytics Process

Input: numpy, pandas, sklearn, matplotlib, pyplot
Output: Best Predictive Model
Data: Train data X^{test}, Test data X^{train}
    Initialize Base List ← [Linear & Ridge Regression, Decision Tree]
    Initialize Ensemble List ← [Random Forest, XGBoost, LightGBM]

Generate Summary Statistics
if *feature engineering* **then**
    if *missing values present* **then**
        Impute missing values
    end
    if *oulier values exist* **then**
        Impute outlier
    end
    if *feature extraction* **then**
        {year, month, day} ← Open Date
        quater ← month
        for *categorical variables* **do**
            One-Hot Encoding for {City, City Group, Type, weekdays}
        end
    end
    analysis
end
Data visualization using Univariate and Bivariate analysis
/* */
/* **Stage 1: Base Modeling** */
for *model in Base List, Ensemble List* **do**
    Y^B_{model} ← Model prediction
    Z^B_{model} ← Evaluate model accuracy (Y^B_{model}) using R^2 Score
end
/* */
/* **Stage 2: Weighted Regression Modeling** */
a_{model} ← $({}^{10}P_1 \times {}^{10}P_1^{10} \times {}^{10}P_1)$ combination considering 3 base models
for *All* a_{model} *combination for Base Model List* **do**
    Y^S_{model} ← $a_{model} \times Y^B_{model}$ (as in figure 7A)
    Z^S_{model} ←Evaluate model accuracy (Y^S_{model}) using R^2 Score
    Save optimal a_{model} combination
end
/* */
/* **Stage 3: Second Stage Modeling** */
Predict Y^G ← from Linear Regression on Y^B_{model} as input (as in figure 7B)
Z^G ←Evaluate model accuracy (Y^G) using R^2 Score
if Z^B_{model} *is better than* (Z^S_{model}, Z^G) **then**
    return Y^B_{model} with maximum Z^B_{model} as the optimal model
    else if Z^S_{model} *is better than* (Z^B_{model}, Z^G) **then**
        return $a_{model} \times Y^B_{model}$ combination as the optimal model
        else
            return Y^G_{model} as the optimal model
        end
    end
end

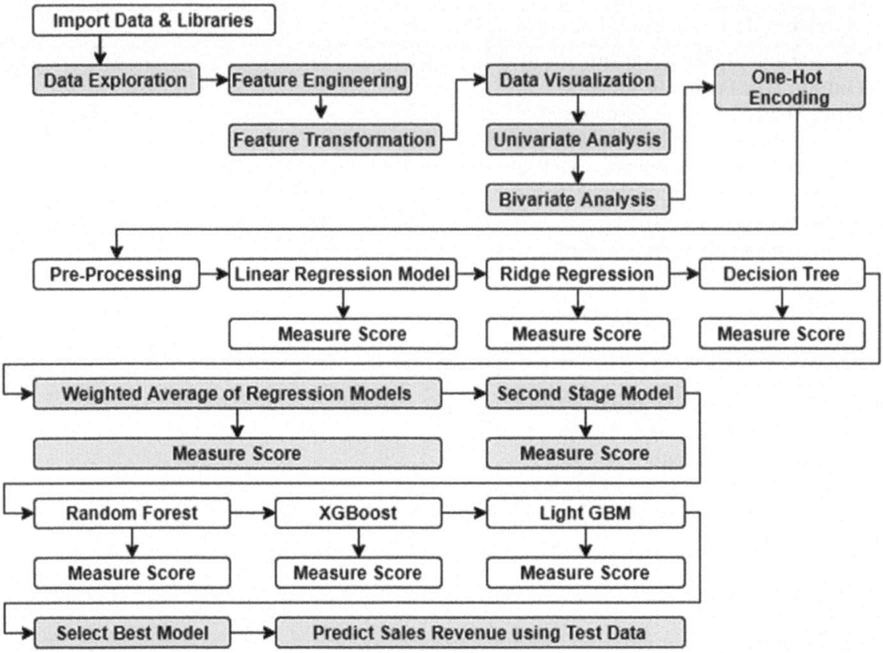

Fig. 4 Methodology followed for restaurant revenue prediction

to change if the model trained is underfit (which result in high bias) or overfit (which result in high variance). This also implies that a moderate training would suffice to make prediction on testing set.

Surprisingly, both training and testing set do not contain any missing values. Thus, the training ans testing set is confirmed to be complete.

4.3.2 Insights Through Univariate and Bivariate Analysis

Impact Annual Revenue over Time: Revenue is reducing with each year, which is a matter of concern. Therefore, TFI should focus on,

1. Existing products that have steady sales to keep business upfloat and mitigate declining sales through offers and promotions,
2. Consider to introduce new products tailored to geography and culture to attract customers and remain competitive,
3. Explore new restaurant sites to expand its outreach to customers, and
4. Make investments to increase brand value by differentiating itself from the competitors.

Impact of City Life on Population Density: Bigger cities attracts more customers and thus generate more revenues. This is obvious, as cities usually have higher population density and ever-evolving lifestyle.

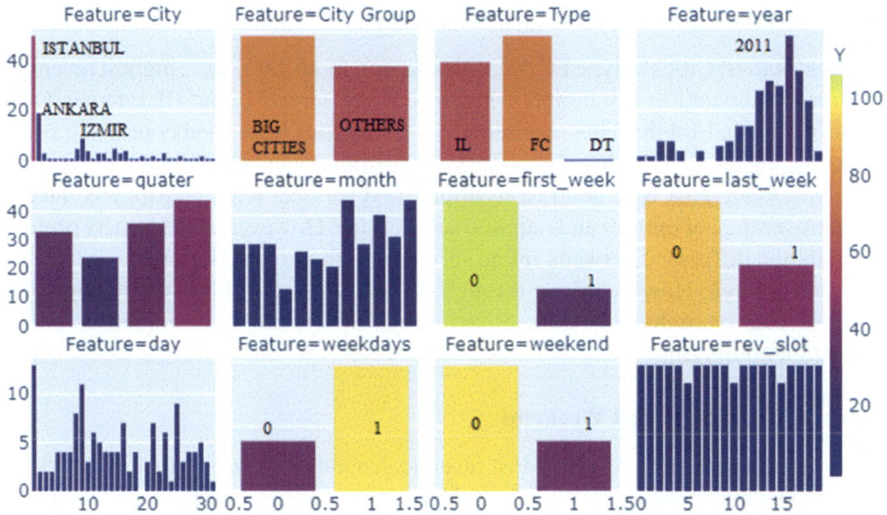

Fig. 5 Data Visualization using **Univariate Analysis** (*for features*: city, city group, type of restaurant, year, quater, month, week (first/last), day, weekdays/weekend, slot reserve), where 1 represents True and vice-versa

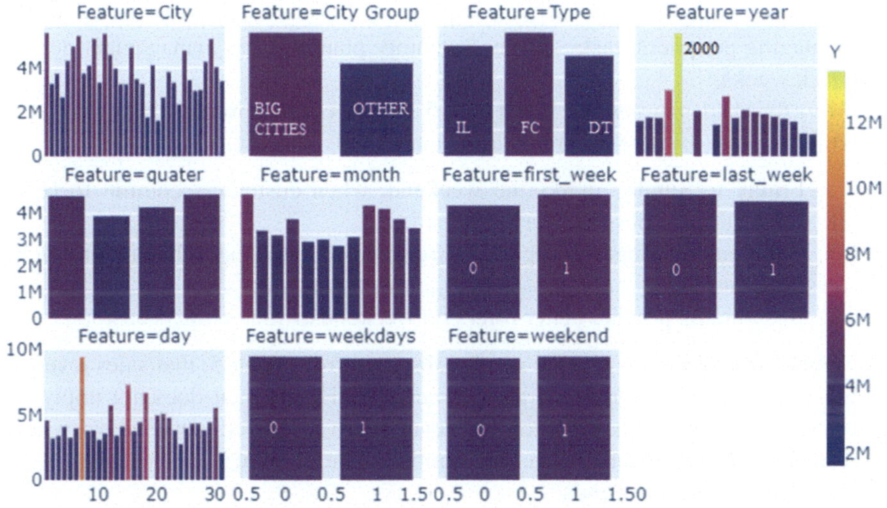

Fig. 6 Data Visualization using **Bivariate Analysis** (*for features*: city, city group, type of restaurant, year, quater, month, week (first/last), day, weekdays/weekend, slot reserve), where 1 represents True and vice-versa

Impact of City Life on Food Court Preference:

1. *Food courts*: Cities of type FC (i.e. cities having food court) have highest revenue, when compared to restaurants where foods are served Inline (IL), Drive Thru (DT), or via Mobile. This is observed as most cities have food courts that cater a diversity of food under one roof where customer can be served.
2. *Inline food*: This is in contrast to inline food service, where customers have a narrow range of options and required self-service. However, as customers prefers both, the difference between inline service and food court is close.
3. *Drive though*: However, drive through are not popular within cities and may be preferable on highways. Thus, drive through fails to attract customers as inline or foot courts can.

Impact of Weekdays and Weekend:

1. *Surge of influx on first week*: Most revenue generation occurs in the first week. It could be due to salary credit that enables customers to spend more as compared to remaining weeks of the month. This can further be confirmed by decrease in revenue on the last week. Therefore, salary/income (reflected spending behavior) influences sales revenue, which is highest in the first week and gradually decreases to lowest in the last week. Therefore, restaurants must strive to fulfill customer demand i.e. by focusing on service level during peak weeks, while managing peripheral tasks such as inventory planning and management during slack weeks.
2. *More footprint on weekend*: From Figs. 5 and 6, it can be observed that restaurants collects most revenue on Friday and Sunday. It is because,

   a. Friday to Sunday marks the weekend, when customers commit to tasks which are not possible during weekdays,
   b. Customers prefers to hangout on weekends due to busy schedule on weekdays, and
   c. Restaurants provide better offers on weekends which attracts customers.

3. *Higher sales revenue ratio*: It can be observed from Table 5, that sales revenue on weekdays is almost same as on weekends. However, it does not imply an increased influx of customers on weekends when compared to weekdays, which can be confirmed from Table 6. This confirms that customers tend to spend more at restaurants on weekends.

Effect of Seasonality: From Fig. 5, it can be observed that the restaurants received more footprints on 1*st* and 4*th* quarter, i.e. around winter. In contrast to this, 2*nd* and 3*rd* quarter received much lesser footprints, i.e. around summer. This observation is also reflected in Fig. 6, which accounts for the sales revenue. Therefore it can stated that the sales revenue increases with increase in footprints.

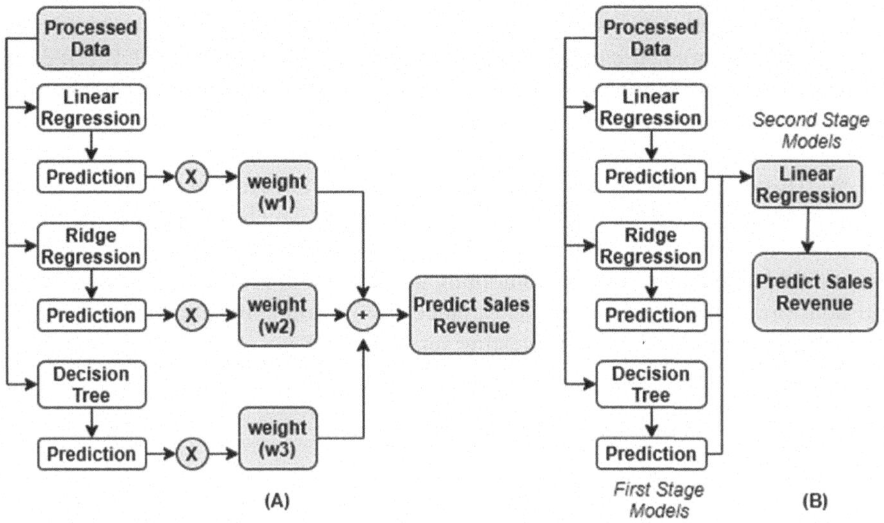

Fig. 7 Model aggregation technique

4.3.3 Modeling

Pre-processing Data Prior to Modeling: Post feature engineering 76 features exist after dropping dependent/target variable/feature in training and testing set, and selecting only the features which are common in training and testing set. All 76 features are selected for model training. However, the feature selection cannot be identified as all-in strategy, but rather customized. This prevents models from training unnecessary set of features, thereby eliminates feature inconsistency improving model performance.

4.3.4 Performance Evaluation

4.3.4.1 Performance of Base Models

The prediction accuracy obtained from (i) linear regression is 55.02%, (ii) ridge regression is 51.33%, and (iii) decision tree (considering 75% feature set with maximum tree depth of 5) is 83.79%. Prediction accuracy of both linear and ridge regression seems to be quite low, where as decision tree provides much better predictions. This implies that the relationship among two or more feature(s) may have non-linear relationship, which decision trees excel in predicting.

4.3.4.2 Performance of Weighted Models

The prediction accuracy of weighted model is shown by Fig. 7A and computed in Table 2. Performance are computed by modifying the value of weights assigned (i.e. by using combination of weight values), however, in an increasing fashion. The best model score by combining linear regression, ridge regression, decision tree models is highlighted in bold. It can be observed that the weight ratio of ($W1$:$W2$:$W3$) is (2:1:10) for best prediction score, which confirms previous observation on presence of non-linear relationship among features.

However, as less than 5%(10%) increase in performance is observed by increasing $W3$ 9 (10) times, the impact of non-linearity among features on dependent/target variable is quite less. This implies that linear models (which in this case are linear regression and ridge regression) cannot be rejected (even with inferior performance), and should be considered when combining models. It is because, combining linear and non-linear models may reduce variance (by discouraging model over-fitting) when predicting sales revenue for testing set.

Table 5 Model performance for weighted modeling approach

Setting	1	2	3	4	5	6	7	8	9	10	11	12
$W1$	1	1	1	1	1	1	1	1	1	2	3	2
$W2$	1	1	1	1	1	1	1	1	1	1	1	2
$W3$	1	2	3	4	5	6	7	8	11	10	10	10
R^2 score (in %)	74.35	80.60	83.10	84.25	84.82	85.11	85.25	85.31	85.29	**85.34**	85.04	84.82

Note: $W1$, $W2$, $W3$ are weights assigned to linear regression, ridge regression, decision tree, value of $W3 = \{9, 10\}$ has been eliminated as it shows no change from $W3 = \{8\}$.

Further, observed values of prediction accuracy indicates that the model suffers neither from under-fit or over-fit, thus reducing the probability of having bias and variance.

4.3.4.3 Performance of Ensemble Models

Ensemble models rules out variance by selecting a fraction of data i.e. sample or features for training and testing. For this purpose, three popular ensemble models are considered which are shown in Table 6 with respective (necessary) hyper-parameter values and prediction scores. Note, most hyper-parameters share similar values. This allows model comparison through accuracy score.

Table 6 Performance of ensemble models

	Random forest	XGBoost	Light GBM
Hyper-parameters	No. of trees = 20 Learning rate = 0.5 Tree depth = 5 Max features = 75%	No. of trees = 20 Tree depth = 5	No. of trees = 20 Learning rate = 0.5
Accuracy score (in %)	72.55	99.99	71.37

It can be observed from Table 6, that both random forest and light GBM provides adequate prediction scores with an accuracy between 70–80%. Therefore, both random forest and light GBM can be considered for predicting sales revenue from testing set. However, accuracy of both random forest and light GBM is quite low, when compared to second stage regression model. XGBoost provides an accuracy of 99.99%. This implies that XGBoost overfits the model, and thus cannot be used to predict sales revenue from testing set.

4.3.4.4 Performance of Second Stage Regression Models

The second stage model shown in Fig. 7B have prediction accuracy of 86.73%, which is an improvement over weighted regression model. However, the improvement seems low i.e. 1.39% over weighted regression model.

Although the increase in performance from weighted approach seems low, the second stage regression model is still preferable. It is because of,

1. *Employed automation*: second stage regression model improves on weighted method by automating weight assignment,
2. *Continuous weights*: second stage regression model takes continuous weights into consideration instead of only integer values considered in weighted approach which provides sub-optimal solutions,
3. *Faster deployment*: weighted approach requires manual intervention (to setup lower and upper bounds on weights based on domain knowledge), which is not desirable as it prevents automated processing while project deployment, and
4. *Multi-stage model extension*: second stage regression model provides window to build multi stage regression model(s) with much less effort and complexity.

4.3.5 Model Selection for Predicting Sales Revenue

To predict sales revenue from test data set, it is imperative to select the model with best prediction accuracy. Therefore, by comparing

1. Base models (linear regression, ridge regression, decision tree),
2. Weighted model (combination of base models through weights),
3. Second stage regression model (combination of base models into another regression model in second stage), and
4. Ensemble models (Random forest, XGBoost, Light GBM),

it can observed that *'second stage regression model'* provides the best accuracy score, without under-fitting or over-fitting the model. Thus, second stage regression model is selected to predict sales revenue from test set.

5 Conclusion and Future Scope

The contribution of this work can be summarized into two folds: the process of operational analytics is discussed in brief yet detailed, and the application of real-world dataset for better understanding. To summarize, in the first fold, data exploration and analysis through feature engineering and selection, data visualization techniques, base modeling and ensembles are discussed, which provides list of data collection methods, type of data used, data analysis tools, modeling approach and tuning, validation measures, reporting. In the second fold, the restaurant revenue prediction dataset is considered where crucial insights into the problem, strength, weakness, opportunities and threats to business, knowledge on model performance is obtained through systematic data analysis and modeling approach.

As operational analytics is still infant in the area of data science, there is ample opportunities for development. Further studies can be pursued on developing novel techniques in data exploration and visualization, efficient feature engineering techniques such as automatic feature identification and transformation techniques, forming feature relevance matrix, feature correlation identifier, and more, and devise modeling approaches such as automated feature selection, model selection and designing structural determinants for ensembles. It should be noted that, operational analytics covers a broad spectrum and thus is not confined to the above discussion. Therefore, existing literature cited in this work provides excellent ground for future work.

References

1. Tufféry, S. (2011). *Data Mining and Statistics for Decision Making*. Hoboken: Wiley.
2. Mosteller, F., Tukey, J. W., et al. (1977). *Data Analysis and Regression: A Second Course in Statistics*. Reading: Addison-Wesley.
3. Hoaglin, D. C., Mosteller, F., & Tukey, J. W. (1983). *Understanding Robust and Exploratory Data Analysis* (Vol. 3). New York: Wiley.
4. Cui, Z., Badam, S. K., Adil Yalçin, M., & Elmqvist, N. (2019). Datasite: proactive visual data exploration with computation of insight-based recommendations. *Information Visualization*, *18*(2), 251–267.
5. Pabinger, S., Dander, A., Fischer, M., Snajder, R., Sperk, M., Efremova, M., Krabichler, B., Speicher, M. R., Zschocke, J., & Trajanoski, Z. (2014). A survey of tools for variant analysis of next-generation genome sequencing data. *Briefings in Bioinformatics*, *15*(2), 256–278.
6. Rautenhaus, M., Böttinger, M., Siemen, S., Hoffman, R., Kirby, R. M., Mirzargar, M., Röber, N., & Westermann, R. (2017). Visualization in meteorology a survey of techniques and tools for data analysis tasks. *IEEE Transactions on Visualization and Computer Graphics*, *24*(12), 3268–3296.
7. Endert, A., Ribarsky, W., Turkay, C., William Wong, B.L., Nabney, I., Díaz Blanco, I., & Rossi, F. (2017). The state of the art in integrating machine learning into visual analytics. In *Computer Graphics Forum* (Vol. 36, pp. 458–486). Wiley Online Library
8. Liu, S., Wang, X., Liu, M., & Zhu, J. (2017). Towards better analysis of machine learning models: A visual analytics perspective. *Visual Informatics*, *1*(1), 48–56.
9. Zhang, M.-L., & Zhou, Z.-H. (2006). Multilabel neural networks with applications to functional genomics and text categorization. *IEEE Transactions on Knowledge and Data Engineering*, *18*(10), 1338–1351.
10. Idreos, S., Papaemmanouil, O., & Chaudhuri, S. (2015). Overview of data exploration techniques. In *Proceedings of the 2015 ACM SIGMOD International Conference on Management of Data* (pp. 277–281).
11. Khan, M., & Khan, S. S. (2011). Data and information visualization methods, and interactive mechanisms: A survey. *International Journal of Computer Applications*, *34*(1), 1–14.

12. Godfrey, P., Gryz, J., & Lasek, P. (2016). Interactive visualization of large data sets. *IEEE Transactions on Knowledge and Data Engineering, 28*(8), 2142–2157.
13. Dey, N., Ashour, A. S., Shi, F., Fong, S. J., & Simon Sherratt, R. (2017). Developing residential wireless sensor networks for ECG healthcare monitoring. *IEEE Transactions on Consumer Electronics, 63*(4), 442–449.
14. Elhayatmy, G., Dey, N., & Ashour, A.S. (2018). Internet of things based wireless body area network in healthcare. In *Internet of Things and Big Data Analytics Toward Next-generation Intelligence* (pp. 3–20). Springer.
15. Das, S. K., & Tripathi, S. (2018). Adaptive and intelligent energy efficient routing for transparent heterogeneous ad-hoc network by fusion of game theory and linear programming. *Applied Intelligence, 48*(7), 1825–1845.
16. Das, S. K., & Tripathi, S. (2019). Energy efficient routing formation algorithm for hybrid ad-hoc network: A geometric programming approach. *Peer-to-Peer Networking and Applications, 12*(1), 102–128.
17. Santosh Kumar, D., Sourav, S., & Nilanjan, D., et al. (2020). Design frameworks for wireless networks. In *Lecture Notes in Networks and Systems, Springer* (pp. 1–439)
18. Tao, F., Qi, Q., Liu, A., & Kusiak, A. (2018). Data-driven smart manufacturing. *Journal of Manufacturing Systems, 48*, 157–169.
19. Gibbs, W. J. (2015). *Contemporary Research Methods and Data Analytics in the News Industry*. Hershey: IGI Global.
20. Bro, R., van den Berg, F., Thybo, A., Andersen, C. M., Jørgensen, B. M., & Andersen, H. (2002). Multivariate data analysis as a tool in advanced quality monitoring in the food production chain. *Trends in Food Science & Technology, 13*(6–7), 235–244.
21. Hey, T., Tansley, S., Tolle, K., et al. (2009). *The Fourth Paradigm: Data-intensive Scientific Discovery* (Vol. 1). RedmondRedmond: Microsoft Research.
22. Panigrahi, S., Kundu, A., Sural, S., & Majumdar, A. K. (2009). Credit card fraud detection: A fusion approach using Dempster-Shafer theory and Bayesian learning. *Information Fusion, 10*(4), 354–363.
23. Zheng, A., & Casari, A. (2018). *Feature Engineering for Machine Learning: Principles and Techniques for Data Scientists*. Sebastopol: O'Reilly Media Inc.
24. Brandt, S. (1976). *Statistical and computational methods in data analysis*. Technical report. Amsterdam: North-Holland Publishing Company.
25. Bengio, Y., Courville, A., & Vincent, P. (2013). Representation learning: A review and new perspectives. *IEEE Transactions on Pattern Analysis and Machine Intelligence, 35*(8), 1798–1828.
26. Coates, A., Ng, A., & Lee, H. (2011). An analysis of single-layer networks in unsupervised feature learning. In *Proceedings of the Fourteenth International Conference on Artificial Intelligence and Statistics* (pp. 215–223).
27. Hinton, G. E., Osindero, S., & Teh, Y.-W. (2006). A fast learning algorithm for deep belief nets. *Neural Computation, 18*(7), 1527–1554.
28. Hinton, G. E., & Salakhutdinov, R. R. (2006). Reducing the dimensionality of data with neural networks. *Science, 313*(5786), 504–507.
29. Sanguansat, P. (2012). *Principal Component Analysis: Multidisciplinary Applications*. Norderstedt: BoD-Books on Demand.
30. Olshausen, B. A., & Field, D. J. (1996). Emergence of simple-cell receptive field properties by learning a sparse code for natural images. *Nature, 381*(6583), 607–609.
31. Yu, H.-F., Lo, H.-Y., Hsieh, H.-P., Lou, J.-K., McKenzie, T.G., Chou, J.-W., Chung, P.-H., Ho, C.-H., Chang, C.-F., & Wei, Y.-H., et al. (2010). Feature engineering and classifier ensemble for KDD cup 2010. In *KDD Cup*.
32. Zhao, Z., Morstatter, F., Sharma, S., Alelyani, S., Anand, A., & Liu, H. (2010). Advancing feature selection research. *ASU Feature Selection Repository*, 1–28.
33. Langley, P., et al. (1994). Selection of relevant features in machine learning. In *Proceedings of the AAAI Fall Symposium on Relevance* (Vol. 184, pp. 245–271).

34. Khotanzad, A., & Hong, Y. H. (1990). Rotation invariant image recognition using features selected via a systematic method. *Pattern Recognition*, *23*(10), 1089–1101.
35. Goltsev, A., & Gritsenko, V. (2012). Investigation of efficient features for image recognition by neural networks. *Neural Networks*, *28*, 15–23.
36. Rashedi, E., Nezamabadi-Pour, H., & Saryazdi, S. (2013). A simultaneous feature adaptation and feature selection method for content-based image retrieval systems. *Knowledge-Based Systems*, *39*, 85–94.
37. Lewis, D. D., Yang, Y., Rose, T. G., & Li, F. (2004). RCV1: A new benchmark collection for text categorization research. *Journal of Machine Learning Research*, *5*, 361–397.
38. Van Landeghem, S., Abeel, T., Saeys, Y., & Van de Peer, Y. (2010). Discriminative and informative features for biomolecular text mining with ensemble feature selection. *Bioinformatics*, *26*(18), i554–i560.
39. Song, Q., Ni, J., & Wang, G. (2011). A fast clustering-based feature subset selection algorithm for high-dimensional data. *IEEE Transactions on Knowledge and Data Engineering*, *25*(1), 1–14.
40. Gibert, J., Valveny, E., & Bunke, H. (2012). Feature selection on node statistics based embedding of graphs. *Pattern Recognition Letters*, *33*(15), 1980–1990.
41. Li, H., Li, C.-J., Xian-Jun, W., & Sun, J. (2014). Statistics-based wrapper for feature selection: An implementation on financial distress identification with support vector machine. *Applied Soft Computing*, *19*, 57–67.
42. Morgan, B. J. T. (2001). Model selection and inference: A practical information-theoretic approach. *Biometrics*, *57*(1), 320.
43. Fleuret, F. (2004). Fast binary feature selection with conditional mutual information. *Journal of Machine Learning Research*, *5*, 1531–1555.
44. Xu, Z., King, I., Lyu, M. R.-T., & Jin, R. (2010). Discriminative semi-supervised feature selection via manifold regularization. *IEEE Transactions on Neural networks*, *21*(7), 1033–1047.
45. Swiniarski, R. W., & Skowron, A. (2003). Rough set methods in feature selection and recognition. *Pattern Recognition Letters*, *24*(6), 833–849.
46. Derrac, J., Cornelis, C., García, S., & Herrera, F. (2012). Enhancing evolutionary instance selection algorithms by means of fuzzy rough set based feature selection. *Information Sciences*, *186*(1), 73–92.
47. Graybill, F. A. (1976). *Theory and Application of the Linear Model* (Vol. 183). North Scituate: Duxbury Press.
48. Segerstedt, B. (1992). On ordinary ridge regression in generalized linear models. *Communications in Statistics-Theory and Methods*, *21*(8), 2227–2246.
49. Singh, S., & Gupta, P. (2014). Comparative study Id3, cart and C4. 5 decision tree algorithm: A survey. *International Journal of Advanced Information Science and Technology (IJAIST)*, *27*(27), 97–103.
50. Rasoul Safavian, S., & Landgrebe, D. (1991). A survey of decision tree classifier methodology. *IEEE Transactions on Systems, Man, and Cybernetics*, *21*(3), 660–674.
51. Viaene, S., Derrig, R. A., Baesens, B., & Dedene, G. (2002). A comparison of state-of-the-art classification techniques for expert automobile insurance claim fraud detection. *Journal of Risk and Insurance*, *69*(3), 373–421.
52. Sebban, M., Mokrousov, I., Rastogi, N., & Sola, C. (2002). A data-mining approach to spacer oligonucleotide typing of mycobacterium tuberculosis. *Bioinformatics*, *18*(2), 235–243.
53. Yong, Z., Youwen, L., & Shixiong, X. (2009). An improved knn text classification algorithm based on clustering. *Journal of Computers*, *4*(3), 230–237.
54. Kanungo, T., Mount, D. M., Netanyahu, N. S., Piatko, C. D., Silverman, R., & Wu, A. Y. (2002). An efficient k-means clustering algorithm: Analysis and implementation. *IEEE Transactions on Pattern Analysis and Machine Intelligence*, *24*(7), 881–892.
55. Bera, S., Chattopadhyay, M., & Dan, P. K. (2018). A two-stage novel approach using centre ordering of vectors on agglomerative hierarchical clustering for manufacturing cell formation. *Proceedings of the Institution of Mechanical Engineers, Part B: Journal of Engineering Manufacture*, *232*(14), 2651–2662.

56. Polikar, R. (2012). Ensemble learning. In *Ensemble Machine Learning* (pp. 1–34). Springer.
57. Liu, Y., Yao, X., & Higuchi, T. (2000). Evolutionary ensembles with negative correlation learning. *IEEE Transactions on Evolutionary Computation*, *4*(4), 380–387.
58. Breiman, L. (2000). Randomizing outputs to increase prediction accuracy. *Machine Learning*, *40*(3), 229–242.
59. Rodriguez, J. J., Kuncheva, L. I., & Alonso, C. J. (2006). Rotation forest: a new classifier ensemble method. *IEEE Transactions on Pattern Analysis and Machine Intelligence*, *28*(10), 1619–1630.
60. Granitto, P. M., Verdes, P. F., & Alejandro Ceccatto, H. (2005). Neural network ensembles: Evaluation of aggregation algorithms. *Artificial Intelligence*, *163*(2), 139–162.
61. Bühlmann, P. (2010). Handbook of computational statistics: concepts and methods, chapter bagging, boosting and ensemble methods.
62. Breiman, L. (2001). Random forests. *Machine Learning*, *45*(1), 5–32.
63. Breiman, L. (1996). Bagging predictors. *Machine Learning*, *24*(2), 123–140.
64. Breiman, L. (1996). Stacked regressions. *Machine Learning*, *24*(1), 49–64.

Application of Machine Learning Algorithm for Anomaly Detection for Industrial Pumps

Nabanita Dutta, Palanisamy Kaliannan, and Umashankar Subramaniam

Abstract Automation technology has brought a pragmatic change in the field of industrial sector, commerce and agricultural sector etc. where machine learning algorithm is one of the pioneers of this. Machine learning has a broad scale application among that anomaly detection is one of the applications. Industrial pumps are essential parts of any kind of industry which requires proper maintenance which is recognized as condition monitoring. The application of machine learning in industrial sectors is going to 65% up to 2018. More will increase in later future. Condition monitoring is the process which cannot prevent failure but can predict the possibility of failure, fault condition by measuring certain machine parameters. If machine learning algorithm can be implemented then the system will be more efficient and it is possible to detect the problems in the ground level which can help to increase the lifespan of the pumping system. Various machine learning algorithms are available among which most of the cases classification and regression analysis are used to detect the anomalies. When there is discrete system generally classification is used and when continuous function is there regression is used. There are more other machine learning algorithms which can analyze the anomalies in the system with the help of predictive control model. The predictive control hybrid model is the new socket of study where the researchers can forecast to shrink the vigor loss of resource and time and can make the system flawless. Thus, there is a big challenge before the researchers regarding the application of machine learning for detecting the anomalies in the pumping system.

N. Dutta (✉) · P. Kaliannan
Department of Energy and Power Electronics, School of Electrical Engineering, VIT Vellore, Vellore 632014, India
e-mail: nabajhilikbarbi@gmail.com

P. Kaliannan
e-mail: kpaanisamy@vit.ac.in

U. Subramaniam
Renewable Energy Lab, College of Engineering, Prince Sultan University, Riyadh, Saudi Arabia
e-mail: shankarums@gmail.com

S. K. Das et al. (eds.), *Machine Learning Algorithms for Industrial Applications*, Studies in Computational Intelligence 907,
https://doi.org/10.1007/978-3-030-50641-4_14

Keywords Machine Learning · Artificial Intelligence · Pumping System · Energy Efficiency · Condition Monitoring

1 Introduction

Pumping system is such kind of machinery by which fluid can be transferred from low level area to high level area or from low pressure area to high pressure area. Basically pump is a mechanical device and world's 20% of the electrical energy is consumed by the pump system and 25–50% of the energy usage has been done by industrial plant operations [1]. The design of the pump affects energy and material used by pump systems. It is used for process operation and heavy duty equipments needs high discharge pressure and low suction pressure [2]. Fluid is lifted from certain depth due to low pressure at suction side of the pump. Importance of pumping system is recognized in modern civilization because it is an important instrument not only for household and agricultural sector, but also for major industries like chemical, manufacturing, heating, mining, air conditioning etc. The energy efficiency can be maximized by life cycle cost (LCC) analysis tool. Pump efficiency also can be improved by life cycle cost analysis shown in Fig. 1. Because maintenance of pumping system is highly essential and continuous monitoring is needed which is recognized as condition monitoring. Manual and automatic are both the ways where the system should be kept under continuous supervision.

The life cycle cost analysis is a management tool by which selection of pump and other information can be operated. LCC process is cost effective solution by which plant designer can able to compare the cost of the materials of the pump and find out the effective solution. Generally a pump life time is 15 to 20 years. Some of the

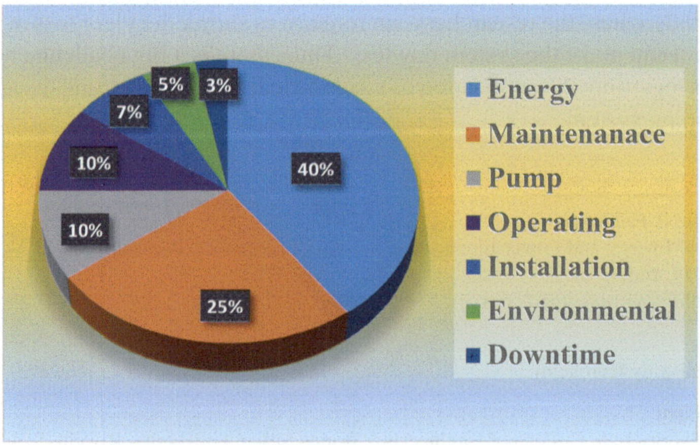

Fig. 1 Life cycle cost of pumping system [3]

elements will be evaluated to present the discounted value of LCC [3]. The equation of L_{CC} has been described in Eq. 1.

$$L_{cc} = C_{ic} + C_{in} + C_e + C_O + C_m + C_S + C_{env} + C_d \quad (1)$$

where

C_{ic} initial costs, purchase price (pump, system, pipe, auxiliary services)
C_{in} installation and commissioning cost (including training)
C_e energy costs (predicted cost for system operation, including pump driver, controls, and any auxiliary services)
C_O operation costs (labor cost of normal system supervision)
C_m maintenance and repair costs (routine and predicted repairs)
C_S down time costs (loss of production)
C_{env} environmental costs (contamination from pumped liquid and auxiliary equipment)
C_d decommissioning/disposal costs (including restoration of the local environment and disposal of auxiliary services).

LCC can be dominated by the energy consumption in larger case and it can happen when pump run more than 2000 h per year. Gathering the system data only energy consumption calculation can be possible and calculation will be simple if output is stable condition.

1.1 Types of Pumps

The pumps are mainly of two types, one is dynamic or centrifugal pump and another is positive displacement pump shown in Fig. 2 [1].

Pumping system is such kind of device by which fluids can be moved or sometimes some slurries by mechanical action and according to the working principle pump can be classified in two categories, one is rotodynamic and another is positive displacement pump. Among all the pumps used in the world 73% are rotodynamic pumps and 27% are positive displacement pump. According to the European Commission report, pump is one of the single largest users of world total electrical energy i.e. 22%. Rotodynamic pumps use centrifugal force to develop velocity in the liquid being handled shown in Fig. 3a. The pressure is increased by decreasing of kinetic energy and velocity is converted to pressure. The system or plant structure depends on the pressure difference in the drives. When pressure changes and decreases when viscosity increases and flow rate also varies. The approach of positive displacement shown in Fig. 3b pump is somewhat different, it uses reciprocating motion to transfer fluid. In a fixed volume it repeatedly transfer the fluid. Figure 3 shows pump types. In general, rotodynamic pumps are preferred among centrifugal pumps as rotodynamic pumps are dominant in pump category. For discharging of the fluid to discharge reservoir centrifugal pump has casing, impeller, and volute. Most of the cases centrifugal pump is preferred as it has high flow rate and medium head and it requires less maintenance cost than positive displacement pump. The main disadvantage of positive displacement pump is continuous maintenance and minimal flow rate. The pumping action is cyclic and can be driven by pistons, screws, gears, rollers, diaphragms or vanes. Generally the average efficiency of the pump is below 40%, and selection of pumping system depends on various factors like right size of the pump, trimming of the impeller, minimization of system pressure drop, implementation of proper control valves, implementation of variable speed drives, maintenance of pumping system, using of proper pumping seal, and reduction of unnecessary usage of pumps. Among all the factors use of variable speed drive (VSD) or VFD is a cost effective solution to achieve significant energy savings.

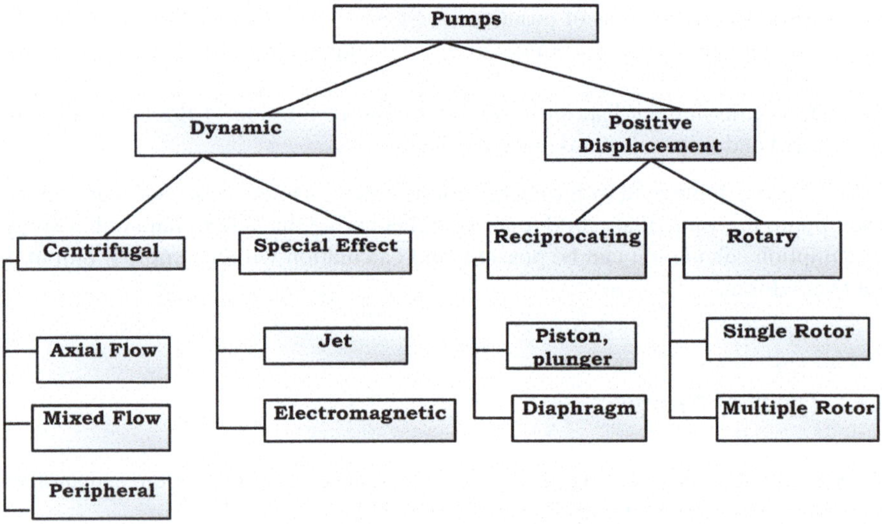

Fig. 2 Types of pumps [1]

a. Centrifugal Pump b. Positive displacement

Fig. 3 Pump types

2 Variable Frequency Drive Based Pumping System

The cost effective solution is the latest model controller and VFD based motor system which is capable to save the energy.

The advance motor controller for VFD drives is simple and cost effective and can achieve significant energy savings. Energy efficiency and therefore reduction of the losses of energy is the main objective for the future research. Throughout the world industry is facing the energy insufficiency problem which is a major challenge. Pumping system is also facing this energy deficiency problem. Increasing the efficiency for the pumping system is highly recommended where larger fluid flow requirement is needed. VFD drives is one of the solutions to increase the efficiency of the system and parallel pumping is the highly recommended solution to increase the efficiency. VFD drives can be applied for parallel pumping also. It can reduce 50% of the energy usage and increase the efficiency up to 80%. Optimal design can improve the pump efficiency and VFD based drives can increase the efficiency from 5 to 50%. Conventional valve can be replaced by VFD which can save the energy by changing the speed of the pump shown in Fig. 4 [5].

2.1 *Pump Characteristics*

The flow rate varies with the change of head value and it has constant relation with pump head. In the large and heavy industry where only one pump in insufficient there two or more than parallel pumps are used. In this situation flow rate of the each pump will be added to get the overall flow rate and head value will be constant like previous. The parallel pump has its variety, it can be similar or different. It follows the affinity law of the pump shown in Figs. 5 and 6.

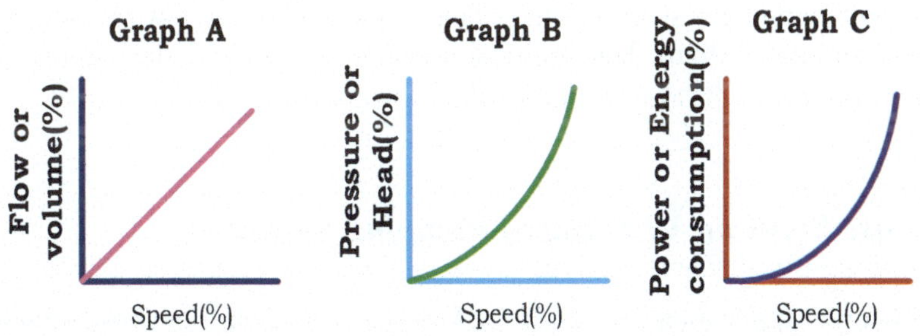

$$\frac{Flow_1}{Flow_2} = \frac{RPM_1}{RPM_2} \qquad \frac{Head_1}{Head_2} = \left(\frac{RPM_1}{RPM_2}\right)^2 \qquad \frac{Power_1}{Power_2} = \left(\frac{RPM_1}{RPM_2}\right)^3$$

Fig. 4 Affinity law [5]

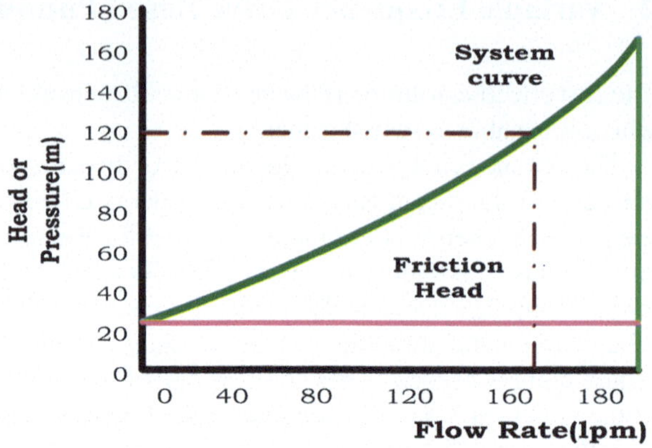

Fig. 5 Elements of a system curve

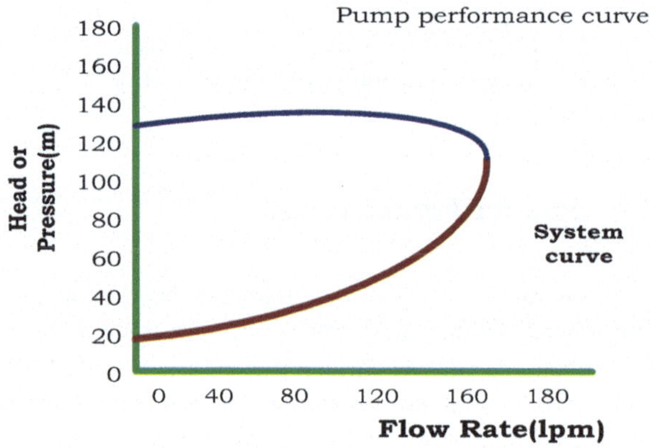

Fig. 6 Combination of the system and pump curves

The static head, friction head, and resulting system curve are shown in the figure for a typical pumping system. In this example, the maximum flow rate required is 160 liters per minute (lpm). This information helps to determine the required pump and impeller size for the system to provide the maximum required flow [6].

2.2 *Case Study on VFD Based Pumping System*

The VFD drive application can help to reduce the cost of the elements required for control valves. For controlling the valve additional piping system is required and valve needs to be adjusted. The valve loss will be 15 kW if piping loss is 10 kW. Because of this loss internal loss also will increase from 50 kW to 90 kW. For VFD system no control valve is required. So smaller pump also can be used with low loss. For 50 kW head 68 kW pump and 68 kW motor load is required. This reduces the costing of the system shown in Fig. 7.

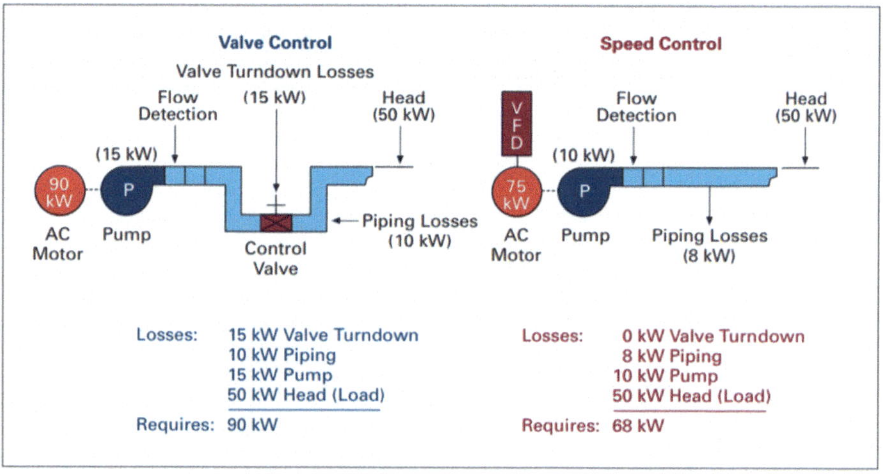

Fig. 7 Computerized energy saving analysis

2.3 *Failures of Pumps and Its Effects*

The flow can be controlled by the application of drive in the pumping system. Variable speed drives help to save the energy [10]. The overall efficiency of the system depends on many factors of the system and constant pressure also be the part of this dependency. The large pump and motor has higher efficiency in generally but when pump operated in low speed the system efficiency also will drop. To operate the water pump in 50 Hz range the overall system efficiency should be the 90% of the system rated operating point. So in 85 Hz drop will be 35%. The efficiency of the water pump generally maximum is 85% and minimum is 35% [11]. The speed changes of the pump affect the pump system curve. But sometimes excessive increment of flow rate in the pipeline causes major failure like water hammering in the system, it causes hydraulic shock, excessive vibration causes bearing failure, stator rotor failure. Instead of bearing failures several problems can occur in the system like sludge, cavitation, dry run, wear, deposits etc. [7–9].

The whole system can be interrupted due to various faults in the pump. Table 1 is the brief description of faulty components and its frequency and Table 2 is the summary of causes of common faults of pump shown in Figs. 8 and 9.

Due to operational problem or continuous usage the efficiency of the pump can be reduced.

The common faults which are seen in pumping system generally that can destroy the whole system. The most common faults can be mechanical, electrical and vibration kind of faults. The common faults are bearing fault, inter turn fault, and other faults like impeller breaking, and single phasing, broken rotor bar faults etc. 42% is the bearing fault and 28% is the stator fault and 22% of the fault is others faults and 8% is rotor fault. These faults directly and indirectly affect the whole system

Table 1 Faulty components and its frequency

Faulty components	Reported frequency (%)
Sliding ring seal	31
Rolling bearing	22
Leakage	10
Driving motor	10
Rotor	9
Sliding bearings	8
Clutch	4
Split pipe	3
Casing	3

Table 2 Malfunctions of the pumping system and their causes

Faults	Causes
Cavitation	Creation of vapor bubbles inside the pump
Gas in fluid	Due to pressure drop some gas mixes with the liquid
Dry run	Due to lack of liquid excessive heating of pump
Sludge	Unnecessary particle enter inside the pipeline
Water hammering	Due to excessive vibration
Wear	Mechanical damage of pipe wall for hard particles
Deposit	Deposit of organic or chemical materials
Oscillation	Unbalancing of rotor

and impeller breaking, casing damage, seal damage can happen. Due to these faults overall system efficiency can be reduced. The faults has adverse effect in the system [11].

2.3.1 Cavitation

Centrifugal pump is used for different operating condition and can be failed in any time. The cavitation is the typical problem of the pump which leads to the total failure of the pumping system and internal parts also doctorates due to this failure. By the help of external measurement method like vibration analysis cavitation phenomenon can be identified. For that condition monitoring is required. Cavitation is the formation of bubbles in the low suction pressure area of the pump. For this reason vapor pressure goes below the suction pressure. Continuous monitoring is needed for industrial pumps to minimize the loss production. The case study based on acoustic emission (AE) plays significant role for detecting cavitation problem in pump. The best

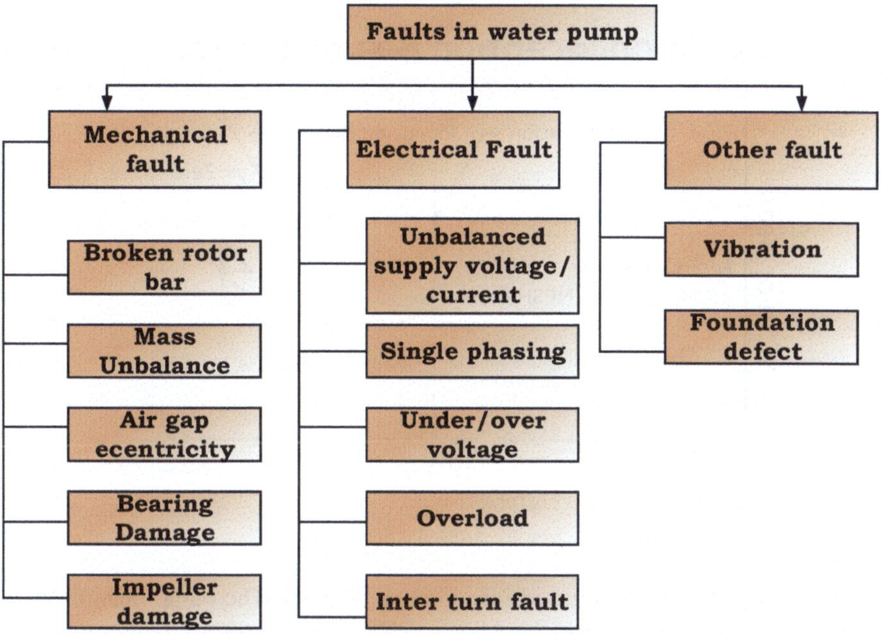

Fig. 8 Various types of faults in pump

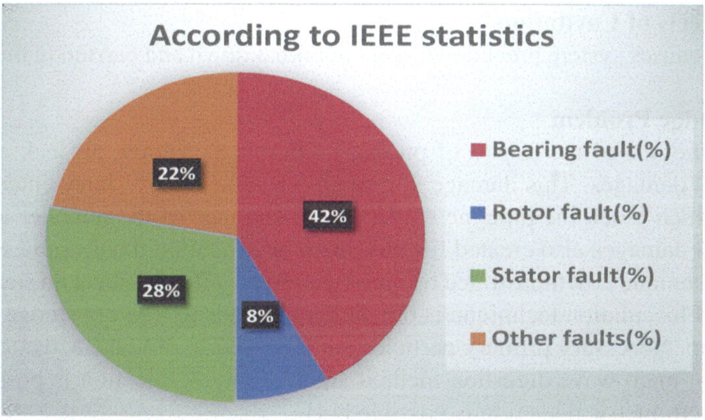

Fig. 9 Statistics of faults

efficiency point (BEP) is the key point of findings the cavitation in pump. AE is the superior tool to detect faults in the system. Cavitation time AE level decreases generally and BEP is another weapon to find the anomalies. But AE required high cost sensors which is costly for the users to detect the faults though condition monitoring method. With the help of 98% of precision cavitation can be detected. Hence

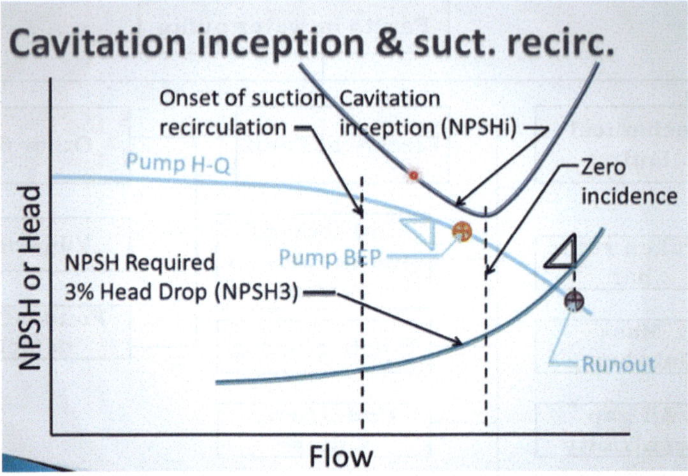

Fig. 10 Effect of cavitation

electrical Current sensors are used for vibration analysis. The pump manufacturer provide a minimum pressure which is known as NPSHA and pump requires minimum suction pressure which is known as NPSHR. The NPSHA should be always higher than NPSHR. If it goes below the NPSHR then cavitation occurs in Fig. 10 [13].

2.3.1.1 Effects of Cavitation

Cavitation causes system breakdown, impeller breakdown and carrion in the system.

2.3.1.2 Sludge Problem

Sludge is the one of the causes of pressure difference creation, air leak and other mechanical damages. This damage is created for unnecessary slurry enter into the pipeline which block the pipeline of the pump, entrance of the impeller and other mechanical damages also created for this. In an investigation the usefulness of flocculation technique is demonstrated by investigating the effect of shear on steady-state floc sizes. Flocculation technique is one of the methods to find out sludge problem in the pump. Necessary primary particles are observed to produce activated sludge problem. In microwave digestion method automatic flow injection is possible and atomic spectrometric detection is possible [14].

2.3.1.2.1 Effects of Sludge

Sludge is also a major problem in the piping system where some unnecessary particles enter into the pipe creates head loss and efficiency degradation.

2.3.1.2.2 Head Loss

For a particular system there is one particular system head but if the unnecessary particles enter into the system then the performance curve will reduce which causes efficiency loss in the system in Fig. 11.

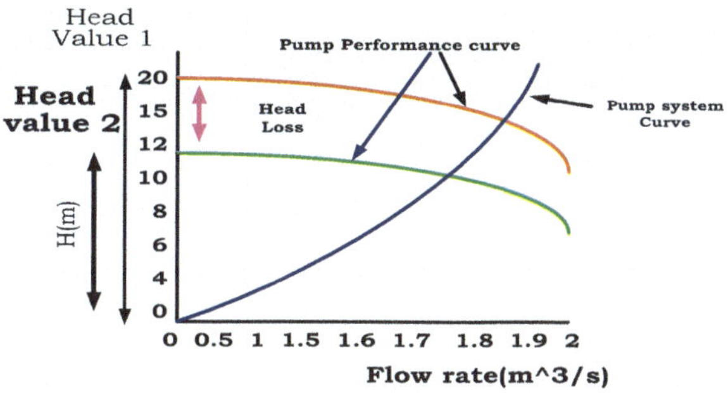

Fig. 11 Effect of Sludge in pumping system

2.3.1.2.3 Cavitation

Sludge indirectly creates cavitation problem also. Due to head loss NPSHA value goes below the NPSHR value which causes cavitation.

2.3.1.2.4 Effects of Sludge

Sludge is also a major problem in the piping system where some unnecessary particles enter into the pipe creates head loss and efficiency degradation.

2.3.1.2.5 Head Loss

For a particular system there is one particular system head but if the unnecessary particles enter into the system then the performance curve will reduce which causes efficiency loss in the system in Fig. 11.

2.3.1.2.6 Cavitation

Sludge indirectly creates cavitation problem also. Due to head loss NPSHA value goes below the NPSHR value which causes cavitation.

2.3.1.3 Water Hammering Problem

Water hammering is one of the problem which occurs when fluid flow changes rapidly [39]. A high pressure has been created in the pipeline and high forces created which causes impeller breaking and leakage in the pipeline. Depending on the condition of the fluid water hammer can occur and causes pipe burst and collapse [15].

2.3.1.3.1 Effects of Water Hammer

It is a vital problem in the field of pump and causes the failure of the system. In the home plumbing system water hammer has its effects. The mathematical discussion has been done to analysis the water hammer effect in the system and some solution can be found out. The paper presents unsteady momentum and continuity equations which has unsteady water distribution problem and steady state energy and continuity equations known as Hardy cross method. It is a shock wave when sudden flow rate of the pump changes water hammer can occur. The fluid velocity changes also causes water hammer. The pressure wave can damage the system if sudden shock wave has been created in Fig. 12 [16].

Fig. 12 Water hammering problem in pumping system in different temperature Vs time

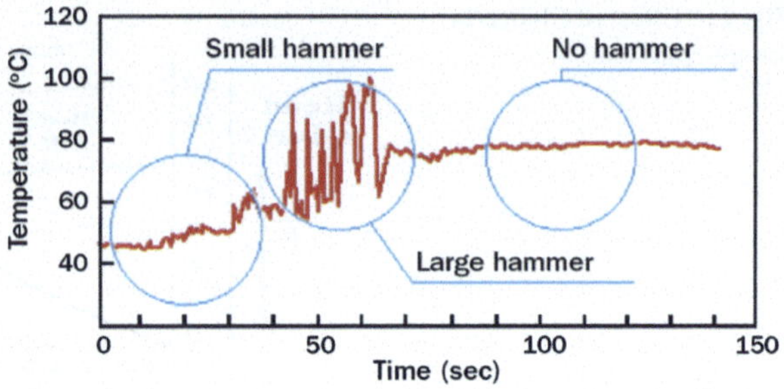

The other problems in the pumping system are bearing problem, seal failure, lubrication failure, excessive vibrations, fatigue etc.

Bearing failure: The mechanical part of the pump is modelled using simple considerations based on Newton's second law. The frictions losses in the bearing and seals are modelled by a simple linear friction term, as the friction losses are very small compared to the torque necessary to drive the pump, and therefore are not important in the model. Mechanical faults are recognized as external faults and categorized in various parts like broken rotor bar, impeller breaking fault, air gap eccentricity, bearing damage, sludge etc. Impeller breaking faults are two parts like cavitation and water hammering. Cavitation and water hammering are the causes of impeller breaking in the pump. Bearing failure frequently happen but ultimately machine has a catastrophic effect for this fault. Installation problems are often caused by improperly forcing the bearing onto the shaft or in the housing. This produces physical damage which leads to premature failure. The electromechanical symmetry can be disturbed by shaft bearing problem and dynamic or static changes also possible for healthy operating conditions [17].

Seal Failure: Mechanical seals can be damaged due to dry run and many applications are there by which mechanical seals are protected and remains lubricated even the system is completely dry run [40].

In non-hazardous applications, a pump sealed with packing that is lubricated from an external source will survive better during dry running, given that the source is compatible with the fluid being pumped [18].

2.4 *Conventional Method of Identifying Pumping Faults*

A fault detection system is said to perform effectively. Mainly two methods are there to detect the faults in pumping system. One is signal based system and another is model based system [19].

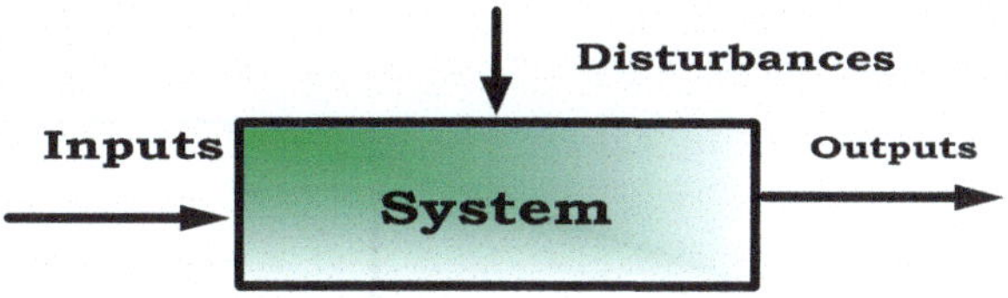

Fig. 13 Signal based fault detection

2.4.1 Signal-Based Fault Detection Method

Based on the raw system measurements signal based fault analysis is possible. Motor vibrations, signature analysis, motor currents can be measured through this method [37]. Signal-based fault detection techniques are based on processing and analyzing raw system measurements such as motor currents, vibration signals and/or other process-based signals [20]. The main advantage of signal based model is it does not require any mathematical model. It requires both model and data which are the major drawback of this system. For this approach it is difficult to ensure robustness of the system. Without model it can solve the problem in certain extent. The signal processing can be done by FFT algorithms, wavelets and higher order statistics. Under signal based analysis there are four types of processes which are limit checking and trend checking, data analysis, and spectrum analysis and pattern recognition. Signal based model is used to detect low frequency vibrations for a specific fault application [38]. Stationary wavelet method has been used in a work. Different effects has been analyzed by over decomposition method. Some controller algorithm like GCMBO for FOPID controller can be used for optimal design and this can be used for optimization of signal analysis in fault detection [41]. The whole process needs three steps, first is the transformation of signal to wavelet coefficient, second is Fuzzification of threshold coefficient, and third is detecting the decision of selection of signal. Fuzzy decision making allows formal modelling of decision-making for imprecise and uncertain conditions shown in Fig. 13 [21].

2.4.2 Model Based System

Analytical redundancy is applied in the process based model. Analytical model helps to make diagnosis of the residual evaluation and actual outputs. The structure can assume the model. When state variable changes faults also changes. The basic principle of a model-based fault detection scheme is to generate residuals that are defined as the differences between the measured and the model predicted outputs [4]. It needs mathematical model for building the system. Only with the help of residual model it can predict the system anomalies. It ensures the robustness of the system. Whole model is required to be simulated then only fault analysis is possible. It is not so helpful for continuous monitoring shown in Fig. 14.

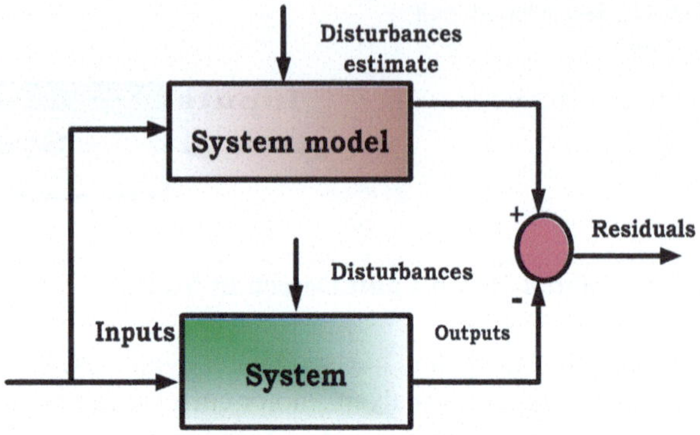

Fig. 14 Model based fault detection

3 Application of AI for Smart Building

Machine learning is a promising technique for many practical applications. In this perspective, we illustrate the development and application for machine learning. It is indicated that the theories and applications of machine learning method in the field of energy conservation and indoor environment are not mature, due to the difficulty of the determination for model structure with better prediction. In order to significant contribution to the problems, we utilize the artificial neural network (ANN) model to predict the indoor cultivable fungi concentration, which achieves the better accuracy and convenience [23]. The proposal of hybrid method further expands the application fields of machine learning method. Further, ANN model was successfully applied for the optimization of building energy system. Building energy and environment are closely related to people's lifestyle. On the one hand, reduction of energy consumption has been considered as the prominent factor for economic growth, since the energy demand of residential and commercial buildings could reach 40% of total energy demand in both US and EU. Each and every country is progressing in the sector of AI, but China is the leading among all shown in Fig. 15.

Many AI algorithms are capable of learning from new data, and this is under the self-learning process such as Bayesian network, decision tree, k- nearest network etc. If infinite data are given to the system, memory can gain the knowledge by self-learning and can learn to approximate any function and considering every possibility it can achieve the best suitable output by hypothesis testing. Based on different strategies learning algorithm work and the work continues for betterment of future. With the help of step by step process early researchers developed algorithms and some reasoning was used to identify the local puzzles [24]. This total system is called combinational explosion but here the process will be slow when data size will be large. Now the concept of machine learning, deep learning and IoT are the more advanced concepts which can handle huge amount of data within very short period of time. The numerous faults can be predicted by condition monitoring in the pumping system with the help of trained data and predict the condition of the fault. The faulty condition can be found out by computational way. The recent world is the

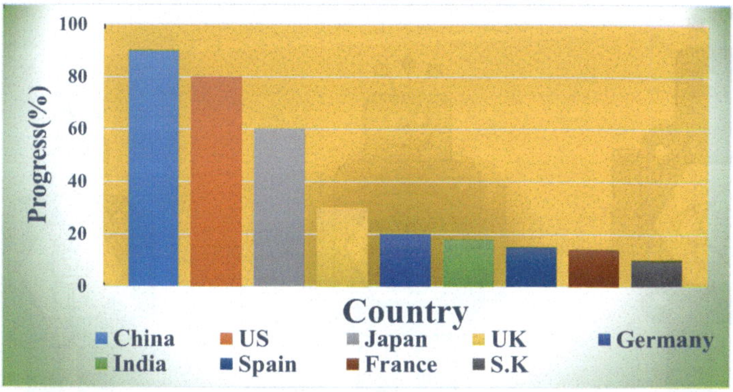

Fig. 15 Progress of AI research for different countries

sign of advance technology like application of deep learning and machine learning algorithm and key advantage of deep learning is the efficiency of the system increase when the size of the data increases. So more accurately problem can be identified [25, 26].

For analysis of data with logical structure deep learning model is required. It acts like human being and deep learning uses layered structure which is popular as artificial neural network (ANN). The design of ANN is inspired by biological network. The machine model becomes more capable than normal machine model.

Best example of deep learning is Alpha Go. Thus both machine learning and deep learning technology can be applied for pumping application along with other applications but it depends on data, pattern and situation. So the main relationship between AI, ML and DL is that AI is the technique by which human intelligence is exhibited by machines, machine learning is the approach to achieve the AI and deep learning is for implementing the machine learning shown in Fig. 16.

In the midst of technological advancement around the globe the privileged generation are bound to live where everything is time bounded. To keep pace with the busy schedule of the daily routine of human civilization, technology also becomes more effective, advance and more predictive shown in Fig. 17 [25].

3.1 Application of AI in Pumping System

The popular algorithms like SVM, KNN, ELM, neural network, linear regression all are the parts of supervised, unsupervised and semi supervised algorithms. With the help of artificial intelligence the devices are more efficient [12, 22]. The computational statistics has close relation with machine learning algorithm and progressive data can be learnt and particular tasks can be analyzed. In the first wave of AI the main challenge was to understand human intelligence but in the second wave the main target is to connect the whole world with cheap computing power, internet and

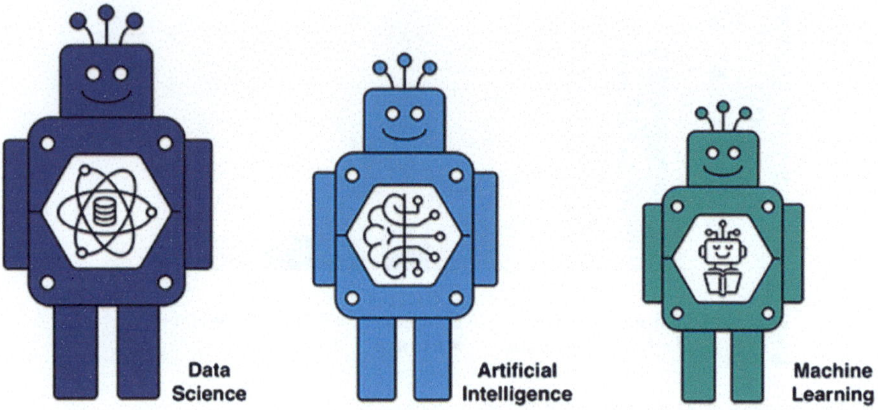

Fig. 16 Steps of AI [25]

Fig. 17 Workflow of AI

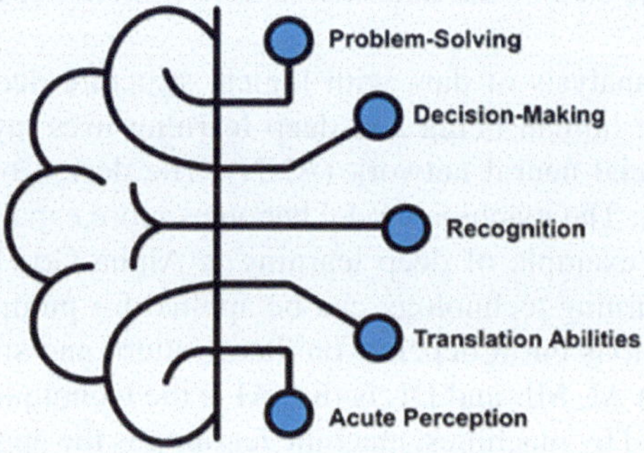

huge amount of training data. AI cannot reproduce the human intelligence but it has the power to adopt human intelligence in its best suited way. AI changed the human's everyday life. It not only reduces the human effort it makes the world more error less, more accurate and faster. In banking sector numerous activities have been done by AI. It has given a big challenge in share market and trending sector also. In the medical sector AI bought the new path, 24 h customer support is possible only for the application of AI. Now for heavy industry incredible development have been done, in air transportation also some control, transportation purpose AI plays the important role. In computer and TV games AI has bought the remarkable changes.

Deep Learning model has state of accuracy which can perform beyond human performance. The technique has been extracted from the machine learning itself and relevant features are extracted with respect to images. Control models are enough flexible and has its wide range of application. It depends on speed, pressure and temperature etc. So artificial intelligence (AI) technology is the key solution to keep

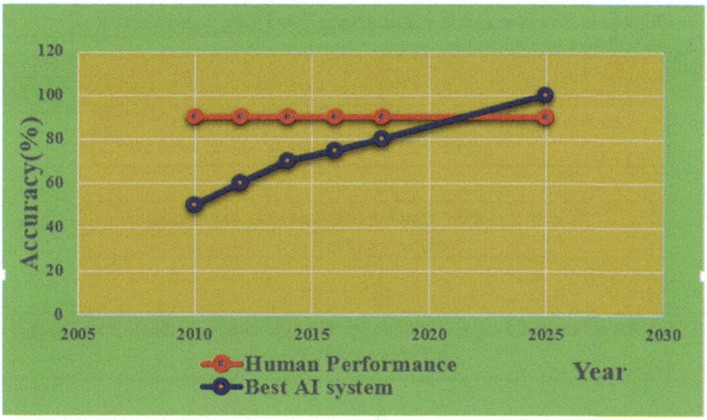

Fig. 18 AI comparison with human performance

the system under continuous monitoring to predict the failure of the system, decaying of the parts of the pumping system which is helpful for sustainable living [27].

Machine learning, deep learning and IoT based technology which are the part of AI are helpful for predictive control technology and help to increase the efficiency of the pumping system by reducing energy loss. Artificial intelligence (AI) is one kind of science by which a machine can achieve the human intelligence by computer programming. It can adopt the human intelligence but it is not limited to the methods which are observed biologically [28]. The main motto of AI is to achieve the human intelligence as far the computer program can adopt. Sometimes it is not possible to achieve the goal but target of AI is to find out the best suited output from the machine compare to human intelligence. In the era of AI it is possible to reduce the human burden by implementing machine oriented work and machine is fed to extra logic for predicting best output shown in Fig. 18 [29, 42].

Monoblock centrifugal pumps are used for various applications and continuous monitoring for pumping system is necessary to reduce the sudden breakdown of the system. In this situation for continuous monitoring vibration based approaches are widely used and particularly fuzzy logic, support vector machine, neural network are used for continuous monitoring of fault diagonise. In the piping system leakage causes economic destruction and energy losses and leakage of the piping should be taken care throughout the lifetime of the pump [30]. For minimization of the damage pressure leak detection is necessary. Real-time Fault Detection for Advanced Maintenance of Sustainable Technical Systems-is most fault detection systems, have proved their efficiency in the detection of anomalies and disruptions in technical systems. In real time applications anomalies and disruptions is time consuming and not applicable. Support vector machine algorithm is one of the machine learning algorithms based on time frequency analysis and suitable parameters fault detection possible [37]. The time domain analysis is statistical feature by which wavelet can be transformed and yearning the population around the world for the purpose of

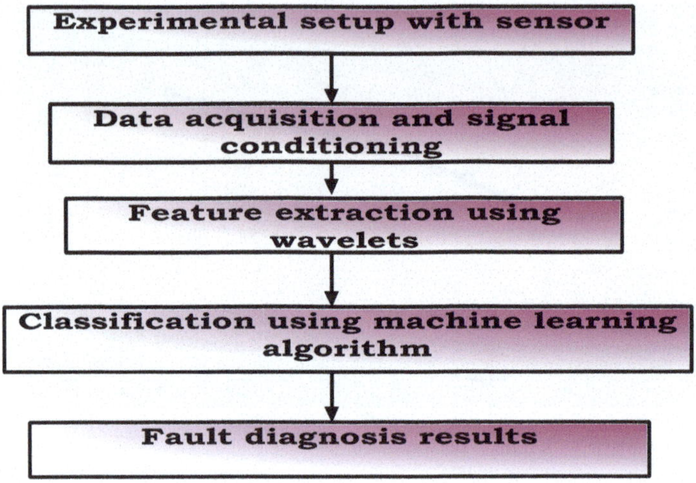

Fig. 19 Flow chart of fault diagnosis of pumping system

energy savings. Motor current signature analysis has a widespread application and mainly rotor faults can be identified with blade passing frequency, vibrations and compressors of pump fan. Rotor faults are dangerous and can be identified by signature analysis. No of blades, fans affects to identify the faults. Machine learning is the best suited method to predict the faults in pumping system which is based on automation technology [33]. The method is computational method and allows the system to learn by itself [34]. So machine learning has wide spread application in automation technology especially in industrial applications. ML has various algorithms like classification, clustering, regression etc. which has more no of algorithms like SVM, KNN, and ELM etc. It is based on data acquisition system and feature can be extracted for training model and testing data. The whole model is known as prediction control model shown in Fig. 19 [31].

3.2 Industrial Application of Machine Learning

The application of artificial intelligence is very popular today and machine learning is one of the tools of AI which works similar as human brain and are used in each and every sector recently. For the application of smart HVAC technology machine learning has kept a big step forward. Alpha Go Zero, the application of reinforcement learning, most of the companies are using this technology for the optimization purpose and handling the data. The Deep mind technology has given the new path to google, siri, Alaska to reduce energy consumption and make it sustainable energy. Google are using automl the part of RL to generate the neural network model for natural language processing and computer vision. For virtual flow meter, thermometer and emission control turbo speed device also machine learning technology is applicable. Danfoss is a manufacturer, product service group which build the product for cooling, heating and refrigeration purposes. They have large scale application in HVAC system. Danfoss has given a unique solution for sea water and river osmosis solutions for saving energy up to 50% of the previous energy consumption. It also provides the cost effective and reliable solution by VLT Aqua drive and pressure transmitter. The variable frequency Drive (VFD) based system can able to give solutions for small sized plant, extensive application, short payback time for high efficiency system, less maintenance cost, and compact footprint. Danfoss App pump has given the solution for fresh water for a decade without any maintenance [35, 36]. In North America previously some axial piston pump was used and after this pump installed by Danfoss its running without any maintenance. The total energy has been saved up to 38% for this solution in Hawaii Island. In this island some RO plant installed which cost and energy is saving and also CO2 emission less. Danfoss Autonomous Vehicle Intergration system (Davis) has given the driverless, self-driving vehicle technology which has given the future world a best solution for energy saving. Danfoss lean heat technology is also the best solution for building energy saving. In most of the industries 5 problems face and these 5 problems can be easily solved by application of machine learning. Transformer and production process for smart manufacturing, predictive maintenance, autonomous vehicle and interactive machines in production, optimized energy management for climate and energy change all are the application of machine learning technology. Hybrid Ad Hoc network can be used for optimization of hybrid and multiple network which can be applicable for condition monitoring technique and energy efficient application [43–45]. In domestic and industrial application water supply quality can be controlled by artificial neural network by cuckoo search method [46].

3.3 Application of Machine Learning in Proposed Work

Various algorithms of machine learning for anomalies detection of pumping system mainly SVM, regression analysis and neural networks are used. All these algorithms are used for prediction of the anomalies in the basic level. In this case the data collection has been done both healthy and faulty condition for comparing the results and after collection of data the analysis should be done by any software like matlab, python or any analytical software to classify the faults through machine learning algorithm. After the causes have been identified the predictive model is made for universal solution to prediction of faults before the system is totally damaged by analysis various parameters like voltage, current, speed, torque etc. limitations. The proposed work has been described in Fig. 20.

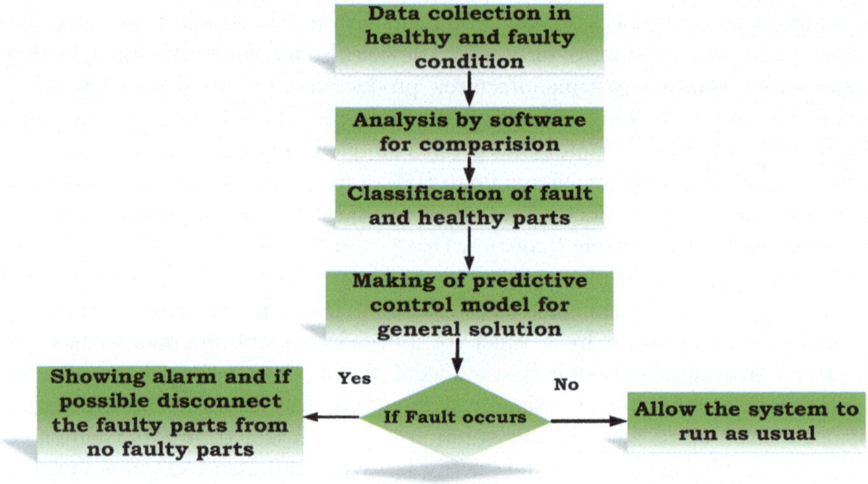

Fig. 20 Flowchart of the proposed work

3.4 Comparison of Various Machine Learning Algorithm

The algorithms can be compared by their accuracy rate and this accuracy depends on training time and prediction time. It is seen that to identify various anomalies in pumping system regression analysis, linear discriminant analysis and neural network are more suitable than other algorithms with respect to their accuracy rate. Though it is overall comparison the accuracy and performance depends on the application and data size for analysis. Use of resampling method, statistical analysis and cross validation can determine the actual performance and accuracy of the algorithm. The comparison of various machine learning algorithms like logistic regression (LR), Logistic Discriminant algorithm (LDR), K-Nearest Neighbor (K-NN), Classification and regression trees (CART), Naïve Bayes (NB), Support Vector machine (SVM) and neural network has been described in Fig. 21 for the proposed work.

The application of artificial intelligence for pumping system cannot helpful to increase the efficiency of the pump when there is no fault it can monitor the pump whether any anomalies is there in the pumping system due to continuous operation. It can predict the anomalies in the basic level and warn the users before the system is totally damaged. It can predict the life cycle of the pumping system also. By predicting the anomalies in can prevent the huge energy loss and efficiency of the system also improved like this way.

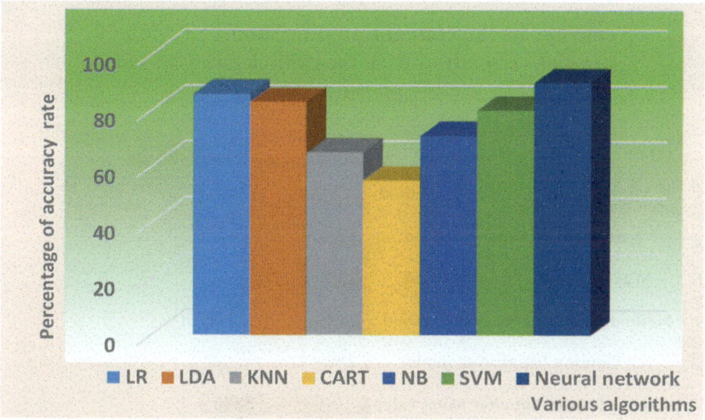

Fig. 21 Comparison of various machine learning algorithm

3.5 *Comparison of Machine Learning Algorithm with Conventional Methods*

Machine learning algorithm based predictive control technique came in the industrial application I n recent time but before that for anomaly detection or continuous monitoring some conventional methods like method of characteristics, wave propagation technique, velocity adjustment method, column separation method were used which are complicated, mathematical model based, time consuming, unable to predict anomalies in the basic level. The overall efficiency of various conventional methods with machine learning based predictive control technology has been compared in Fig. 22.

3.6 *Simulation Based Work*

In the model based work a dynamic model can be built and output data can be measured [32]. The good model accurately predict the response of the system and predictions is not good all time if residuals are large and there is a aspect to detect the failures. Different types of vibrational analysis depending upon the system such as wind gusts, contact with running engines and turbines, or ground vibrations. The impacts are a result of impulsive bump tests on the system that are added to excite the system sufficiently [33]. The excitation comes from periodic bumps as well as ground vibrations modeled by filtered white noise. The output of the system is collected by a sensor that is subject to measurement noise. The model is able to simulate various scenarios involving the structure in a healthy or a damaged state. Here the system behaviour has been compared with healthy and faulty condition of the pump with the help of vibration data shown in Fig. 23. Figure shows the time frequency analysis of

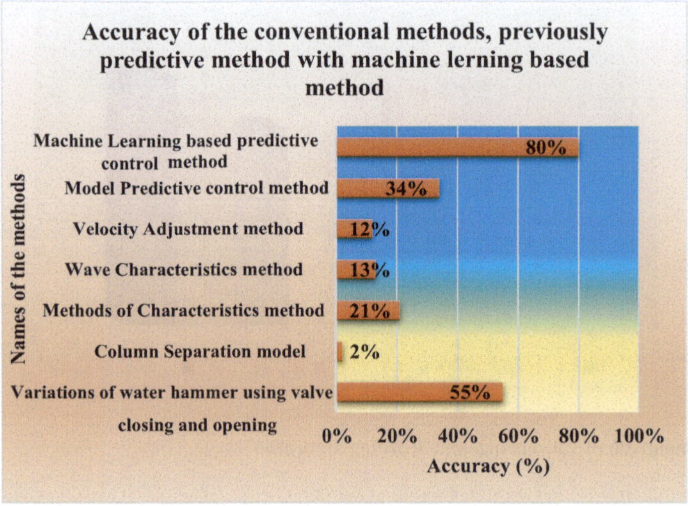

Fig. 22 Comparison of various conventional and predictive control method with machine learning-based predictive control method

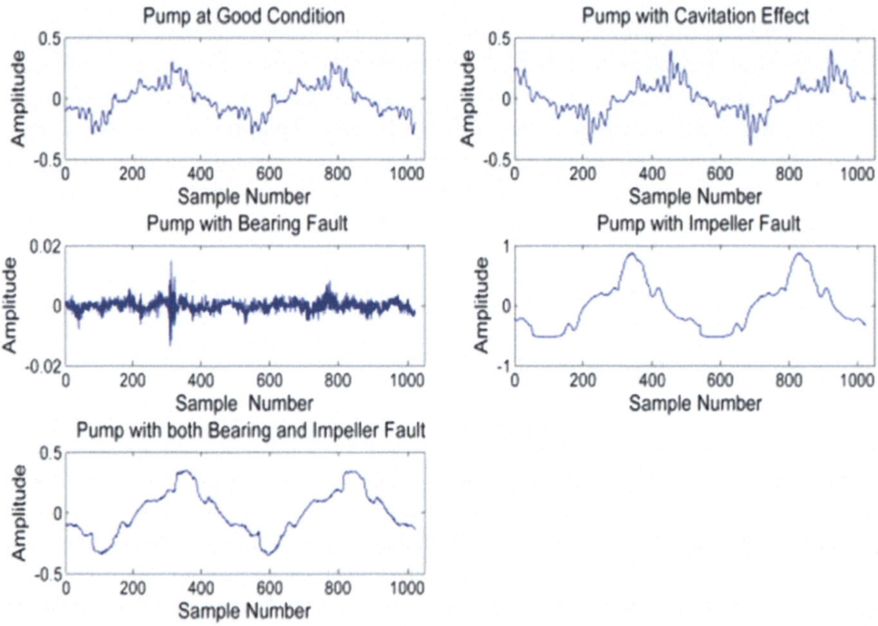

Fig. 23 Healthy and faulty condition of the pump

pumping system shown in Fig. 24. Figure shows the classification of faults shown in Fig. 25.

After analysis of the simulation results it is decided that when fault can be occurred then some noise will be seen in the faulty signal and signal peak can cause the danger zone limitation for which alarm will be shown.

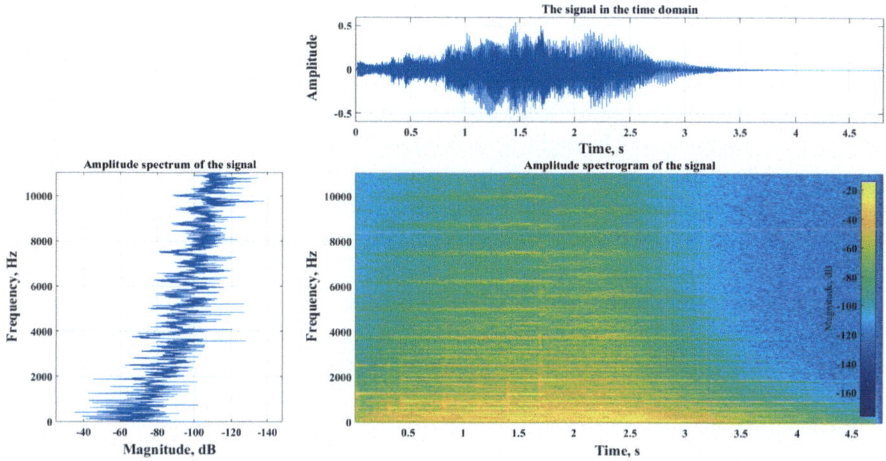

Fig. 24 Time frequency analysis of normal and faulty condition

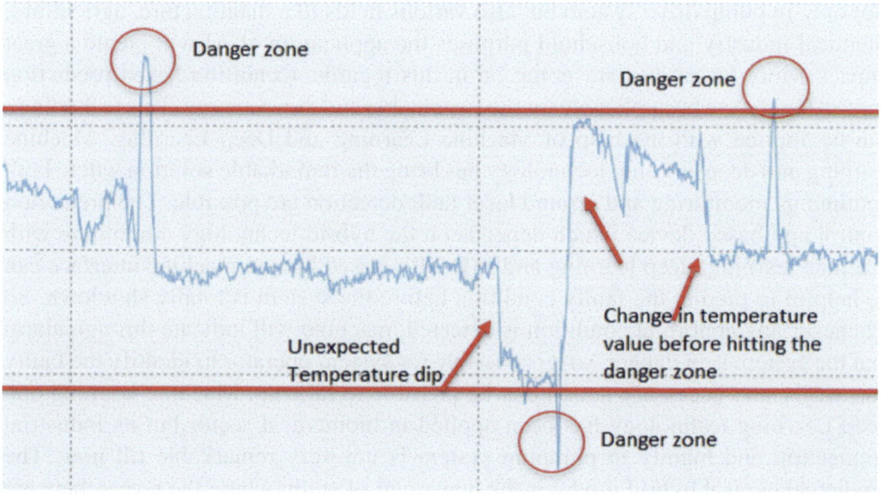

Fig. 25 Classification of faults

4 Conclusion

Pumping system is an important part for major industries. According to the International Energy Agency electric motors consume 46% of the electricity generated in the world. They account for nearly 70% of the total industrial electricity consumption in industries. As per the report made by the European Commission the pumping system accounts for nearly 22% of the energy supplied by electric motors. So continuous monitoring and increment of energy efficiency is essential. Machine learning and IoT based technology can be able to build up the predictive control model to achieve the target of continuous monitoring and save the systems from huge energy loss. Artificial intelligence has brought a revolutionary change in the field of machine learning and deep learning with the help of data and images. Condition based prediction control theory can be pertinent to the sustainable and green energy. Thus there is a big challenge before the researchers regarding the application of artificial intelligence in detecting the anomalies in the pumping system for smart building and other industrial sectors also. In the analysis it is seen that logistic regression, logistic discriminant algorithm and neural network are more suitable for the proposed work of classification of faults in industrial pumping system.

5 Future Scope

Artificial intelligence has bring a momentary changes in the industrial application. Not only in pump drive system but also various fields like manufacture, agriculture, chemical industry and household purposes the application of AI will create a great impact. More researches are going on in this regards. Condition based prediction control theory can be applicable for the sustainable and green energy. This technology can be applied with the help of Machine Learning and Deep Learning. Machine learning and deep learning technology has bring the remarkable solution where both continuous monitoring and ground level fault detection are possible. The prediction control app based device which depends on the hybrid technology combining with machine learning, deep learning and IoT, GPU based human machine interface can be helpful to predict the faulty condition before the system is totally shutdown. So whenever any abnormal condition is detected, machine will indicate through alarm that the system is in danger, so it is possible for system operator to identify the faulty part easily and necessary action can be possible to take up. Machine learning and deep Learning technology has been applied in biomedical sector but its industrial application and mainly in pumping system is not very remarkable till now. The predictive control hybrid model is the new point of study where the researchers are planning to reduce the energy loss and time of the process and trying to make the system flawless.

Acknowledgements The authors like to thank VIT Vellore Advance drives lab to give the opportunity of this type of research in their labs and Danfoss Industries Pvt ltd. Chennai to share their knowledge regarding the topic and continuous monitoring about the research.

Conflict of Interest The authors don't have any conflict of interest.

References

1. Shankar, V. K. A., Umashankar, S., Paramasivam, S., & Hanigovszki, N. (2016). A comprehensive review on energy efficiency enhancement initiatives in centrifugal pumping system. *Applied Energy, 181*, 495–513. https://doi.org/10.1016/j.apenergy.2016.08.070.
2. Thumati, B. T., Halligan, G. R., & Jagannathan, S. (2012). A novel fault diagnostics and prediction scheme using a nonlinear observer with artificial immune system as an online approximator. *IEEE Transactions on Control Systems Technology, 21*(3), 569–578. https://doi.org/10.1109/TCST.2012.2186635.
3. Ahonen, T., Ahola, J., Kestilä, J., Tiainen, R., & Lindh, T. (2007). Life cycle cost analysis of inverter-driven pumps.
4. Kallesoe, C. S., Izaili-Zamanabadi, R., Rasmussen, H., & Cocquempot, V. (2004, September). Model based fault diagnosis in a centrifugal pump application using structural analysis. In *Proceedings of the 2004 IEEE International Conference on Control Applications, 2004.* (Vol. 2, pp. 1229–1235). IEEE. https://doi.org/10.1109/CCA.2004.1387541.
5. Gao, Z., Cecati, C., & Ding, S. X. (2015). A survey of fault diagnosis and fault-tolerant techniques—Part I: Fault diagnosis with model-based and signal-based approaches. *IEEE Transactions on Industrial Electronics, 62*(6), 3757–3767. https://doi.org/10.1109/TIE.2015.2417511.
6. Alfayez, L., Mba, D., & Dyson, G. (2005). The application of acoustic emission for detecting incipient cavitation and the best efficiency point of a 60 kW centrifugal pump: case study. *NDT and E International, 38*(5), 354–358. https://doi.org/10.1016/j.ndteint.2004.10.002.
7. Stopa, M. M., Cardoso Filho, B. J., & Martinez, C. B. (2013). Incipient detection of cavitation phenomenon in centrifugal pumps. *IEEE Transactions on Industry Applications, 50*(1), 120–126. https://doi.org/10.1109/TIA.2013.2267709.
8. Hernandez-Marin, M., & Burbey, T. J. (2012). Fault-controlled deformation and stress from pumping-induced groundwater flow. *Journal of Hydrology, 428*, 80–93. https://doi.org/10.1016/j.jhydrol.2012.01.025.
9. Dutta, N., Umashankar, S., Shankar, V. A., Padmanaban, S., Leonowicz, Z., & Wheeler, P. (2018, June). Centrifugal pump cavitation detection using machine learning algorithm technique. In *2018 IEEE International Conference on Environment and Electrical Engineering and 2018 IEEE Industrial and Commercial Power Systems Europe (EEEIC/I&CPS Europe)* (pp. 1–6). IEEE. https://doi.org/10.1109/eeeic.2018.8494594.
10. Muralidharan, V., & Sugumaran, V. (2012). A comparative study of Naïve Bayes classifier and Bayes net classifier for fault diagnosis of monoblock centrifugal pump using wavelet analysis. *Applied Soft Computing, 12*(8), 2023–2029. https://doi.org/10.1016/j.asoc.2012.03.021.
11. Muralidharan, V., Sugumaran, V., & Indira, V. (2014). Fault diagnosis of monoblock centrifugal pump using SVM. *Engineering Science and Technology, an International Journal, 17*(3), 152–157. https://doi.org/10.1016/j.jestch.2014.04.005.
12. Farokhzad, S., Ahmadi, H., Jaefari, A., Abad, M. R. A. A., & Kohan, M. R. (2012). 897. Artificial neural network based classification of faults in centrifugal water pump. *Journal of Vibroengineering, 14*(4).
13. Dutta, N., Subramaniam, U., & Padmanaban, S. (2020, January). Mathematical models of classification algorithm of machine learning. In *International Meeting on Advanced Technologies in Energy and Electrical Engineering* (Vol. 2019, No. 1, p. 3). Hamad bin Khalifa University Press (HBKU Press).

14. Park, Y., Jeong, M., Lee, S. B., Antonino-Daviu, J. A., & Teska, M. (2017). Influence of blade pass frequency vibrations on MCSA-based rotor fault detection of induction motors. *IEEE Transactions on Industry Applications, 53*(3), 2049–2058. https://doi.org/10.1109/TIA.2017.2672526.
15. Pham, T. T., Thamrin, C., Robinson, P. D., McEwan, A. L., & Leong, P. H. (2016). Respiratory artefact removal in forced oscillation measurements: A machine learning approach. *IEEE Transactions on Biomedical Engineering, 64*(8), 1679–1687.
16. Siryani, J., Tanju, B., & Eveleigh, T. J. (2017). A machine learning decision-support system improves the internet of things' smart meter operations. *IEEE Internet of Things Journal, 4*(4), 1056–1066. https://doi.org/10.1109/JIOT.2017.2722358.
17. Telford, R. D., Galloway, S., Stephen, B., & Elders, I. (2016). Diagnosis of series DC arc faults—A machine learning approach. *IEEE Transactions on Industrial Informatics, 13*(4), 1598–1609. https://doi.org/10.1109/TII.2016.2633335.
18. Williamson, R., & Andrews, B. J. (2000). Gait event detection for FES using accelerometers and supervised machine learning. *IEEE Transactions on Rehabilitation Engineering, 8*(3), 312–319. https://doi.org/10.1109/86.867873.
19. Jain, S., Bajaj, V., & Kumar, A. (2016). Efficient algorithm for classification of electrocardiogram beats based on artificial bee colony-based least-squares support vector machines classifier. *Electronics Letters, 52*(14), 1198–1200. https://doi.org/10.1049/el.2016.1171.
20. Bouboulis, P., Theodoridis, S., Mavroforakis, C., & Evaggelatou-Dalla, L. (2014). Complex support vector machines for regression and quaternary classification. *IEEE Transactions on Neural Networks and Learning Systems, 26*(6), 1260–1274. https://doi.org/10.1109/TNNLS.2014.2336679.
21. Bufler, T. D., & Narayanan, R. M. (2016). Radar classification of indoor targets using support vector machines. *IET Radar, Sonar and Navigation, 10*(8), 1468–1476. https://doi.org/10.1049/iet-rsn.2015.0580.
22. Zhang, Y. (2012). *Support vector machine classification algorithm and its application* (pp. 179–186). Berlin, Heidelberg: Springer. https://doi.org/10.1007/978-3-642-34041-3_27.
23. Yao, L., Tang, J., & Li, J. (2009). Entire solution path for support vector machine for positive and unlabeled classification. *Tsinghua Science and Technology, 14*(2), 242–251.
24. Davy, M., Gretton, A., Doucet, A., & Rayner, P. J. (2002). Optimized support vector machines for nonstationary signal classification. *IEEE Signal Processing Letters, 9*(12), 442–445. https://doi.org/10.1109/LSP.2002.806070.
25. Demir, B., & Erturk, S. (2010). Empirical mode decomposition of hyperspectral images for support vector machine classification. *IEEE Transactions on Geoscience and Remote Sensing, 48*(11), 4071–4084. https://doi.org/10.1109/TGRS.2010.2070510.
26. Doumpos, M., Zopounidis, C., & Golfinopoulou, V. (2007). Additive support vector machines for pattern classification. *IEEE Transactions on Systems, Man, and Cybernetics, Part B (Cybernetics), 37*(3), 540–550. https://doi.org/10.1109/TSMCB.2006.887427.
27. Foody, G. M., & Mathur, A. (2004). A relative evaluation of multiclass image classification by support vector machines. *IEEE Transactions on Geoscience and Remote Sensing, 42*(6), 1335–1343. https://doi.org/10.1109/TGRS.2004.827257.
28. Abdulshaheed, A., Mustapha, F., & Ghavamian, A. (2017). A pressure-based method for monitoring leaks in a pipe distribution system: A review. *Renewable and Sustainable Energy Reviews, 69,* 902–911. https://doi.org/10.1016/j.rser.2016.08.024.
29. Bansal, S., Sahoo, S., Tiwari, R., & Bordoloi, D. J. (2013). Multiclass fault diagnosis in gears using support vector machine algorithms based on frequency domain data. *Measurement, 46*(9), 3469–3481. https://doi.org/10.1016/j.measurement.2013.05.015.
30. Bordoloi, D. J., & Tiwari, R. (2014). Support vector machine based optimization of multi-fault classification of gears with evolutionary algorithms from time–frequency vibration data. *Measurement, 55,* 1–14. https://doi.org/10.1016/j.measurement.2014.04.024.
31. Carunaiselvane, C., & Chelliah, T. R. (2017). Present trends and future prospects of asynchronous machines in renewable energy systems. *Renewable and Sustainable Energy Reviews, 74*(1028–1041), 2017. https://doi.org/10.1016/j.rser.2016.11.069.

32. Zhao, R., Yan, R., Chen, Z., Mao, K., Wang, P., & Gao, R. X. (2019). Deep learning and its applications to machine health monitoring. *Mechanical Systems and Signal Processing, 115,* 213–237. https://doi.org/10.1016/j.ymssp.2018.05.050.
33. Shankar, V. A., Umashankar, S., Sanjeevikumar, P., Mihet-Popa, L., Fedák, V., & Ramachandaramurthy, V. K. (2018). Power quality performance analysis of grid tied PV fed parallel pumping system under normal and vibrating condition. *Energy Procedia, 145,* 497–503.
34. Shankar, V. A., Umashankar, S., Padmanaban, S., & Paramasivam, S. (2017). Adaptive neuro-fuzzy inference system (ANFIS) based direct torque control of PMSM driven centrifugal pump. *International Journal of Renewable Energy Research (IJRER), 7*(3), 1436–1447.
35. Shankar, V. A., Umashankar, S., Paramasivam, S., Sanjeevikumar, P., & Venkatesh, K. (2018). Investigation of direct torque control-based synchronous reluctance motor drive for pumping. In *Advances in Systems, Control and Automation* (pp. 319–327). Singapore: Springer.
36. Shankar, V. A., Umashankar, S., Sanjeevikumar, P., Viliam, F., Ramachandaramurthy, V. K., & Mihet-Popa, L. (2018). Investigations of power quality disturbances in a variable speed parallel pumping system with grid tied solar PV. *Energy Procedia, 145,* 490–496.
37. Shankar, V. A., Umashankar, S., Padmanaban, S., Bhaskar, M. S., Ramachandaramurthy, V. K., & Fedák, V. (2017, October). Comparative study of photovoltaic based power converter topologies for pumping applications. In *2017 IEEE Conference on Energy Conversion (CENCON)* (pp. 174–179). IEEE.
38. Shankar, V. A., Umashankar, S., & Paramasivam, S. (2017, April). Investigations on performance evaluation of VFD fed PMSM using DTC control strategies for pumping applications. In *2017 Innovations in Power and Advanced Computing Technologies (i-PACT)* (pp. 1–8). IEEE.
39. Arun Shankar, V. K., Subramaniam, U., Padmanaban, S., Holm-Nielsen, J. B., Blaabjerg, F., & Paramasivam, S. (2019). Experimental investigation of power signatures for cavitation and water hammer in an industrial parallel pumping system. *Energies, 12*(7), 1351.
40. Shankar, V. K. A., Umashankar, S., Paramasivam, S., Sanjeevikumar, P., & Sailesh, K. D. (2018). Experimental investigation of VFD-fed scalar control of induction motor for pumping application. In *Advances in Smart Grid and Renewable Energy* (pp. 287–295). Singapore: Springer.
41. Kaliannan, J., Baskaran, A., Dey, N., & Ashour, A. S. (2016). Ant colony optimization algorithm based PID controller for LFC of single area power system with non-linearity and boiler dynamics. *World Journal of Modelling and Simulation, 12*(1), 3–14.
42. Jagatheesan, K., Anand, B., Dey, N., & Ashour, A. S. (2018). Effect of SMES unit in AGC of an interconnected multi-area thermal power system with ACO-tuned PID controller. In: *Advancements in Applied Metaheuristic Computing* (pp. 164–184). IGI Global.
43. Das, S. K., & Tripathi, S. (2017). Energy efficient routing formation technique for hybrid ad hoc network using fusion of artificial intelligence techniques. *International Journal of Communication Systems, 30*(16), e3340.
44. Das, S. K., & Tripathi, S. (2018). Adaptive and intelligent energy efficient routing for transparent heterogeneous ad-hoc network by fusion of game theory and linear programming. *Applied Intelligence, 48*(7), 1825–1845.
45. Das, S. K., & Tripathi, S. (2019). Energy efficient routing formation algorithm for hybrid ad-hoc network: A geometric programming approach. *Peer-to-Peer Networking and Applications, 12*(1), 102–128.
46. Chatterjee, S., Sarkar, S., Dey, N., Ashour, A. S., Sen, S., & Hassanien, A. E. (2017). Application of cuckoo search in water quality prediction using artificial neural network. *International Journal of Computational Intelligence Studies, 6*(2–3), 229–244.

Decision Making System

Fast Accessing Non-volatile, High Performance-High Density, Optimized Array for Machine Learning Processor

Divya Mishra, Abhishek Kumar, Vishwas Mishra, Shobhit Tyagi, and Shyam Akashe

Abstract Traditional memory technologies for example; FLASH memory, SRAM (Static Random Access Memory) & DRAM (Dynamic Random Access Memory) are not able to cope with machine learning processor because of high density & low power requirement. FLASH memories have already attained their physical limit and therefore can't be scaled further due to its bounded tenacity. Non-volatile memories have gained tremendous popularity after being in theoretical study for more than 30 years. Among all various types of Nonvolatile memories, Memristors are the one which are compact, highly dense, fast and nonvolatile. It combines nonvolatile nature of flash memories, speed of SRAMs and high dense nature of DRAMs. Earlier one transistor one memristor based array have been designed but it consisted of CMOS transistor that now has attained fundamental limits and nothing new can be improvised in it because it can't be scaled further. So, in place of transistor FinFET have been used to form the memory element 1F1M, which fulfill the requirement of machine learning processors. FinFET is the most promising transistor known for its superior controllability of short channel effects and robust threshold voltage (V_{th}) through a double gate. The main challenge in this time is chip design with scaling of Integrated circuits (IC's) at Low Power. The thinner W_{fin} FinFET shows less performance loss than the thicker W_{fin} FinFET. In VLSI, the Power, Area and

D. Mishra · A. Kumar · V. Mishra (✉) · S. Tyagi
Swami Vivekanand Subharti University, Meerut 250005, India
e-mail: vishwasmishra88@gmail.com

D. Mishra
e-mail: mishradivya1311@gmail.com

A. Kumar
e-mail: abhishekec02@gmail.com

S. Tyagi
e-mail: shobhittyagi.1857@gmail.com

S. Akashe
ITM University, Gwalior, Gwalior 474001, India
e-mail: shyam.akashe@itmuniversity.ac.in

S. K. Das et al. (eds.), *Machine Learning Algorithms for Industrial Applications*, Studies in Computational Intelligence 907,
https://doi.org/10.1007/978-3-030-50641-4_15

Delay are key factors for improving circuit functionality, when any of these can be decreased then the circuit output can be improved. Due to different threshold voltage, memory types in FinFET offer Data Retention Voltage variables compared to CMOS based memory.

Keywords Very Large Scale Integrated Circuit · Static Random Access Memory · Dynamic Random Access Memory · One FinFET One Memristor · Integrated Circuit

1 Introduction

Electronic industry is going through a major revolutionary phase in designing each and every part of anodic device as there is a boom for low power, high performance consumer products with whirl winding developing market. Power dissipation has been a customary neglected constraint but as the scaling of integration has improved by the number of transistors mounted on a single chip is rising as a result of which power and factors like chip density, energy consumption has taken dominant place. The increasing demand of electronic products has led to the booming of data storing devices to newer level. Memory devices are one of the significant components for storing past, present and future information. Design of memory is an active area of research in the machine learning space. In recent years, techniques such as Neural Turing Machines (NTM) have made significant progress setting up the foundation for building human-like memory design in machine learning systems. So instead, we like to approach the subject from a different angle and attempt to answer fundamental questions that we should have in mind when thinking about memory in machine learning models. They play a key role from very simple and small device that is mobile to large devices such as radars, satellites, military applications etc. Depending on the storage capacity i.e. permanent or temporary, there is various classification of memory that is used for multiple uses [1, 2].

Any conventional memory is found to have these three fundamental requisites:

- ON (Logic 1) and OFF (Logic 0) states
- Means by which state can be controlled
- Means by which the state can be read

The limitations of CMOS based memories have forced the designers to design devices that are not only handy, have high capacity, high speed but also low leakage in all respects [3]. With the commencement of semiconductor aural and video players vogue for nonvolatile depot has escalated. Voltaic memories come in plenty of configurations and styles. The character of memory unit that is desired for a particular employment is a function of size, accessing time, patterns, applications and entities requirements [4].

After a CMOS a new emerging device came into scene that is FinFET. FinFET stands for Fin-Field Effect Transistor discovered by the University of California,

Berkeley researchers describing a non-planner system, DG-FET based on a SOI (Silicon on Insulator) substrate. The distinctive feature of the FinFET is that the channel formed between the source to drain and that channel enfolded by a thin "fin" of silicon and the double-sided protected layer acts as a gate terminal [5]. DG-FETs are extensions of CMOS and the double gate FinFET is proposed for future memory circuits [6].

The multi-gate structure offers improved charge drift and diffusion power over the channel and thus helps to reduce the latch-up effect and counteract the other short channel effect. Since the FinFET gate drain and source are doped with the same kind of dopant, there is no p-n junction forming along the length of the channel and the leakage current is minimized. The Higher mobility in the FinFET is attributed to the un-doped channel, which prevents the coulombs scattering. The mobility fraction of n-type to p-type is greater in FinFET [7]. The difference in source-body voltage did not affect the threshold voltage. A powerful & mathematical concept of natural length:

$$\lambda = \sqrt{\frac{\varepsilon_{si}}{2\varepsilon_{ox}} t_{si} t_{ox}} \tag{1}$$

where λ is the short-channel effect, t_{si} is thickness of device body, t_{ox} is thickness of gate oxide. In FinFET the side walls of the fins known as Channel. The Channel width is described by:

$$FinW_{eff} = Fin_{WIDTH} + 4\frac{\varepsilon_{si}}{\varepsilon_{ox}} t_{ox} \tag{2}$$

The relation between the threshold voltage of Gate1 (ΔVt_{G1}) and Gate2 biasing can be:

$$\frac{\Delta Vt_{G1}}{\Delta V_{G2}} = -\frac{C_{Fin} C_{ox2}}{C_{ox1}(C_{Fin} + C_{ox2})} \approx -\frac{3t_{ox1}}{3t_{ox2} + Fin_{WIDTH}} \tag{3}$$

Trapping the body's buried oxide potential when the depletion between the source and the body is lower than that of the electrons injected into the body and the drain region collects it. If the electrical field is high, it causes impact ionization in the drain to occur and leads to a current runaway causing a snapback. Basic structure of FinFET is shown in Fig. 1.

For low power application threshold is lowered which also contributes leakage currents. There are many common ways to design some circuits. One of these approaches is to reduce power supply around or below threshold voltage [8]. Yet lowering the nominal voltage has to compromise to some degree its working and the effects of lowering the supply voltage are embarrassment in cell stability, noise margin, current-to-off current ratio and high PVT (Process Voltage Temperature) sensitivity variations [9]. There are different approaches for leakage reduction during both standby mode and run time, when the circuit is in operation.

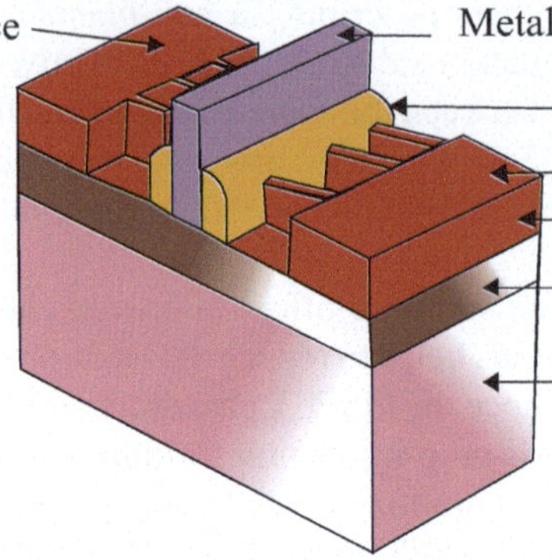

Fig. 1 3D FinFET structure

1.1 Relation Between the Short Channel and Drain Induced Barrier

4 × 4 array uses the combination of 1F1M in a previous arrays the basic components we needed is decoder precharge, write driver and sense amplifier but as we get the emerging device Memristor they work like a combination of precharge, write driver and sense amplifier because it behaves as a fourth fundamentals of electronics or electrical circuitry [10].

Memristor linked with all the fundamental elements the link equation is shown below-

$$\emptyset = \int_0^t (V_{out} - V_{in})\,dt \tag{4}$$

$$M = \frac{d\emptyset}{dq} = \frac{\int_0^t (V_{out} - V_{in})\,dt}{dq} \tag{5}$$

So here flux is related to time it help to link Memristor with flux but also to avoid link between the flux by using voltage as well as time functionality these way the Memristor linked with flux but also not get linked with magnetic quality of flux so the Memristor is not affected with the magnetic flux.

Now the second component is resistance the doping of titanium dioxide helps to reduce the resistance of the semiconductor. Using GaAs or silicon we can employ Memristor for different applications. The resistance behaves as-

$$R = \frac{V}{I} = \frac{dV}{dI} \tag{6}$$

$$M = \frac{d\emptyset}{dq} = \frac{d\emptyset}{dt}\,X\,\frac{dt}{dq} \tag{7}$$

$$M(q(t)) = \frac{\frac{d\emptyset}{dt}}{\frac{dq}{dt}} = \frac{V}{I} \approx R \tag{8}$$

The 1F1M helps to behave and work on various thresholds as well to decrease the sub threshold drawbacks and also the short channel effect (SCE) and drain induced barrier lowering (DIBL) to reduce the threshold we have been towards the DIBL short channel effect decreases because-

$$V_{TH} = V_{TH\infty} - SCE - DIBL \tag{9}$$

where, $V_{TH\infty}$ is the threshold voltage of long channel device,

$$\text{SCE} = 0.64\frac{\varepsilon_{si}}{\varepsilon_{ox}}\left[1 + \frac{x_i}{L_{el}^2}\right]\frac{t_{ox}}{L_{el}}\,X\,\frac{t_{dep}}{L_{el}}V_{bi} \tag{10}$$

$$\text{SCE} = 0.64\frac{\varepsilon_{si}}{\varepsilon_{ox}}EIV_{bi} \tag{11}$$

where L_{el} is the electrical channel length and V_{bi} is the source or drain built in potential and t_{ox} gate oxide thickness and t_{dep} is the Penetration depth of gate field in channel region and EI is the electrostatic integrity factor.

$$\text{DIBL} = 0.8\frac{\varepsilon_{si}}{\varepsilon_{ox}}\left[1 + \frac{x_i^2}{L_{el}^2}\right]\frac{t_{ox}}{L_{el}}\,X\,\frac{t_{dep}}{L_{el}}V_{DS} \tag{12}$$

$$\text{DIBL} = 0.8\frac{\varepsilon_{si}}{\varepsilon_{ox}}EIV_{DS} \tag{13}$$

In Memristor and FinFET, the SCE reduces using the different condition-

$$\text{SCE} = 0.64\frac{\varepsilon_{si}}{\varepsilon_{ox}}EIV_{bi} \tag{14}$$

$$M(q(t)) = \frac{\frac{d\emptyset}{dt}}{\frac{dq}{dt}} = \frac{V(t)}{I(t)} \tag{15}$$

$$V(t) = M(q(t))I(t) \tag{16}$$

$$\text{SCE} = 0.64\frac{\varepsilon_{si}}{\varepsilon_{ox}}EI\{V_{bi}(\approx V(t))\} \tag{17}$$

$$\text{SCE} = 0.64\frac{\varepsilon_{si}}{\varepsilon_{ox}}EI\{M(q(t))i(t)\} \tag{18}$$

If we have to decrease the SCE then the $M(q(t))$ helps also i(t) to reduce the SCE. Now the SCE and DIBL also get reduce due to the doping of the titanium dioxide dopant due to that the permittivity of the Memristor is vary from ε_{si} to $\varepsilon_{si} + \varepsilon_{TiO2-x}$

$$\text{SCE} = 0.64 \frac{(\varepsilon_{si} + \varepsilon_{TiO2-x})}{\varepsilon_{ox}} EI\{M(q(t))i(t)\} \tag{19}$$

EI is the electrostatic Integrity.

$$\text{EI} = \left[1 + \frac{x_i^2}{L_{el}^2}\right] \frac{t_{ox}}{L_{el}} X \frac{t_{dep}}{L_{el}} \tag{20}$$

$$\text{EI} = \left[1 + \frac{x_i^2}{L_{el}^2}\right] \frac{(t_{ox} + t_{TiO2-x})}{L_{el}} X \frac{t_{dep}(\approx t_{TiO2-x})}{L_{el}} \tag{21}$$

So the SCE is-

$$\text{SCE} = 0.64 \left[\frac{(\varepsilon_{si} + \varepsilon_{TiO2-x})}{\varepsilon_{ox}}\right] \left[1 + \frac{x_i^2}{L_{el}^2}\right] \left[\frac{(t_{ox} + t_{TiO2-x})}{L_{el}}\right] \frac{t_{TiO2-x}}{L_{el}} \tag{22}$$

$$\text{SCE} = 0.64 \left[\frac{(\varepsilon_{si} + \varepsilon_{TiO2-x})}{\varepsilon_{ox}}\right] (EI)_{TiO2}\{M(q(t))i(t)\} \tag{23}$$

Similarly in DIBL

$$\text{DIBL} = 0.8 \left[\frac{(\varepsilon_{si} + \varepsilon_{TiO2-x})}{\varepsilon_{ox}}\right] (EI)_{TiO2}\{M(q(t))i(t)\} \tag{24}$$

So we control the threshold by

$$V_{TH} = V_{TH\infty} - SCE - DIBL \tag{25}$$

If we get the proportionality then

$$V_{TH} \,\alpha\, \varepsilon_{TiO2-x} \tag{26}$$

$$V_{TH} \,\alpha\, M(q(t))i(t) \tag{27}$$

So for any improvement we want in V_{TH} then we can control it by ε_{TiO2-x} and here X is the length of doped area. So by varying the doped area we can control the threshold voltage [11] (Fig. 2).

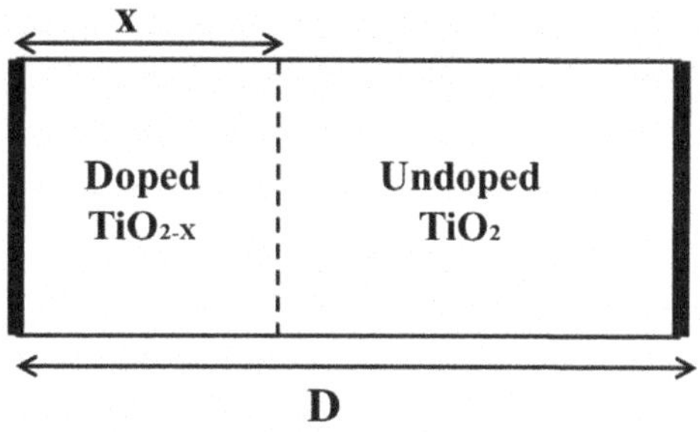

Fig. 2 Representation of Memristor as per HP labs

2 Common Effect of FinFET and Memristor over Channel Length

Starting with the Poisson's equation-

$$\frac{d^2\Phi(x,y,z)}{dx^2} + \frac{d^2\Phi(x,y,z)}{dy^2} + \frac{d^2\Phi(x,y,z)}{dz^2} = \frac{qN_{sub}}{\varepsilon_{si}} \tag{28}$$

For single or double gate $\frac{d\Phi}{dz} = 0$ then

$$\frac{d^2\Phi(x,y)}{dx^2} + \frac{d^2\Phi(x,y)}{dy^2} = \frac{qN_{sub}}{\varepsilon_{si}} \tag{29}$$

By implanting over 1 FinFET 1 Memristor, we include the ε_{TiO2-x} &

$$M = \frac{\int_0^t (V_{out} - V_{in})}{dq} = \frac{dv}{dq} \tag{30}$$

$$dq = \frac{dv}{M} \tag{31}$$

By integrating these equation-

$$q = \frac{(V_{out} - V_{in})}{M(q(t))} = \frac{dv}{M(q(t))} \tag{32}$$

So we can put the q value in the Poisson equation that is-

$$\frac{d^2\Phi(x,y)}{dx^2} + \frac{d^2\Phi(x,y)}{dy^2} = \frac{dv}{M}\frac{N_{sub}}{(\varepsilon_{si} + \varepsilon_{TiO2-x})} \tag{33}$$

We are now applying the boundary condition-

$$\frac{d^2\Phi(x,y)}{dy^2} = 0 \text{ at } y = \frac{t_{si}}{2} \tag{34}$$

$$\Phi(x,y) = \Phi_f(x) + \left(\frac{\varepsilon_{ox}}{\varepsilon_{si} + \varepsilon_{TiO2-x}}\right)\left(\frac{\Phi_s(x) - \Phi_{gs}}{t_{ox} X t_{TiO2-x}}\right) - \left(\frac{1}{t_{si} X t_{TiO2-x}}\right)\left(\frac{\varepsilon_{ox}}{(\varepsilon_{si} + \varepsilon_{TiO2-x})}\right)\left(\frac{\Phi_s(x) - \Phi_{gs}}{t_{ox}}\right)y^2 \tag{35}$$

$\Phi_s(x)$ is front and back surface potential

$$\Phi_{gs} = V_{gs} - V_{FB} \tag{36}$$

where V_{gs} the front is gate voltage and V_{FB} is flat band voltage. Center of fin of FinFET $\Phi_c(x)$ is mostly good for SCE, $y = \frac{t_{si}}{2}$

Now discuss relation between $\Phi_c(x)$ and $\Phi_s(x)$ is-

$$\Phi_s(x) = \left(\frac{1}{1 + \frac{\varepsilon_{ox}}{4(\varepsilon_{si} + TiO2)} X \frac{(t_{si} X t_{TiO2-x})}{t_{ox}}}\right)\left[\Phi_c(x) + \frac{\varepsilon_{ox}}{4(\varepsilon_{si} + \varepsilon_{TiO2-x})}\Phi_{gs}\right] \tag{37}$$

Expressing $\Phi(x,y)$ as a function of $\Phi_c(x)$-

$$\Phi(x,y) = \left[1 + \left(\frac{\varepsilon_{ox}}{\varepsilon_{si} + \varepsilon_{TiO2-x}} X \frac{y}{t_{ox}} - \right)\left(\frac{\varepsilon_{ox}}{\varepsilon_{si} + \varepsilon_{TiO2-x}}\right) X \frac{y^2}{t_{ox}}\right](t_{si} X t_{TiO2-x}) \tag{38}$$

$$\left[\frac{\Phi_c(x) + \frac{\varepsilon_{ox}}{4(\varepsilon_{si} + \varepsilon_{TiO2-x})} * \frac{(t_{si} X t_{TiO2-x})}{t_{ox}} * \Phi_{gs}}{1 + \frac{\varepsilon_{ox}}{4(\varepsilon_{si} + \varepsilon_{TiO2-x})} - \frac{(t_{si} X t_{TiO2-x})}{t_{ox}}}\right] - \left[\frac{\varepsilon_{ox}}{\varepsilon_{si} + \varepsilon_{TiO2-x}} X \frac{y}{t_{ox}} \Phi_{gs} - \left(\frac{\varepsilon_{ox}}{\varepsilon_{si} + \varepsilon_{TiO2-x}}\right)\frac{y^2}{t_{ox}(t_{si} X t_{TiO2-x})}\Phi_{gs}\right] \tag{39}$$

Substitute Eqs. (39) in (28),

$$\frac{d^2\Phi_c(x)}{dx^2} + \frac{\Phi_{gs} - \Phi_c(x)}{\lambda^2} = \frac{dv}{M} * \frac{N_{sub}}{(\varepsilon_{si} + \varepsilon_{TiO2-x})} \tag{40}$$

Let us assume good SCE control can be achieved with $L_g = n\,\lambda$, n is some arbitrary number,

$$\frac{L_g}{n} = \lambda = \sqrt{\left(\frac{(\varepsilon_{si} + \varepsilon_{TiO2-x})}{2\varepsilon_{ox}}\left[1 + \frac{\varepsilon_{ox}}{4(\varepsilon_{si} + \varepsilon_{TiO2-x})} X \frac{(t_{si} X t_{TiO2-x})}{t_{ox}}\right]\right) t_{si} t_{ox} t_{TiO2-x}} \tag{41}$$

In the limiting cases of $L_g = 0$ due to bounded result in channel length less value then,

$$\frac{\varepsilon_{si}}{2\varepsilon_{ox}}\left[1+\frac{\varepsilon_{ox}}{4(\varepsilon_{si}+\varepsilon_{TiO2-x})}\right]t_{si}t_{ox}t_{TiO2-x}=0 \tag{42}$$

This implies

$$t_{si} = -\frac{4(\varepsilon_{si}+\varepsilon_{TiO2-x})}{\varepsilon_{ox}}t_{ox}t_{TiO2-x} \tag{43}$$

$$t_{TiO2-x} = -\frac{4(\varepsilon_{si}+\varepsilon_{TiO2-x})}{\varepsilon_{ox}}t_{si}t_{ox} \tag{44}$$

So the FinFET width becomes

$$F_{in}W_{eff} = F_{inwidth} + 4\frac{\varepsilon_{si}}{\varepsilon_{ox}}t_{ox} \tag{45}$$

By neglecting ε_{TiO2-x} in FinFET device and channel width in Memristor is-

$$M_{width} = W_{eff} + \frac{4(\varepsilon_{si}+\varepsilon_{TiO2-x})}{\varepsilon_{ox}}t_{si}t_{ox} \tag{46}$$

3 4 x 4 Array Using 1 FinFET and 1 Memristor

Array is represented as $2^N \times M$, where N represents the number of address bits or inputs to a decoder or depth and M are the data bits; it also represents the width of array [12]. 4×4 array can be written as $2^2 \times 4$ which shows that there will be two (here $N = 2$) address bits or input to the decoder and 4 data bits that is the width of bits is 4 bit (Fig. 3).

3.1 Working Principle of 1F1M

In Write mode, there are A0, A1 goes to 00 then the first row is getting high and when the B0, B1 goes 00 then the first column goes high it means that the first cell is get selected and the value that one is stored in an cell 1. So this way the 1 FinFET and 1 Memristor stores the single bit value. Similarly, in different cells of the 4×4 array, A0A1B0B1 combination helps to select the different cells to store the data in an array [13] (Fig. 4).

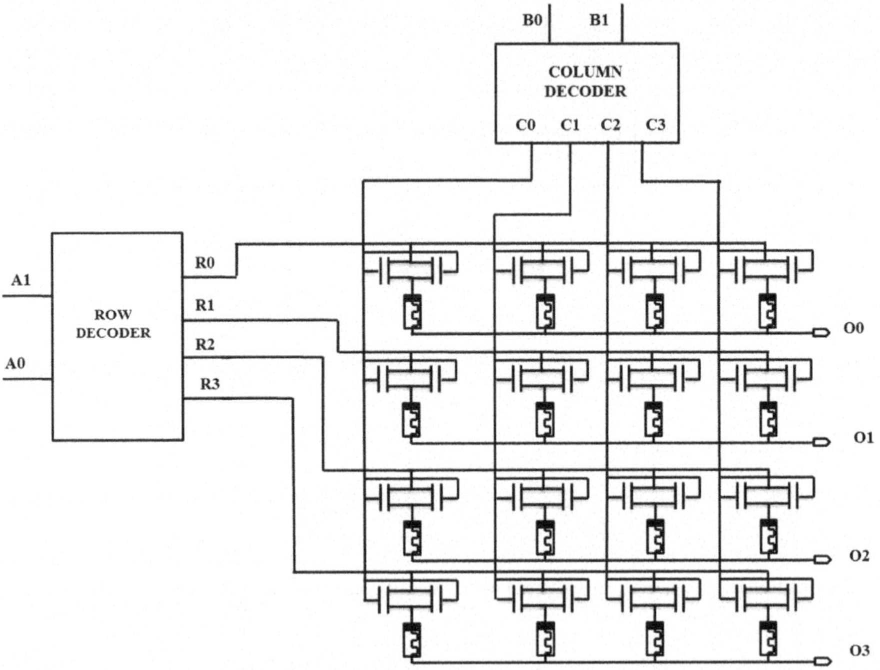

Fig. 3 Schematic of a 1F1M array

In read mode, there are A0, A1 goes to 00 then the first row is getting high and when the B0, B1 gives nothing then the first column goes neither high nor low, it means that the first row is get selected and the value that is stored in an first cell is can be read out at the output pin O0. Similarly, in different cells of the 4 × 4 array, A0A1 combination helps to select the different cells to read the data from an array [14] (Fig. 5).

4 Simulation Result

The proposed 1F1M i.e. One FinFET One Memristor is memory topology are accomplished using the Virtuoso Tool of Cadence IC 6.1 version simulator of cadence is used for simulation of the output. The simulations are performed using the 45 nm technology and Table 1 describe the simulation results which are performed using the various parameters. The design of proposed One FinFET One Memristor and also called FinFET based Resistive Random Access Memory or in short FRRAM memory and that type of memory device are based on a non-volatile element called "Memristor". Due to its simple structure, low power, Nano size and non-volatile in nature, it is embedded with the FinFET to get a combination of FinFET & Memristor

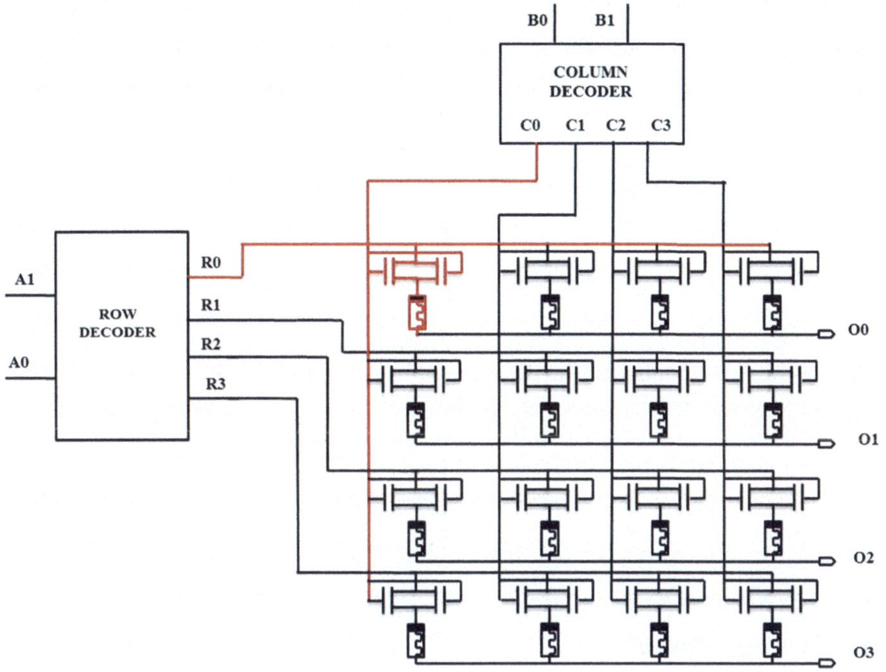

Fig. 4 Write mechanism of 1F1M array

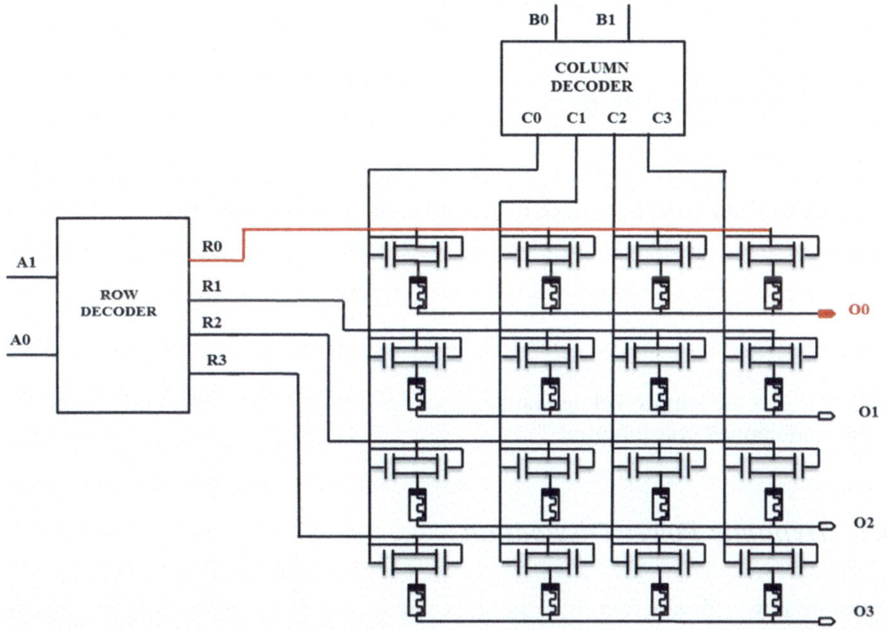

Fig. 5 Read mechanism of 1F1M array

Table 1 Simulation parameters

S. no.	Parameters	Notation	Value
1	Process technology	–	45 nm
2	Supply voltage	*Vdd*	0.7 V
3	Rise time	*tdr*	1 ns
4	Fall time	*tdf*	1 ns
5	Delay	–	0 ns
6	Temperature	*T*	27 °C

design. These hybrid designs boost up the memory design with their unique features. To verify the functionality of the device supply voltage of 0.7 V is given for 45 nm technology and transient response are shown in the form of waveforms and the results of FinFET based Resistive Random Access Memory like Average Power or also called as Total Transient Power and Leakage Current and Leakage Power are determined on cadence virtuoso tool as shown in Table 1. Now we discuss definition of these parameters.

4.1 Average Power

The Average Power or Total Transient Power dissipation occurs due to the static and dynamic power dissipation and unit of power is in watts [14].

4.2 Static Power Dissipation

Static power dissipation occurs in a circuit due to the existing of direct path between V_{DD} to V_{SS} that is supply voltage to ground node, region behind that sub-threshold condition in this condition transistors are in off state and mathematically it is given as [15]-

$$P_{Static} = V_{DD} * I_{leakage} \tag{47}$$

where, V_{dd} is the supply voltage and $I_{leakage}$ is the quiescent supply current and P_S is the static power consumption.

4.3 Dynamic Power Dissipation

Dynamic power dissipation in a circuit is due to the switching activity of capacitances. It consist two phenomena one is charging and other is discharging the load capacitances and mathematically it is given as:

$$PD = \frac{1}{2} C_L V_{DD}^2 f \tag{48}$$

where, switching capacitance is denoted by C_L and switching activity of output node is denoted by α and operating frequency of the system is denoted by f and supply voltage is denoted by V.

4.4 Leakage Current

Leakage current is the current that passes through the protective ground conductor to the ground [16]. In the absence of a grounding contact, it is the current that could flow from any conductive component or the surface of non-conductive parts to the ground if a conductive path is open. The main source of leakage current is sub-threshold leakage and is characterized as the condition of low inversion conduction current when $V_s < V_{th}$ is in the CMOS transistor, and is mathematically described as-

$$\mathrm{I} = \mathrm{I_0} \,.\, \exp\left(\frac{\mathrm{Vgs} - \mathrm{Vth}}{\eta \mathrm{kT/q}}\right).\left[1 - \exp\left(\frac{1 - \mathrm{Vds}}{\mathrm{kT/q}}\right)\right] \tag{49}$$

$$\mathrm{I_0} = \mu_0 \mathrm{C_{ox}} (\mathrm{W/L})(\mathrm{kT/q})^2 \left(1 - \mathrm{e}^{1.8}\right) \tag{50}$$

where, low field mobility is denoted by μ_0, transistor channel width & length are denoted by W & L, the gate oxide capacitance is denoted by C_{ox}, the electronic charge is denoted by q and the Boltzmann's constant is denoted by k [17, 18].

4.5 Parameters of One FinFET One Memristor Memory Array

The functionality of the one FinFET one Memristor memory array and show transient response of one FinFET one Memristor memory array in the below:

4.6 Transient Response

The transient response shows the response of a system in the form of output by giving different inputs to it. Above figure indicates the input and output waveform of one FinFET one Memristor memory array with the help of a virtuoso simulator with schematic result analysis (Fig. 6).

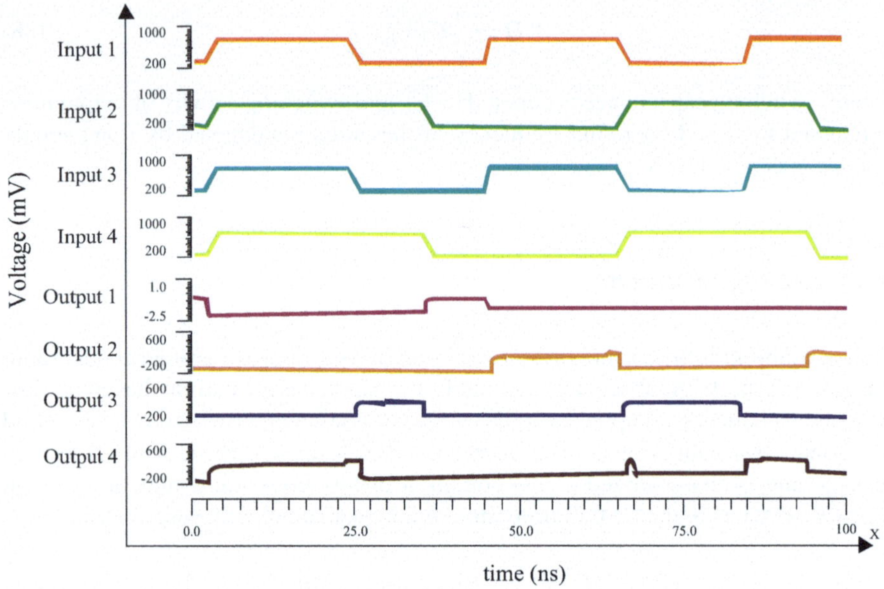

Fig. 6 Transient response of 1 FinFET 1 Memristor (1F1M) memory array

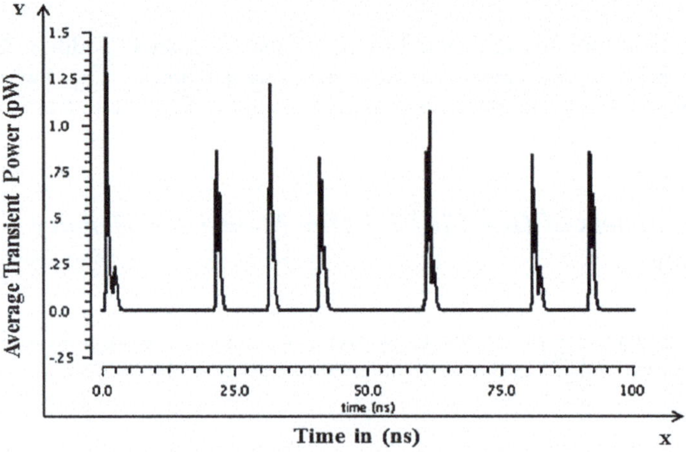

Fig. 7 Average power of one FinFET one Memristor memory array

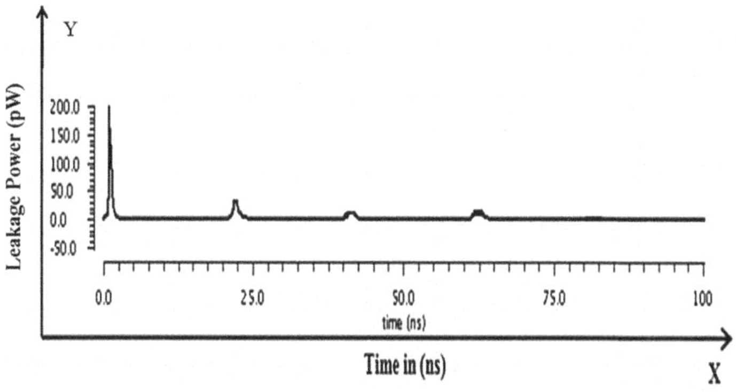

Fig. 8 Leakage power of 1 FinFET 1 Memristor memory array

4.7 Average Power of One FinFET One Memristor Memory Array

Figure 7 waveform shows that Average power that use in one FinFET one Memristor memory. In this waveform we shows that Y-axis represent the power that major in nanometer and X-axis represent the time scale that major in nano-seconds.

4.8 Leakage Power of One FinFET One Memristor Memory Array

Figure 8 waveform shows the leakage power that is generated by one FinFET one Memristor memory array. In this waveform we shows that Y-axis represent the power that major in nanometer and X-axis represent the time scale that major in nano-seconds.

4.9 Leakage Current One FinFET One Memristor Memory Array

Figure 9 waveform shows the leakage current that is generated during read write mechanism or operation by one FinFET one Memristor memory array. In this waveform we shows that Y-axis represent the current that major in microampere and X-axis represent the time scale that major in Nano-seconds. The simulation results of different topologies of one FinFET one Memristorarray with different powers is shown below in the form of Table 2.

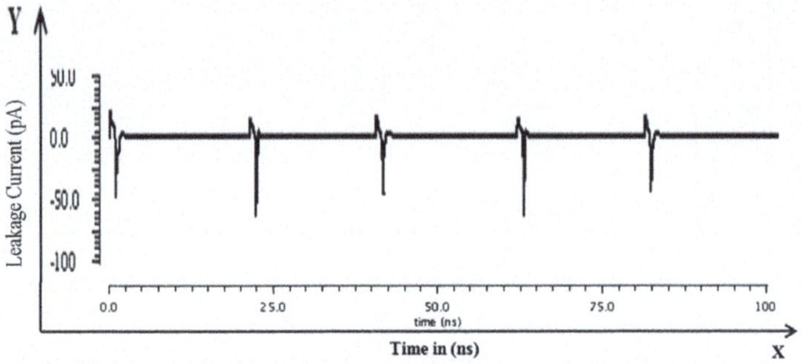

Fig. 9 Leakage power of one FinFET one Memristor memory array

Table 2 Simulation results of comparison of memory arrays

Comparison of memory arrays			
Parameters	Voltage	1F1M array	Conventional array
Average power	0.5	43.56 nW	83.59 nW
Leakage power	0.5	479 pW	619 pW
Leakage current	0.5	89.28 pA	28.93 nA

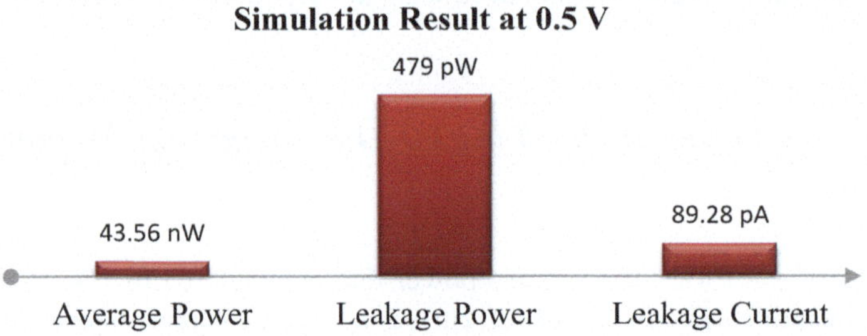

Fig. 10 Graphical representations of simulation results of 1F1M array

Table 2 shows the simulation results of one FinFET one Memristor in this table shows that three parameters are calculated and these parameter are total power, leakage power and leakage current. Graphical Representation of One FinFET One Memristor Memory Array is shown in Fig. 10.

In the above figure shows the graphs of one FinFET One Memristor memory Array and these graphs namely Average Power, Leakage Power and Leakage Current.

5 Conclusion

1F1M, that is, One FinFET One Memristor is a memory element that is used to store multiple-bit decimal numbers, as well as binary values which will useful for machine learning processor. This is a novel approach to creating a high-density network that can be used to store information from handheld small devices, internal storage, hard disks to large satellite, radars, various military uses and machine processors. In the last few years, the improvement in computer competence has been closely linked to the reduction of CMOS technology. Electrical interconnection has become both a barrier to performance and one of the major sources of power dissipation that has recently become the most significant factor that acts as a major barrier to technological growth. The main reasons for developing 1F1M memory feature are to build very high density storage devices with very low power consumption, low leakage and very little delay in accessing data. Such 1F1 M systems are also CMOS compliant.

Acknowledgements This work was supported by ITM University Gwalior, with collaboration Cadence Design System Bangalore. The authors would also like to thank to Professor Shyam Akashe for their enlightening technical advice.

References

1. Gavaskar, K., & Priya, S. (2013). Design of efficient low power stable 4-bit memory cell. In *International Journal of Computer Applications International Conference on Innovations In Intelligent Instrumentation, Optimization And Signal Processing* (pp. 0975–8887).
2. Kim, N., Flautner, K., Blaauw, D. & Mudge, T. (2004). Circuit and micro architectural techniques for reducing cache leakage power. *IEEE Transactions on Very Large Scale Integration (VLSI) Systems 12*(2), 167–184.
3. Vontobel, P. O., Robinett, W., Kuekes, P. J., Stewart, D. R., Straznicky, J. & Williams, R. S. (2009) Writing to and reading from a nanoscale crossbar memory based on memristors. *Nanotechnology, 20*(42), 425204.
4. Strukov, D. B., Snider, G. S., Stewart, D. R., & Williams, R. S. (2008). The missing memristor found. *Nature, 453*, 80–83.
5. Yu, S., Liang, J., Wu, Y., & Wong, H. S. P. (2010). Read/write schemes analysis for novel complementary resistive switches in passive crossbar memory arrays. *Nanotechnology, 21*(46), 4652.
6. Kozicki M. N., et al. (2004). Nonvolatile memory based on solid electrolytes. In *IEEE Proceedings of Non-Volatile Memory Technology Symposium.*
7. Vourkas, I., & Sirakoulis, G. Ch. (2014) Memristor-based combinational circuits: A design methodology for encoders/decoders. *Microelectronics Journal, 45*, 59–70.
8. Chua, L. O. (1971). Memristor – the missing circuit element. *IEEE Transactions on Circuit Theory, 18*(5), 507–519.
9. Zidan, M. A., Fahmy, H. A. H., Hussain, M. M., & Salama, K. N. (2013). Memristor-based memory: The sneak paths problem and solutions. *Microelectronics Journal, 44,* 176–183.
10. Ehiro, M., Inoue, K. & Awaya, N. (2005, May). Nonvolatile memory device. U.S Patent 6 888 745, 3 May 2005.

11. Rabaey, J. M., Chandrakasan, A., & Nikolic, B. (2002). *Digital Integrated Circuits: A Design perspective*. Upper Saddle River: Prentice Hall.
12. Singh, S., & Mishra, V. (2018). Enhanced Static Noise Margin and Increased Stability SRAM Cell with Emerging Device Memristor at 45-nm Technology. *Radio Electronics and Communication System*, *61*(5), 200–206.
13. Dayal, A., Pandey, S. P., Khandelwal, S., Akashe, S.: Multiple-gate silicon on insulator (SOI) MOSFETs: Device design and analysis. In *(IEEE Conference) Annual International Conference on Emerging Research Areas and 2013 International Conference on Microelectronics, Communications and Renewable Energy" (AICERA/ICMiCR)*, Kanjirapally, India (pp. 1–6) (2013).
14. Saha, D., & Sarkar, S. K. (2014). High-speed reduced-leakage SRAM memory cell design techniques for low-power 65 nm FD-SOI/SON CMOS technology. *Elsevier Microelectronics Journal, 45,* 848–856.
15. Garg, A., & Kim, T. T.-H. (2013). SRAM array structures for energy efficiency enhancement. *IEEE Transactions on Circuits and Systems—II, 60*(6), 351–355.
16. Khan, S., Hamdioui, S., Kukner, H., Raghavan, P., & Catthoor, F. (2013). Bias temperature instability analysis in SRAM decoder. In: *18th IEEE European Test Symposium (ETS)* (pp. 4673–6377).
17. Wen, Liang, Li, Zhentao, & Li, Yong. (2013). Single ended, robust 8T SRAM cell for low-voltage operation. *Elsevier Microelectronics Journal, 44,* 718–728.
18. Bellerimath, P. S., & Banakar, R. M. (2013). Implementation of 16x16 SRAM memory array using 180 nm technology. *IJCET International Journal Of Current Engineering and Technology*, 2277–4106.

Long Term Evolution for Secured Smart Railway Communications Using Internet of Things

Shweta Babu Prasad and P. Madhumathy

Abstract The railroads have consistently met hurdles due to maintenance and upkeep issues. The length and scale of the network impedes maintenance. Likewise executing and utilizing new age innovation is not feasible still changes should be made on current frameworks. Internet of things is new age innovation that can be executed inexpensively. This can be actualized by simulating the IoT network in small-scale by utilizing Android application and also 4G LTE systems for real time execution. The proposed system uses various sensors to detect distinctive parameters which are regularly checked via the IoT network. And in case any parameter is triggered the system will automatically stop. Although derailment is imminent due to enormous mass of trains, the effect of the crash can be definitely improved by immediate mediation of IoT network. These systems will likewise lessen the overload on government on maintenance and safety related issues of existing framework with lower expenses. These systems also vulnerable to hacking by external factors and one time password system is also used for security. Real time tracking of parameters from base station allows systems to find train journey details such as average speed and performance parameter.

Keywords Internet of Things · Sensors · Embedded System · Security · Privacy

1 Introduction

The recommended system uses various sensors for railway operations. An IR sensor is utilized to detect crack in the railway track, ultrasonic sensor to detect the obstacle in front of the train, flame sensor to detect the fire in the train and an emergency

S. B. Prasad · P. Madhumathy (✉)
Department of Electronics and Communication, Dayananda Sagar Academy of Technology and Management, Bangalore, India
e-mail: sakthi999@gmail.com

S. B. Prasad
e-mail: shwetabp89@gmail.com

S. K. Das et al. (eds.), *Machine Learning Algorithms for Industrial Applications*, Studies in Computational Intelligence 907,
https://doi.org/10.1007/978-3-030-50641-4_16

switch which stops the train when the user presses it. If any of the faults are detected, the train slows down and stops. The current situation and parameters from the sensors are amended to the specialized IoT website. By making use of IoT, it is possible to bring the train to a stop. In order for manual stoppage, the system sends an OTP after dispensing the stop command. When the user issues an authentic OTP, the system stops the train. The light and fan are under the control of a touch sensor and in case it is not able detect the presence of a user, the light and the fan remain turned off thereby conserving energy.

The framework centers primarily on the effectiveness of the observation procedure of the railway operations by utilizing wireless links that disposes of the necessity of human intervention to check for faults which is unreliable and requires high maintenance.

1.1 Security and Privacy

Additionally, the security and privacy aspects also need to be considered. As the sensitive data travels between the sensory networks to the IoT network, it is important that only authorized and authentic entities have access to it. Hence, in the proposed system, for manual stoppage of the train in times of emergency, an OTP is required. The OTP is issued only by the authentic user. In this way, in the IoT consisting of various hardware and software components linked together, there is liability of sensitive information being exposed to adversaries who can utilize the information to their advantage. Hence, security and privacy has been introduced to avoid any vulnerability and unauthorized and unauthenticated access to data.

1.2 Chapter Organization

This chapter is divided into nine sections.

The chapter begins with the introduction which includes the background information and the different components used in making the system.

Section 2 presents the literature survey performed for the chapter which is the study of different papers related to the topic.

Section 3 describes the existing system.

Section 4 describes the proposed system.

Section 5 presents the block diagram of the system.

Section 6 gives a detailed description of the various hardware and software components used to build the system.

Section 7 covers the results obtained and also comprises of discussions made in this regard.

Section 8 presents the details of the user interface.

Section 9 describes the summary of the chapter including the conclusions drawn and also suggests several ideas for future work.

2 Literature Survey

Chen and et al. propose a vision of IoT where an insight is given about the applications, challenges, and opportunities from the perspective of China [1]. As there is progressive increase in requirement of IoT, there is also a growth in sensors, network, radio access and other platforms for mass production. In this paper, the authors introduce a profitable IoT solution which has IoT based network device platform, server and gateway for new age railway framework which is assessed to establish the appropriateness by a thorough analysis about how IoT systems can be utilized for maintenance by realizing a PoC and carrying out experiments. Network framework of IoT solution is proposed to check the utilization and goal attainment of the Radio Access Technologies (RATs) for conveying data generated by IoT for power usage by carrying out an comprehensive field test with system level experiments. By checking the results in different disciplines, the establishment of benefits of IoT is done.

Martinez and et al. propose the modeling power consumption in IoT devices [2]. In this paper the authors examine the issue of criterion evaluation for railway wagon suspensions to cater data to reinforce condition-based support. A dynamical model developed Rao–Blackwellized particle filter (RBPF) is obtained on which simulations are performed to correlate the performance of parameter estimation with multiple sensor values stability concerning the ambiguity in the data of the irregular track input values. This is essentially tested from a Coradia Class 175 railway vehicle having just a bogie and onboard sensors, and some basic outcomes are obtained.

Kim and et al. propose the design and performance evaluation of automatic train control over LTE [3]. With the growth of train technology, control and wireless communications, automatic train operation have become prominent. Further, if any glitches occur regarding taking care of the QoS from the train traffic control leads to adverse effects. Accordingly the operators aspire to investigate the specs in order that wireless communications framework is competent enough of assuring the required QoS. Here, the authors introduce a beneficial QoS administrative design in order to control the traffic from the train established on the technique utilized in a traditional LTE system. As per the recommended proposal, the authors gauge the usefulness of the LTE system by testing in a commercial railway area. The main arguments backing the train control functions by the LTE system are the composition of a QoS policy established on inspecting the attributes of the train control traffic and the convenient alteration of the cell specification throughout the cell devising and development measures so as to solve any network problems having adversaries with data pause.

Song and et al. propose the tests to be carried out and performance assessment of long duration growth for wireless railway communications [4]. This research paper introduces a system in which groups of infrared sensors are utilized to record the path of the train along with direction. Aforementioned report is utilized to close/open the railway blockade automatically by means of a motor connected to a microcontroller unit. It notifies the driver of a potential encounter with a train coming from the opposite direction by an SMS sent from the GSM module connected to the microcontroller. The exact spot of the train is transferred to a webpage through the GSM module for tracking.

Fragma-Lamas and et al. propose an analysis on Industrial IoT (IIoT)-Connected Railways [5]. It is stated by the authors that the railway is able to utilize conveniences generated by the IIoT and facilitate transmission automation concealed by the archetype of Internet of Trains. This analytic report contains details of how the growth in technologies of communication took place since GSM-R and also throwing light over the different revolutionized parameters. The preference of the new age broadband transmission schemes along with the rise of Wireless Sensor Networks (WSNs) to be utilized in the railways are also illustrated along with how the migration from GSM-R took place. Additionally, this detailed report identifies situations and frameworks by using which the railways can perform better with commercial IIoT. Further the short and medium-term IIoT set up duties for smart railways are assessed. A study is also performed on the current probe on predictive conservation, smart infrastructure, advanced monitoring of assets, video surveillance systems, railway operations, Passenger and Freight Information Systems (PIS/FIS), train control systems, safety assurance, signaling systems, cyber security and energy efficiency. Hence it can be said that the detailed study throws light on various technologies and services that will reform the railway industry and make it easier to defy the challenges.

Jat and et al. propose an inventive wireless QoS technology for delivery of Big Data video in WLAN [6]. The results obtained in the study undertaken by the authors prove that the intelligent wireless QoS technology in terms of big data video communication over a WLAN is efficient. Structural Similarity Index SSIM and Video Quality Metric VQM video quality matrixes are utilized to measure the received video. The results claim that dynamics frame aggregation mechanism are better than the big data video delivery for SSIM and VQM compared to frame aggregation mechanism defined by the draft of IEEE802.11n WLAN.

Mukherjee and et al. suggest about the nature motivated computer applications for wireless sensor networks [7]. They throw light on different constraints like usage of battery, low speed of communication along with security. They elaborate on the economical and ideal solutions to the issues regarding the WSN nature-inspired algorithms.

Das and et al. elaborate on the present condition of wireless networks globally [8]. The study is centered on making use of AI and soft computing to construct models for wireless networks. The approaches hold crucial position in order to develop powerful algorithm. These techniques play a vital role in developing a more robust algorithm suitable for the dynamic and heterogeneous environment.

Singh and et al. determine the frequently adopted unsecure patterns in MANETs [9]. The discussion elaborates that MANETs are more prone to attacks than the wired networks. These attacks take place during the transmission of the messages and the suitable solutions are provided for the same.

Jayakumar and et al. suggest the suitable QoS parameters for basic web services and applications so that they can be deployed over the cloud [10]. The exhaustive study proves that there is no fixed rule for the acceptable standard or values for the quality attributes of the web services and web applications. Hence this study facilitates on finding the best of web services and web applications for their suitability or deploy ability in the cloud.

3 Existing System

The existing system comprises of an IR, temperature and touch sensor along with Zigbee and buzzer. IR sensor is utilized to identify the cracks in the tracks by obstacle detection principle. If the IR detects the changes, the buzzer expends the alert signal to notify the neighboring areas and the motor automatically shuts down. Infrared (IR) transmitter is a form of LED from which infrared rays are given out. In the same way, IR receiver is utilized to detect the IR rays transmitted by the IR transmitter and these two are placed in a straight line. If they are positioned correctly, both the sensors constantly deliver the sensed output. A temperature sensor is interfaced with the controller to detect the temperature. Again if the temperature exceeds the normal, the buzzer expends the alert signal to notify the neighboring areas and the motor automatically shuts down.

4 Proposed System

The issues with the existing model is that the signal from the base station has to penetrate into the vehicle, and undergoes a loss of up to 24 dB which has to be taken care by increasing the transmission power along with the sensitivity of the receiver. The structure can be combined into present structures without considerable infrastructural modifications. Aforementioned structures are implemented in several countries and are extremely effective in ensuring safety of passengers. Government agencies need to implement such systems to undertake special measures of safety, support and reliability. The main objective is to identify cracks on railway track, obstacle in front of the train, fire emergencies and added functionality is that the touch sensor senses the presence of the user and turns the light and fan ON or OFF. Manual stoppage of the train can also be implemented to avoid any accidents and ensure safety of the passengers. But this can be performed only by issuing an OTP which is sent to an authorized and authentic user. This way adversary cannot issue manual stoppage and security and privacy of the system is preserved. Additionally the entire information needs to be updated to a specific IoT website.

5 Block Diagram

The recommended system comprises of the following components (i) data collector (ii) data processor and (iii) communication unit as shown in the Fig. 1. The sensors collect data from the environment which are processed by the system and communicated to the third party app and the IoT server as shown in the receiver block diagram in Fig. 2.

The recommended system utilizes the IR, ultrasonic, flame sensors and an emergency switch for railway operations. The ultrasonic sensor detects the obstacle in front of the train. For detection of cracks in the tracks an IR sensor is used. The flame sensor detects the fire in the train. In case of fire detection, the train slows down and stops. When the user press emergency button, the system stops the train and IoT can also be used for this purpose. The values generated by the sensors are written into the IoT website. To stop the train manually, the system sends an OTP once the stop command is given. When the user enters the valid OTP, the system stops the train.

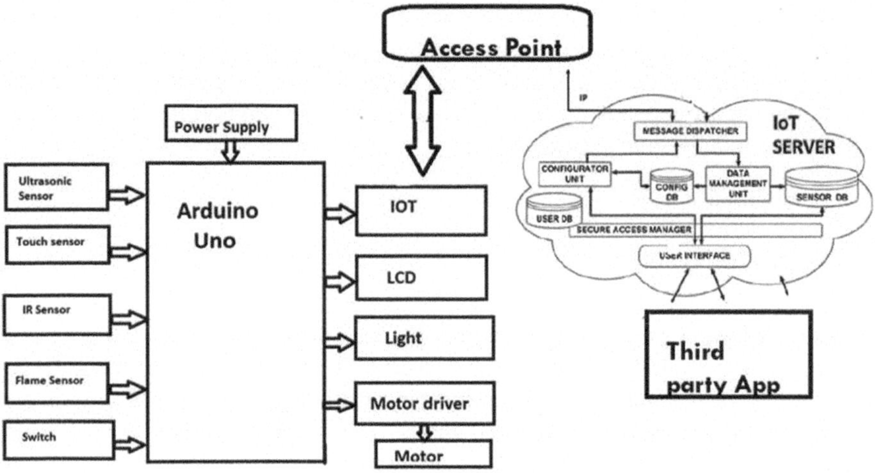

Fig. 1 Block diagram of the system

Fig. 2 Receiver block diagram

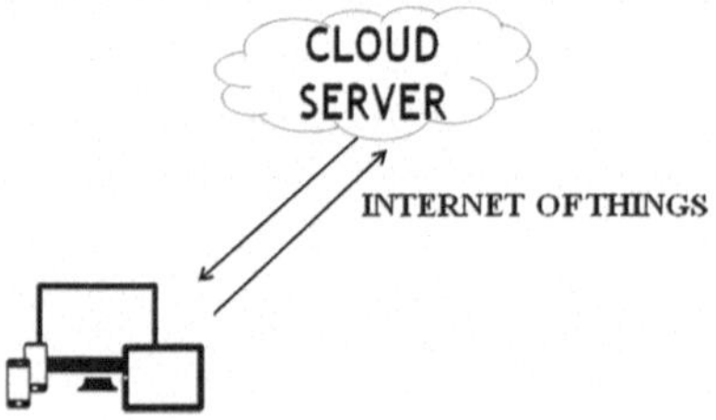

The light and fan are controlled by means of a touch sensor. In case the touch sensor does not detect the presence of the user, it switches off the light and fan thereby conserving power.

6 Details of Hardware and Software

In this section the different hardware components used are mentioned along with their functionality.

1. **Ultrasonic Sensor:** Ultrasonic sensors can recognize mobility of objects and figure out the range of separation. They can also identify the edge of material in web guiding system.
2. **Touch Sensor:** A touch sensor is a device that can identify a physical touch. It allows a device or object to detect touch, by a human user or operator.
3. **IR Sensor:** A passive infrared sensor (PIR) is installed in detectors to gauge infrared (IR) light emitted by objects.
4. **Flame Sensor Module:** This device perceives flame and radiation along with light wavelength between 760 nm and 1100 nm. The range of perception is 100 cm.
5. **Microcontroller:** The microcontroller used is Atmega 328p which is top notch performer Atmel 8-bit AVR RISC-based microcontroller.
6. **Geared DC Motor:** They are an extension of DC motor which and they come with a gear assembly attachment to the motor.
7. **L293D DC Motor Driver:** L293D is a well known motor driving IC. It is a 16 pin IC.
8. **DC Motor Switching and Control:** Small DC motors can be turned "On" or "Off" via switches, relays, transistors or MOSFET circuits with the most basic motor control known as "Linear" control.
9. **Power Supply Circuit: A** power supply unit (PSU) is one which provides electrical and other forms of energy to an output or a set of different loads.
10. **LCD Display (JDH162A):** A 16X2 LCD (Liquid Crystal Display) screen is a widely used and simplest electronic display unit used in different operations.
11. **ESP8266 WI-FI MODULE:** It is a low cost Wi-Fi IC with complete TCP/IP stack and microcontroller abilities manufactured by Espress if Systems.

The figure below depicts the interfacing diagram (Fig. 3).

Pseudo code

The pseudo code is given below and flowchart shown in Fig. 4. Depict the connection of the various components of the model and also the flow of control is shown.

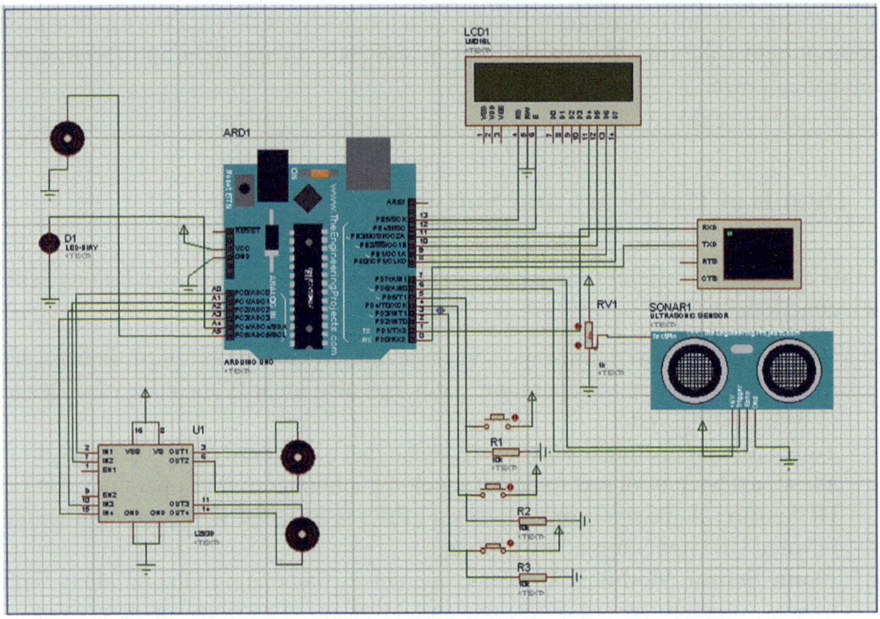

Fig. 3 Interfacing diagram

Step 1: Start

Step 2: The conditions are read from various sensors.

Step 3: Depending on the conditions of values read by the sensors, the various sensors either take appropriate action by utilizing the motor, update the IoT website and display to the LCD or they just display condition on the LCD and update to the IoT website.

Step 4: Stop

7 Results and Discussions

7.1 Hardware Section of the Model

The picture below shows the final model (Fig. 5).

The model is tested by exposing it to different conditions to make sure that the complete set of parameters are tested and satisfactory results are obtained and additionally the values are updated to the network. This has to be compulsorily satisfied

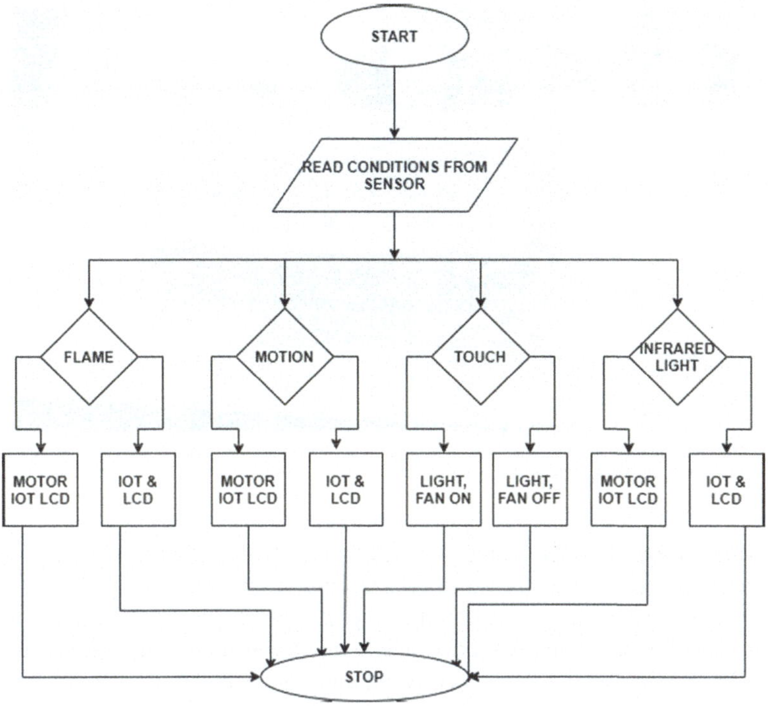

Fig. 4 Flowchart to depict component connection and flow control

Fig. 5 Hardware section of the model

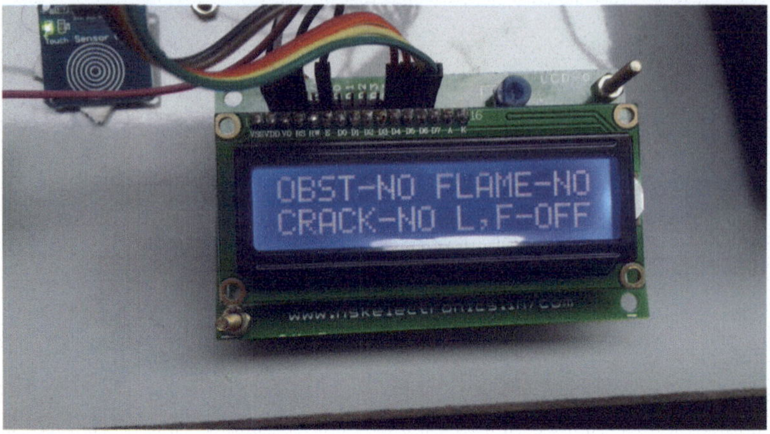

Fig. 6 Start condition prerequisites

for the train's forward motion and likewise this is also notified to the base station. The dynamic parameters can be adjusted according to the range (Fig. 6).

The different sensors have considerable importance and placed to guide the motion of the robot and hence they play a major role in administering an impressive safety solution. They monitor the parameters in real time and provide safety by stopping the train which is in a remote location without. These ensure remote stoppage of real time parameters without the interference of operators/human. They can be deployed and implemented in the existing modules with not much alteration to the infrastructure. In the model IoT network is simulated by using a third party application. The values of the dynamic parameters can be set in the working model, the ultrasonic sensor is set at a range of 90 cm to work with the 5 s delay.

7.1.1 Obstacle Detection

To test the working of the model in the case of obstacle detection, the model is exposed to an obstacle. As a result, the LCD screen reads that the obstacle is in fact detected and is true and shows that the other parameters are not detected or their condition met is false. In real time the sensor model can be placed at the front and rear end of the train. The ultrasonic sensor in the model, sense and avoid the collision by emitting waves to detect object distance, thereby notifying the user of the obstacle and to bring the train to a stop (Fig. 7).

7.1.2 Crack Detection

To test the working of the model in the case of crack detection, the model is exposed to a crack. As a result, the LCD screen reads that the crack is in fact detected and is

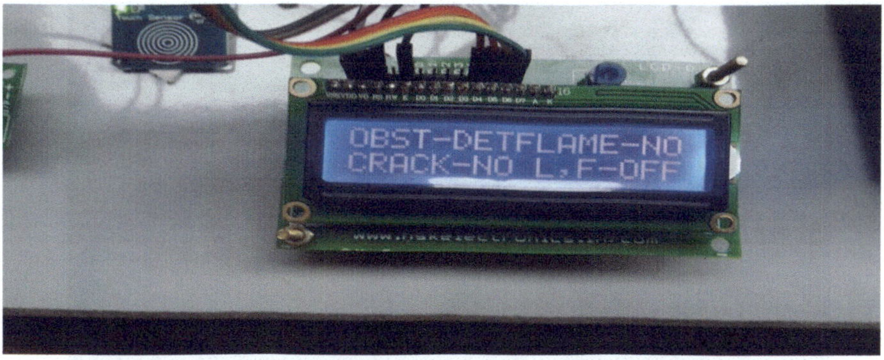

Fig. 7 Obstacle detection

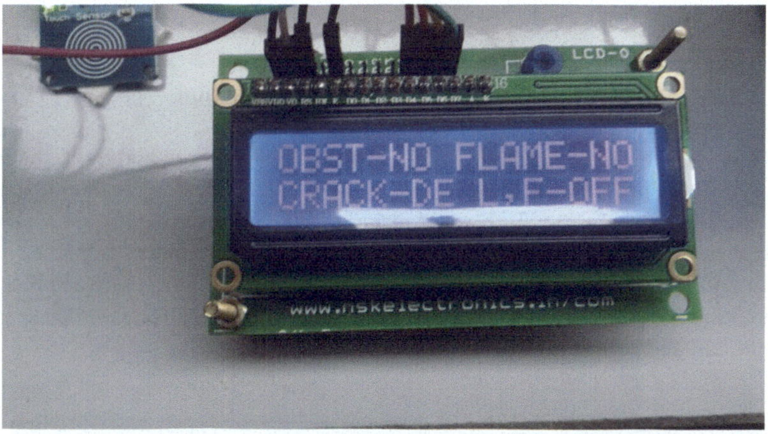

Fig. 8 Crack detected

true and shows that the other parameters are not detected or their condition met is false. In real time, the sensor model is placed at the track, where the IR sensor detects the crack and sends a notification to the user thereby stopping the train (Fig. 8).

7.1.3 Flame Detection

To test the working of the model in the case of flame or fire detection, the model is exposed to a flame. As a result, the LCD screen reads that the flame or fire is in fact detected and is true and shows that the other parameters are not detected or their condition met is false. In real time, the sensor model is placed inside the train. The flame sensor module present as a part of the system detects a fire in case of an emergency and notifies the user for further action (Fig. 9).

Fig. 9 Flame detected

7.1.4 Light and Fan on/off Depending on User Detection

To test the working of the model in order to switch ON/OFF fan and light, the model is exposed to the presence of a user. Here the touch detector detects the presence of the user and switches ON the light and fan and hence this is shown in the LCD which reads that the light and fan are ON. In real time the sensor model is placed inside the train and is exposed to the movements of the user. The touch sensor present on the model then detects the motion and performs appropriate action in order to switch off/on the fan and/or light (Fig. 10).

Fig. 10 Touch sensor detected

Fig. 11 Login page

8 User Interface

8.1 Security Process Results

Figure 11 shows the initial login page of the user interface for the IoT device control. This is where the users are supposed to register themselves by providing unique username and password.

(i) Secure access to the network is ensured by using an OTP. The use of unique OTP which is sent only to authorize users prevents unauthorized access to the network.
(ii) Once the user enters the necessary details, the registration process is completed.
(iii) Now the user is led to a third-party application to facilitate access to the device.
(iv) After access the user is taken to the third party application for accessing the device. Although the automated system allows for effective safety measures, it leaves the system wide open to attackers who can combine cyber attacks with physical attacks.

Figure 12 shows the details filled by the user which includes the name, contact info, email id along with the username and password.

Figure 13 shows that the user needs to give command to generate OTP and when the user receives the OTP, he must enter it into the space provided to verify it (Fig. 14).

IoT Site Connected
It is possible to utilize IoT site connected to control the device in an isolated manner as the values of the parameters are being written. A security credential is used as

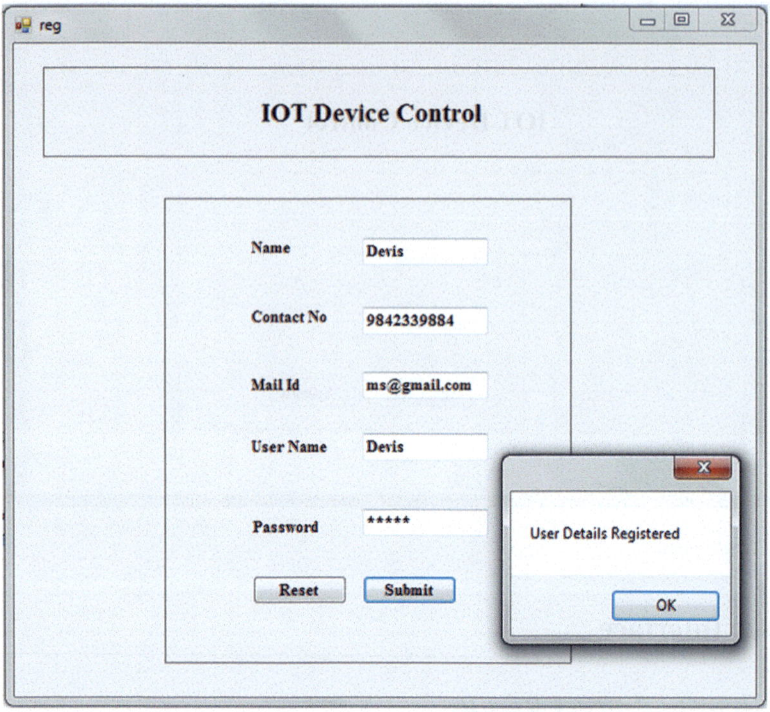

Fig. 12 User registration

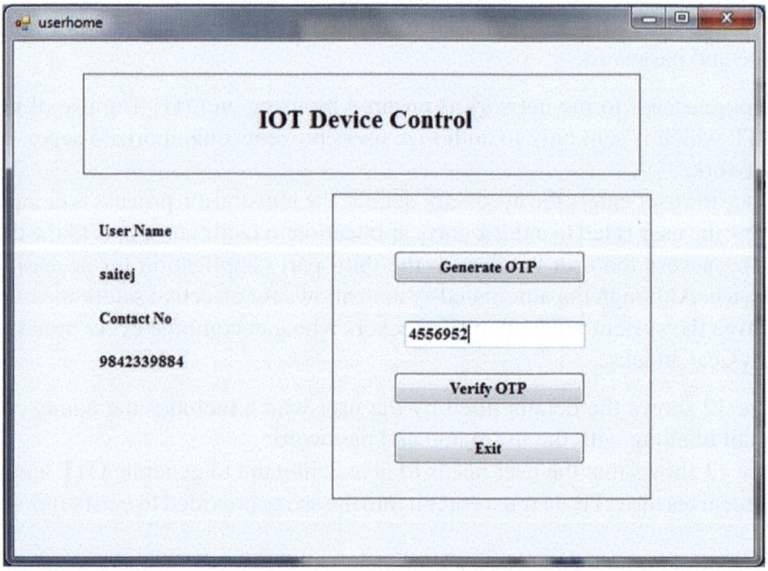

Fig. 13 OTP generation

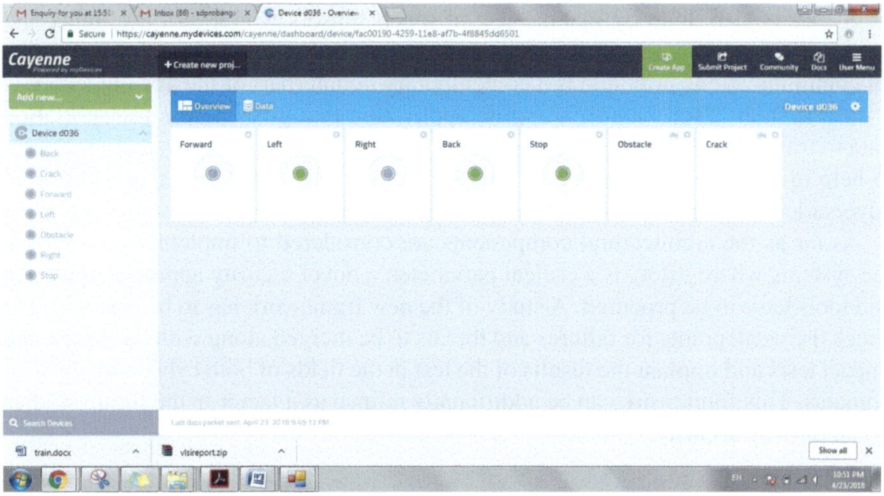

Fig. 14 IoT site connected

robust authentication system instead of outdated static passwords paving way for a temporary one-time password (OTP). An additional layer of security is provided this way which makes unauthorized access to confidential data, accounts and networks difficult. A security credential is extremely important to protect sensitive data to avoid damage attack to systems. In IoT, devices are interconnected with various hardware and software, chances of sensitive information leaking through unauthorized manipulation. Hence it is crucial to enforce various security and privacy measures in order to safeguard sensitive information.

9 Conclusion

The primary goal of the project is to detect the railway crack, obstacle in front of the train, fire emergencies and depending on the touch sensor the system will turn ON and turn OFF the light and fan and it can be safely concluded by observing the results that this has been achieved. All the readings of the values and parameters are updated to a specific IoT website. The model comprises of IR sensor, ultrasonic sensor, flame sensor and an emergency switch to perform their respective function. If any parameter exceeds the predetermined value or when faults are observed in tracks by the sensors, the train is controlled to slow down and ultimately stop. Additionally, the status and values of the parameters are updated to IoT website on a timely basis and hence the train can also be controlled to stop through the IoT. In this case, the user needs to issue a valid OTP so that the system is able to stop the train. Here, authentication, identity management and authorized access to data is accomplished as a part of implementing security and privacy measures between the sensory and IoT networks. Conservation of energy is also done by making use of touch sensors which turn the lights and fans OFF in the absence of the user. The train control system which is established on communications principle is a contemporary counterpart of the age old safety—critical system. There is also a mobile application that gives access transportation data to travelers. This study focuses on the demand for a cyber- physical viewpoint to help in acquiring knowledge of cross-domain attack and defense, in addition to adversaries of cyber discontinuity in physical realm.

As far as the architectural components are considered to implement security in the systems where safety is a critical parameter, a novel security appraisal approach and tools have to be procured. A study of the new framework has to be performed to check the weak points for failures and threats to be merged along with the attack and impact tests and implant the results of the test in the fields of both cyber and physical domains. This framework can be additionally refined as a factor in the future studies urban railway security.

References

1. Chen, S., Xu, H., Liu, D., Hu, B., & Wang, H. (2014). A vision of IoT: Applications, challenges, and opportunities with China perspective. *Internet of Things Journal, 1,* 349–359.
2. Martinez, B., Montn, M., Vilajosana, I., & Prades, J. D. (2015). The power of models: Modelling power consumption for IoT devices. *IEEE Sensors Journal, 15,* 5777–5789.
3. Kim, J., Choi, S. W., Song, Y. S., Yoon, Y. K., & Kim, Y. K. (2015). Automatic train control over LTE: Design and performance evaluation. *IEEE Communications Magazine, 53*(10), 102–109.
4. Song, Y. S., Kim, J., Choi, S. W., & Kim, Y. K. (2016). Long term evolution for wireless railway communications: Test-bed deployment and performance evaluation. *IEEE Communications Magazine, 54*(2), 138–145.
5. Fraga-Lamas, P., Fernández-Caramés, T.M., & Castedo, L. (2017). Towards the internet of smart trains: A review on industrial IoT-connected railways. *IEEE Sensors Journal, 15.*
6. Jat, D. S., Bishnoi, L. C., & Nambahu, S. (2018). An intelligent wireless QoS technology for big data video delivery in WLAN. *International Journal of Ambient Computing and Intelligence (IJACI), 9*(4), 1–14.
7. Mukherjee, D., Das, A.K., Dey, S. (2020). *Nature Inspired Computing for Wireless Sensor Networks.* Singapore: Springer.
8. Das, S.K., Samanta, S., Dey, N., Kumar, R. (2020). Design Frameworks for Wireless Networks. Singapore: Springer.
9. Jayraj, S., Singh, A., & Shree, R. (2011). An assessment of frequently adopted unsecure patterns in mobile ad hoc network: Requirement and security management perspective. *International Journal of Computer Applications, 24*(9), 0975–8887.
10. Jayakumar, S. K. V., Singh, J., & Joseph, K. S. (2014). Suitable QoS parameters survey for standard web services & web applications to understand their cloud deployability. *International Journal of Computational Intelligence Systems, 4,* 1–18.

Application of Flower Pollination Algorithm to Locate Critical Failure Surface for Slope Stability Analysis

Jayraj Singh, Ravishankar Kumar, and Haider Banka

Abstract Analysis and design of the earth slopes has been an essential preface in the area of geotechnical science and engineering for all the times. In a certain moment of time, the geo-hazards or any geological phenomenon may come across to make the slope failures. It may provide extensive loss of life, great economics and environment damage. To reduce the enormous destruction from the slope failures, slope stability analysis can play a necessity role in evaluation of stability factor. In this chapter, a novel nature inspired algorithm based on pollination process of plants is used for locating the critical surface. The quantitative evaluation of stability analysis in terms of factor of safety demonstrated the performance of the approach. The findings indicate the appropriate performance over current methods and declare the optimum solution.

Keywords Meta-Heuristic Algorithms · Optimization · Flower Pollination Algorithm · Critical Failure Surface · Slope Stability Analysis

1 Introduction

Slope stability can be defined as resistance of inclined surface to failure by sliding or collapsing. They either occur naturally or engineered by humans. The problems associated with slope stability emerged in our lives because of the instability and imbalance produced by humans. The slope stability analysis by locating critical slip surface associated with minimum safety factor is rather challenging task due to the lots of imprecise, uncertainty and other dynamic decision-making variables in the actual scenario [1, 2]. Apart from this, the geographical and seismological or environmental change results to decrease the slope stability. Therefore, the chances of

J. Singh (✉) · R. Kumar · H. Banka
IIT (ISM) Dhanbad, Dhanbad, India
e-mail: jayrajsinghit@gmail.com

R. Kumar
e-mail: ravisk265@gmail.com

H. Banka
e-mail: haider.banka@gmail.com

S. K. Das et al. (eds.), *Machine Learning Algorithms for Industrial Applications*, Studies in Computational Intelligence 907,
https://doi.org/10.1007/978-3-030-50641-4_17

occurrence for slope failure has increased. Thus, the growing demand for engineered cutting, filling slopes on construction projects and other geotechnical activities has expanded the need to recognize analytical methods, research tools and techniques for solving the slope stability problems [3]. A knowledge of the geological, hydrological and soil properties is essential to proper implementation of the principles of slope stability [4, 5]. The study must be based on a model that accurately describes the conditions of the subsurface, ground activity and loads applied. This cycle includes the designers, engineers, geologists, contractors, technicians, and maintenance staff. In general, the slope stability analysis is conducted to refer to the safe and economical layout of excavations, embankments, dams, landfills, bridges, and heaps of spoil. Slope stability assessments deal with the identification of important geological, material, environmental, and economic parameters affecting the project, as well as the type, magnitude and frequency of potential slope problems. Proper study and analysis of slope stability enable safe establishment of construction structures [6]. The goal of slope stability analysis is to identify endangered areas, investigate potential failure mechanisms, evaluate slope sensitivity on different triggering mechanisms, optimum slopes design in terms of protection, reliability and economy to possible remedial measures etc. Successful design of a slope includes geological details and position characteristics [7, 8].

Many deterministic and probabilistic based strategies were expanded rapidly to form a optimal design for this vital issue of slope stability by identifying critical failure surface [1, 4, 5, 9, 10]. In general, to finding the critical slip surface is found a NP-hard type problem, which also can be described as a problem of unconstrained global optimization. Some of analytical methods such as rigid element method, limit equilibrium methods, finite element, distinct element methods are used for stability calculation. The most commonly used approach for geotechnical assessment is the approach of limiting the equilibrium (LEM) [11–13, 15]. The LEM method aims to split the entire moving mass into a specified quantity of vertical slices and to measure the slip surfaces for calculating the margin of the safety factor [16, 17]. Many authors have effectively used meta-heuristic techniques in conjunction with various LEM methods to evaluate slope stability [18–21]. Some of the researchers such as Yamagami and Ueta [22] adopted distinct stochastic technique such as method BFGS & DFP to examine the safety factor of the various slopes. Greco [19] used techniques such as Monte-Carlo method. The authors McCombie and Wilkinson, Chen and Morgenstern, Zolfaghari, Das, Jianping and Sengupta, and Upadhyay utilized various search techniques such as grid strategy and genetic algorithm [2, 23–27]. Kahatadeniya *et al.* [28] employed optimization using Ant colony search (ACO). Khajehzadeh *et al.* [29] introduced a new metaheuristic approach namely as GSA approach, which works on the basis of the rules of gravity and motion. Kashani explored the ICA-based stability evolution for complex slope problem [30]. Due to elegancy and flexibility, these meta-heuristic methods are attracting a lot of coverage. This chapter is presented, the flower pollination algorithm (FPA) introduced by Yang [31]. A benchmark slope failure problem from literature were investigated using FPA approach. The findings were examined and compared with other methods of optimisation to show the method's usefulness and supremacy.

1.1 Motivation and Contribution

The problem for analysing slope stability by locating critical surface is usually a quiet complex, unconstrained problem of global optimization and becomes a very challenging task. This is also considered an NP-complete problem. Exploring m-possible solutions in the form of failure surfaces to find a critical surface is an optimisation problem. It requires a high amount of processing time and storage. This chapter relates primarily to identify critical failure surface by using an application of the Flower pollination algorithm (FPA) [32, 33]. The suggested technique offers an empirical analysis for defining a critical surface for evaluating slope stability. Using this approach, engineers and scientists might identify the susceptible slopes, including the preventive remedy to save from severe and extreme loss of life. To mitigate the enormous disruption caused by the slope failures or the possibility of landslides, this study may play a vital role in assessing slope stability in the geotechnical industry.

2 Slope Stabilty Analysis

Slope stability analysis is a popular concern in civil, mining, and geo-technical engineering domain. The Identification of the safe configuration of human-made or natural slopes is accomplished by slope stability analysing [14, 15]. The slope stability research involves the examination of a variety of gravitationally driven slope movements. By owing to landslides, floods, earthquakes and many other geo-hazards, can arise. Such failures may also be devastating and entail the significant loss of damage to the economy, culture and the environment. Due to the geological phenomenon of slope failures and certain lithologies, the slopes may possess specific types of failure. This section introduces the different key parameters which significantly affect the slope stability analysis.

2.1 Slope Failure Types

Slopes in soils and rocks are ubiquitous in nature. Their movements are a common geological phenomenon under the influence of gravity. There are four common type of slope movements due to failures found in nature which are described as follows:

Plane Failure
As compositions of discontinuities in the rock mass block are free to move, a slope undergoes this mode of failure. Amount of rock slipped down a fairly planar failure surface in a slope. The surface of the failure is typically structural discontinuities, including faults and joints.

Wedge Failure
This might happens in rock mass with two or more sets of discontinuities whose intersection line is approximately perpendicular to the slope strike and dip towards the slope planes.

Toppling Failure
It occurs when rock columns created by steeply dipping discontinuities in the rock rotate at or near the base of the slope at an essentially fixed point, followed by slippage between the layers. Removing the pressure and the confining material, as is the case in mines, may proceed to partial relief of the limiting stresses inside the material structure, results in a topple failure.

Rotational Failure
The rotational failure surface can be a form of a circular or non-circular curve in the region. It involves of a rock or waste movement, along an axis parallel to the slope's contours, which includes shear displacement along a concave upward.

2.2 Factors Affecting Slope Stability

The factor of natural forces of wind, water, snow, etc., usually change the topography of earth, which often creates unstable slopes. Apart, the external triggers such as heavy rains, construction projects, earthquakes, and others are likewise the reasons for the instability of slopes. The factors which affects the slope stability of the earth can be described in details as follow:

Geological Discontinuity
A geological discontinuity in a rock or soil mass indicates a difference in the physical or chemical characteristics. The shape of a foliation, joint, bedding plane, fracture, cleavage, fissure, crack or fault plane may be described as a discontinuity. It regulates the different type of soil or rock slope mass failure. A Jointed rock demonstrates greater permeability along discontinuity planes and decreased shear strength. A rock mass that includes several joints is often seen as more fragmented. The positioning of neighboring joints influences in large portion the size of individual blocks regulating the failure mode. A wide spacing of joints allows weak rock mass stability and is accountable for failure in circular or perhaps even waste flow. It also has an effect on permeability of the rock mass.

Water
The impact of water can be taken into two aspects in the slopes. One is surface water or Water table below ground surface that causes the pore pressure emerged from water and the other is the accumulation of rain water that passes through the soil and creates water pressure along the slope. In the medium to hard rock, Water can reduce a rock's surface strength considerably by picking up the fractures within the rock mass. Within a discontinuity, the water pressure acting decreases the effective stress acting on the surface and thus, the shear strength along the plane is also reduced.

Additionally, the pressure of the pores rises if a load is kept at the top of the surface. Such a load would cause the slope to collapse immediately if it exceeds the shear strength of its slope. Water filling in discontinuities can lead to a lowering of the stability situation for natural or artificial slopes. Rainfall water supplement and snow melt adds more slope weight. In addition to this, the slope instability is often caused by ground water. The water occurs almost anywhere under the surface of the earth. These water fills the gaps of pore in the rock between the grains or fractures. Such water will flow into discontinuity within the rock mass replacing the air in the pore space, which increases the soil weight. This contributes to increased effective stress

Strength
When the unconsolidated soil is insaturated or dry, the air in the pore spaces in the soil and mine dumps slope would be compressed by a increase in load. Deforming the mass and assembling fragments of grain or rock that mostly enhance the shear force. However, when a rock mass is porous, an increase in external pressure contributes to a raise in pore density, because water is comparatively heavier. This rise in pore pressure described a buoyant effect, That can be highly enough to withstand the mass of the underlying rock, thereby reduce shear force and friction.

Temperature
The temperature effects also impact a rock slope's strength. Large changes in temperature may cause the rocks to break or fracture. Owing to the resulting contraction and extension, further the rock splits into smaller parts. Cooling of water in discontinuities induces further disruption by loosening the rock mass. Repeated freezing cycles can lead to gradual loss of strength.

Failure Due to Erosion
In the viewpoint of slope stability, there is need to address two aspects of erosion. The first one is a large-scale erosion, such as the flooding of the water at the base of a slope. Here erosion changes the composition of rock mass which is technically unstable. The removal material at the bottom of a possible fall decreases the stress which can stabilize the slope. The second is a typically localized erosion induced by groundwater or surface run-off. This erosion of joints filled soil or weathered rock regions will ultimately weaken the interlocking of adjacent blocks of rock. The lack of such interlocking therefore significantly reduces shear force in rock mass. This reduction in shear strength can allow movements into stable mass of rock. Besides, the localized erosion can also lead to enhance permeability and flow of groundwater which directly affect the stability of rock slope.

Geometry of Slope
The important slope geometry parameters such as slope height and angle are directly affect the slope stability. This stability ultimately depends upon slope height, density, bearing and shear strength of slope foundation. Stability of the slope usually decreases with height increase of the slope. As the height of the slope rises, the shear stress of the slope within the toe enhances due to the addition of weight. The shear stress is correlated with the material mass and the angle of the slope. The tangential stress

grows with increasing slope angle resulting in increased shear stress thereby reducing its stability.

Earthquake Effect

If the earthquake is sudden, big enough or repeated over time until the time required to drain the soil, water can not evacuate the soil. The higher the level of water in the soil, the lower the soil's mechanical strength and the more likely the slope is to crack. In recent decades, various methods have been adopted for accessing the failure system and identifying volatility due to spatial, ecological, climatological and geological exploration.

Effect of Reducing Bench Angle

Considering to have berm is not enough to guide and satisfy the slope stability analysis criteria. Decreasing overall slope is also considered to achieve better stability of the slope. By decreasing angle of slope with few degrees can provide an increasing value in shear strength and decreasing mobilized stress, which causes to give a better factor of safety. Thus, for designing more stable dump slopes, it is required to optimize bench height, berm width, bench angle, and volume of dump.

3 The Methodology Used for FoS Calculation in Circular Failure Surface

The Flower pollination algorithm (FPA) is used in the current study to assess the critical surface. This algorithm has been fully implemented because of its ease of operation and its ability to address highly complicated, non-linear optimization problems. In this study, the Fellenius method is taken from LEM procedure as an objective function for calculating the safety factor. To evaluate the safety factor, this method of slice splits the entire slope into many slices [21]. Thereafter, based on moment equilibrium criteria, the safety factor (FoS) is determined as follow:

$$F = \sum_{i=1}^{n} \frac{F_r}{F_d} = \frac{\sum_{i=1}^{n} [c_i' l_i + (W_i \cos\alpha_i - \mu_i l_i)\tan(\phi_i')]}{\sum_{i=1}^{n} W_i \sin\alpha_i} \tag{1}$$

where ϕ_i and c_i are internal friction angle and cohesion of the material. W_i is weight of the slice due to self-gravity. l_i is the circular arc length of the slice. α_i is the angle between horizontal line and tangent line along the slip surface [35]. In order to evaluate the circular failure slope, the failure slip surface is initially generated at random. The generated surface is then derived based on center point *(a, b)* and radius *(r)*. Further, the intersection points (as shown by 'AD' in Fig. 1) that intersect the circular surface with the line of slope, is determined. The distance from bottom to top of the slip surface determines the slice width. Lastly, the intersections point at slice mid line with slip surface is determined to derive the base slice's angle.

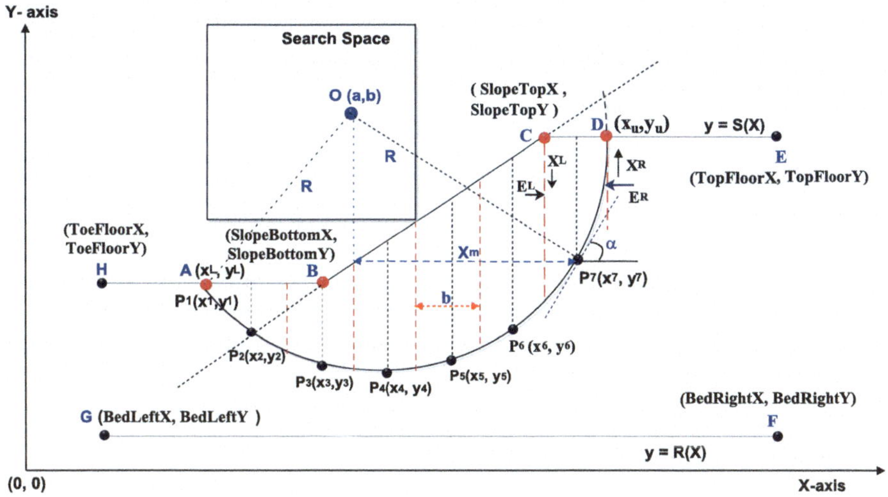

Fig. 1 A pictorial view of a failure slip surface with its nth—slices

From the basic of coordinate geometry, the circle equation can be expressed as:

$$(x-a)^2 + (y-b)^2 = R^2 \tag{2}$$

The points of intersection (x_l, y_l) and (x_u, y_u) of a surface with the line type $y = mx + d$ are as follow:

$$x_{l,u} = \frac{a + bm - dm \pm \sqrt{\partial}}{1 + m^2}, \text{ where } \partial = r^2(1 + m^2) - (b - ma - d)^2 \tag{3}$$

$$y_{l,u} = \sqrt{(R^2 - (x_{l,u} - a)^2)} + b \tag{4}$$

If these points coincide far from the slope geometry region, then these points are correct based on coordinate geometry principles, so that a proper surface with reasonable points on the slip surface is generated. Eventually, the parameters which play a important role in stability calculation are then determined as below:

$$\text{Slice width (b)} = \frac{|x_u - x_l|}{n} \tag{5}$$

here, 'n' indicates the number of slices divided in slope region. Now, the base angle (α) at the mid point (Xmi, Ymi) of the slice computed as:

$$\text{Slice base angle } (\alpha) = sin^{-1}\left(\frac{X_{mi}}{R}\right) \; where, \; x_{mi} = x_i - b/2 \tag{6}$$

Now, the remaining factors that have important role in computing of safety factor (FoS), will be derived as follows:

$$\text{Length of each slice } (L) = \frac{b}{\cos\alpha} \tag{7}$$

$$\text{Tengential force (T) at each slice} = w \times sin\alpha = b \times h \times \gamma \times sin\alpha \tag{8}$$

$$\text{Normal force (N) at each slice} = w \times cos\alpha = b \times h \times \gamma \times cos\alpha \tag{9}$$

By substituting all of the above terms related to each slice in factor of safety equation, the FoS value associated with the surfaces will be derived. The process is further repeated to n—consecutive number of surfaces to explore the search space and observe critical failure surface.

4 Flower Pollination Algorithm

Flower pollination algorithm (FPA) is a newly developed metaheuristic approach which has been widely used to solve several constraint and unconstraint optimization problem [31, 33, 34]. In recent years, the approach is propagated through its various applications in research industries due to its capability to solve complex NP-hard problems. The algorithm works based on pollen transmission, which is performed by pollinators including wind or insects, bats, birds, other animals. For good pollination some forms of flower have different pollinators. There are four important rules of pollination which were inspired from flowering plants and influences the essential equations for updating the FPA algorithm [32, 33].

1. Cross-pollination exists from the pollen of various plants. Pollinators by jumping or flying separate moves, follow the laws of a Levy distribution. That is recognized as the global cycle of pollination.
2. Self-pollination is carried out from pollen from a certain flowers. It is pollinated locally.
3. Constancy of flower is the combination of pollinators and types of flowering. It is an upgrade to the cycle of flower pollination.
4. Local and global pollination are regulated by a probabilty value generated between 0 and 1. This probability is referred to a switch probability.

A plant has many flowers in the real scenario, and the flower patches produce plenty of gametes of pollen. For simplicity, one assumes that every plant has one flower that produces a single game of pollen. Because of this simplicity the solution (xi) presented in problem is equal to a flower or pollen gamete. There are two main steps in the FPA algorithm which include global and local pollination. The first and third rules are used together in the global pollination step to obtain the solution forward step $(x_i,\ t+1)$ using the values described as x_i and t. The formulation of global pollination is depicted in Eq. 10.

$$x_i^{t+1} = x_i^t + L(x_i^t - g*) \tag{10}$$

here, subscript i represents the ith flower which is added to the flowers' pollen. g* is the best available alternative. L is the pollination strength, which is extracted from a Levy distribution. The second rule of local pollination is extended with third flower constancy rule. The latest approach comes with random runs, as seen in Eq. 11.

$$x_i^{t+1} = x_i^t + \epsilon(x_i^t - x_i^k) \tag{11}$$

Where, x_i^t and x_i^k are separate plant solutions. ϵ is randomized between 0 and 1. In accordance with the fourth rule a switch probability (p) is used to pick the type of pollination that will regulate the optimization process throughout iterations. It can be seen the specifics of the optimisation method in the pseudocode given for the FPA algorithm.

Algorithm 1. Pseudocode for the flower pollination algorithm.

```
procedure MINIMIZE THE FITNESS FUNCTION (f(z))
    Initialize a randomly assigned population of n flowers or pollen gametes
    Find the appropriate (g *) solution i.e. best for the initial population
    Defines the probability of a switch (p)
    while ( t < number of iterations) do
        for (i= 1: n) do  /* n is the number of pollens or flowers in the population */
            if rand < p then
                Use global pollination by Eq.10
            else
                Use local pollination by Eq.11
            end if
            Appraising new solutions
            Update population's approaiate solutions
        end for
        Obtain best solution (g*)
    end while
end procedure
```

5 A Numerical Case Study Investigation for Performance Validation

In this section, a slope problem is considered for homogeneous soil material for validating the performance. Niu's problem with the slope is abstracted from the litre [36], where the geotechnical properties are discribed as: unit weight = 20 KN/m^3, friction angle = 26.6°, and cohesion is 10 KPA. The geometrical view of the slope model is depicted in Fig. 2. The problem is solved using FPA algorithm to search the critical failure surface within the specified search area. In addition to this, grid search and GA algorithm are also implemented to investigate the performance of the proposed algorithm. The results of the comparative analysis indicates that the

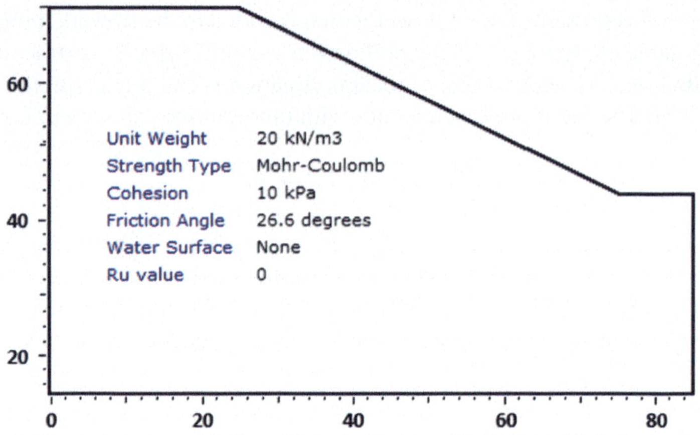

Fig. 2 Geometrical view with soil properties of homogeneous slope model

Table 1 Slice data using fellenius method for calculating FoS with Grid search method

Slice no.	c [kPa]	μ [kPa]	ϕ [deg]	b [m]	l [m]	h [m]	w [m]	α [deg]	Resisting moment (F_r) cl+(wcosα -μl)tanϕ	Deriving moment (F_d) wsinα
1	10	0	26.6	6.427	6.454	1.900	186.351	5.212	186.351	22.189
2	10	0	26.6	6.427	6.432	5.287	404.403	−2.147	404.403	−25.460
3	10	0	26.6	6.427	6.517	7.840	562.860	−9.542	562.860	−167.073
4	10	0	26.6	6.427	6.725	9.525	653.244	−17.105	653.244	−360.113
5	10	0	26.6	6.427	7.091	10.25	668.974	−24.997	668.974	−556.838
6	10	0	26.6	6.427	7.703	9.843	605.722	−33.448	605.722	−697.424
7	10	0	26.6	6.427	8.768	7.952	462.895	−42.859	462.895	−695.283
8	10	0	26.6	6.427	10.970	3.611	245.892	−54.145	245.892	−376.250
Factor of safety (FoS) = $\sum_{i=1}^{n} \frac{F_r}{F_d}$									$FoS = 1.327$	

grid search and GA approaches would have assumed 1.327 and 1.323 to be a lowest factor of safety. But in the FPA algorithm, the safety factor stabilizes at 1.309 (at 200 iteration) as the number of generations increases, That describes the global minimum. The critical failure surface observed during different generations are demonstrated in Fig. 3. The Tables 1, 2 and 3, shows all slice parameter's value for finding the critical failure surface using Grid approach, GA and FPA algorithm, where the fellenius method of slice's is used as fitness function to derive the factor of safety.

6 Result Discussion and Comparison

The efficacy of the proposed FPA algorithm was tested by solving a case study from literature, where the slip surface model simulation was validated using a geotechnical software tool named 'Slide'. To evaluate the critical surface correlated with the

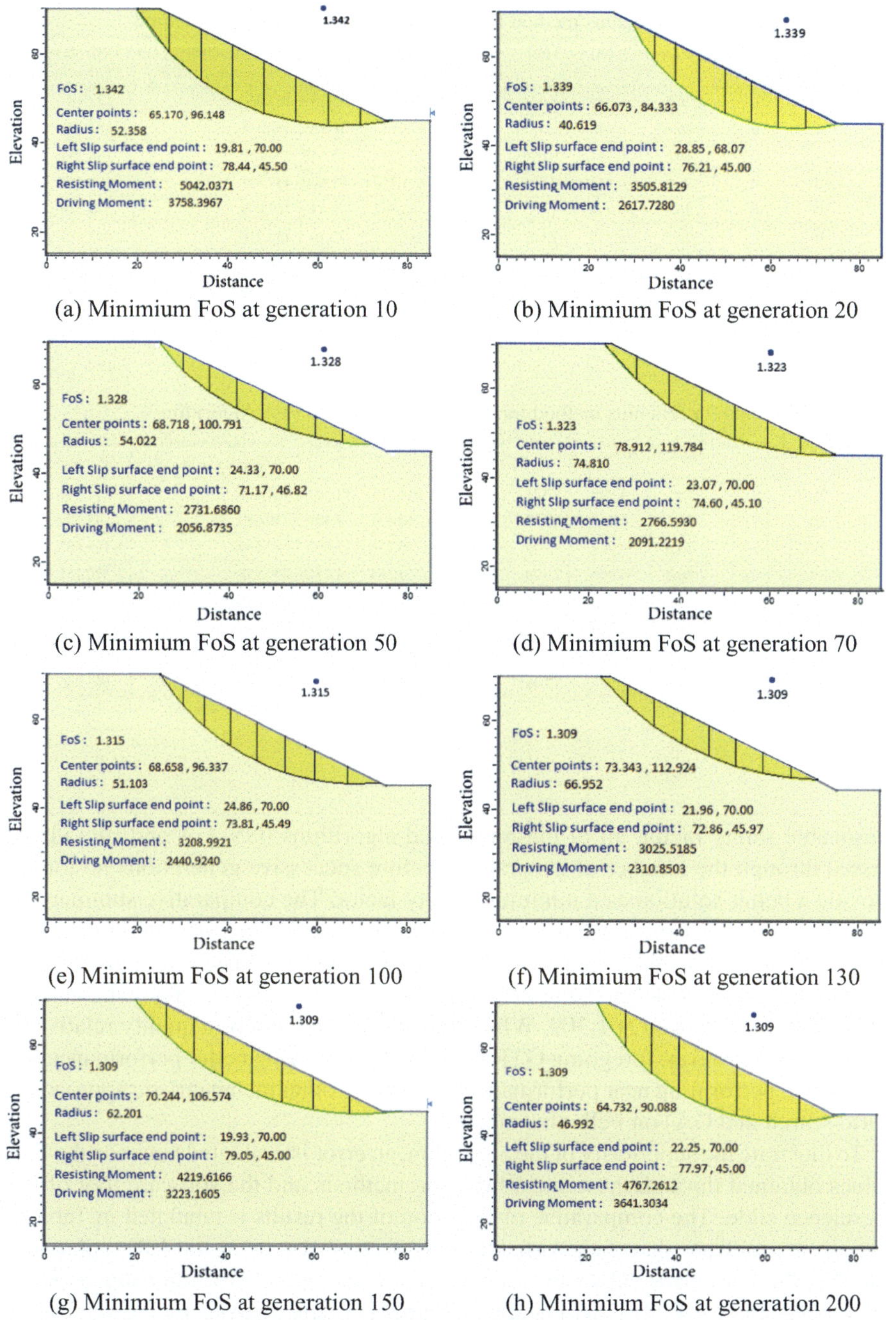

(a) Minimium FoS at generation 10

(b) Minimium FoS at generation 20

(c) Minimium FoS at generation 50

(d) Minimium FoS at generation 70

(e) Minimium FoS at generation 100

(f) Minimium FoS at generation 130

(g) Minimium FoS at generation 150

(h) Minimium FoS at generation 200

Fig. 3 Interpreted view for critical failure surface using FPA algorithm

Table 2 Slice data using fellenius method for calculating FoS with Genetic method

Slice no.	c [kPa]	μ [kPa]	ϕ [deg]	b [m]	l [m]	h [m]	w [m]	α [deg]	Resisting moment (F_r) cl+(wcosα -μl)tanϕ	Deriving moment (F_d) wsinα
1	10	0	26.6	6.441	6.474	1.361	175.359	−5.779	152.060	−17.648
2	10	0	26.6	6.441	6.557	3.659	471.343	−10.768	297.311	−88.020
3	10	0	26.6	6.441	6.696	5.370	691.794	−15.842	400.040	−188.754
4	10	0	26.6	6.441	6.901	6.456	831.772	−21.046	457.547	−298.566
5	10	0	26.6	6.441	7.193	6.858	883.542	−26.44	467.909	−393.226
6	10	0	26.6	6.441	7.603	6.484	835.293	−32.10	430.218	−443.674
7	10	0	26.6	6.441	8.187	5.190	668.650	−38.136	345.152	−412.733
8	10	0	26.6	6.441	9.062	2.743	353.443	−44.721	216.356	−248.601
Factor of safety (FoS) = $\sum_{i=1}^{n} \frac{F_r}{F_d}$									$FoS = 1.323$	

Table 3 Slice data for fellenius method for calculating FoS with FPA Algorithm

Slice no.	c [kPa]	μ [kPa]	ϕ [deg]	b [m]	l [m]	h [m]	w [m]	α [deg]	Resisting moment (F_r) cl+(wcosα -μl)tanϕ	Deriving moment (F_d) wsinα
1	10	0	26.6	6.965	7.121	1.035	144.215	11.991	141.807	29.946
2	10	0	26.6	6.965	6.978	5.473	762.441	3.409	450.680	45.309
3	10	0	26.6	6.965	6.993	8.867	1235.256	−5.097	685.694	−109.687
4	10	0	26.6	6.965	7.170	11.210	1561.684	−13.718	830.998	−370.151
5	10	0	26.6	6.965	7.548	12.419	1730.052	−22.670	874.486	−666.491
6	10	0	26.6	6.965	8.235	12.298	1713.132	−32.256	807.530	−913.898
7	10	0	26.6	6.965	9.52	10.429	1452.781	−42.998	627.157	−990.345
8	10	0	26.6	6.965	12.485	5.761	802.478	−56.118	348.909	−665.986
Factor of safety (FoS) = $\sum_{i=1}^{n} \frac{F_r}{F_d}$									$FoS = 1.309$	

reasonable safety factor, all of the-mentioned algorithms (GS, GA and FPA) have passed through the number of generations before successive generations no longer provide a better solution as a minimum safety factor. The comparative summary of derived FoS values from current study with previous studies are tabulated in Table 4. As can be seen from the result that, the minimum FoS obtained using grid search and GA approach is stabilized at nearly 1.327 and 1.323 on fellenius, whereas the FPA yields to stabilized at 1.309, Which indicates better solution quality relative to GS and GA. However, foregoing GTRM method [36] having better performance but FPA is also performing near performance like wise to other metaheuristic approaches (Grid search and GA) on Fellenius method.

To illustrate the superiority of the algorithm, an error is evaluated between the FoS values obtained through these meta-heuristic methods and the software tool of the rocscience slide. The comparative observation of the results is tabulated in Table 5. The results indicate that approximately 6% of errors occur in the FPA approach, which is the least error values in respect to Grid search and GA algorithm as measured 6.8% and 6.4%. This will declare a higher stability analysis by the algorithm. Additionally, it uses good convergence rate and better quality of solution with local minima avoidance. Therefore, based on these benefits, the proposed algorithm is comparatively suitable for getting better solution quality.

Table 4 Comparative analysis of the findings for the case study safety factor

Optimization algorithms	LEMs-method	Minimum FoS
W. J. Niu [36] (Genetic-traversal random method)	Fellenius	0.648
W. J. Niu [36] (GA method)	Fellenius	1.325
W. J. Niu [36] (Slope/W)	Fellenius, Bishop	1.325
W. J. Niu [36] (Slope/W)	Janbu	1.325
W. J. Niu [36] (Slope/W)	Spencer, M-P	1.325
FPA method (Current study)	Fellenius	1.309

Table 5 The statistical test for maximum error (%) over different methods

LEM's methods	Generations	Meta-heuristcs methods			Rocscience slide tool			Error (in %)		
		Grid search	GA	FPA	Grid search	GA	FPA	Grid search	GA	FPA
Fellenius	10	1.352	1.364	1.345	1.548	1.547	1.426	12.661	11.829	5.680
	50	1.344	1.342	1.315	1.459	1.526	1.418	10.100	12.057	7.263
	100	1.331	1.303	1.312	1.506	1.511	1.401	11.620	13.765	6.352
	200	1.327	1.323	1.309	1.424	1.414	1.393	6.811	6.435	6.030

7 Conclusion

The careful analysis of the research work has aimed to investigate the effects of slope stability by examine the critical failure surface in homogeneous earth slopes. Specifically, a novel meta-heuristic method called FPA has been adopted to investigate the critical surface for slope stability analysis. The strategy was successfully compared the performance with Grid search and GA. Based on the results incurred from the comparative evaluation, It has demonstrated that, the FPA approach has obtained a least value of factor of safety and significantly outperform than Grid search and GA. FPA approach may also explore more in future as an appropriate tool for various slope stability analysis problem.

References

1. Fredlund, D. G., & Krahn, J. (1977). Comparison of slope stability methods of analysis. *Canadian Geotechnical Journal, NRC Research Press, 14*(3), 429–439.
2. Chen, Z. Y., & Shao, C. M. (1988). Evaluation of minimum factor of safety in slope stability analysis. *Canadian Geotechnical Journal, NRC Research Press, 25*(4), 735–748.
3. Abramson, L. W., Lee, T. S., Sharma, S., & Boyce, G. M. (2001). *Slope stability and stabilization methods*. Wiley, Hoboken.
4. Dodagoudar, G. R., Venkatachalam, G., & Srividya, A. (2000). Reliability analysis of slopes using fuzzy sets theory. *Computers and Geotechnics, Elsevier, 27*(2), 101–115.
5. Rubio, E., Hall, J. W., Anderson, M. G., & Srividya, A. (2004). Uncertainty analysis in a slope hydrology and stability model using probabilistic and imprecise information. *Computers and Geotechnics, Elsevier, 31*(7), 529–536.

6. Fellenius, W. (1936). Calculation of stability of earth dam. *Transactions. 2nd Congress Large Dams, Washington, DC* (pp. 445–462), 4(7-8).
7. Singh, J., Banka, H., & Verma, A. K. (2018). Analysis of slope stability and detection of critical failure surface using gravitational search algorithm. *Fourth International Conference on Recent Advances in Information Technology (RAIT)* (pp. 1-6). IEEE.
8. Singh, J., Banka, H., & Verma, A. K. (2019). A BBO-based algorithm for slope stability analysis by locating critical failure surface. *Neural Computing and Applications, Springer, 31*(10), 6401–6418.
9. Mathada, V. S., Venkatachalam, G., & Srividya, A. (2007). Slope stability assessment-a comparison of probabilistic, possibilistic and hybrid approaches. *International Journal of Performability Engineering, Springer, 3*(2), 231–242.
10. Zhang, Z., Liu, Z., Zheng, L., & Zhang, Y. (2014). Development of an adaptive relevance vector machine approach for slope stability inference. *Neural Computing and Applications, Springer, 25*(7–8), 2025–2035.
11. Malkawi, A. I. H., Hassan, W. F., & Sarma, S. K. (2001). Global search method for locating general slip surface using Monte Carlo techniques. *Journal of geotechnical and geoenvironmental engineering, American Society of Civil Engineers, 127*(8), 688–698.
12. Aryal, K. P. (2006). Slope Stability Evaluations by Limit Equilibrium and Finite Element Methods, PhD thesis, Norwegian University of Science and Technology.
13. Khajehzadeh, M., Taha, M. R., El-Shafie, A., & Eslami, M. (2012). Locating the general failure surface of earth slope using particle swarm optimisation. *Civil engineering and environmental systems, Taylor & Francis, 29*(1), 41–57.
14. Yamagami, T. (1988). Search for noncircular slip surfaces by the Morgenstern-Price method. *Proceedings 6th International Conference Numerical Methods in Geomech* (pp. 1335–1340).
15. Singh, J., Verma, A. K., Banka, H. (2020, 22 April). A comparative study for locating critical failure surface in slope stability analysis via meta-heuristic approach. *Handbook of Research on Predictive Modeling and Optimization Methods in Science and Engineering* (pp. 1-18). IGI Global.
16. Bishop, A. W. (1955). The use of the slip circle in the stability analysis of slopes. *Geotechnique, Thomas Telford Ltd., 5*(12), 7–17.
17. Janbu, N. (1973). Slope stability computations: In Embankment-dam Engineerin, Textbook. Eds. RC Hirschfeld and SJ Poulos. *John Wiley and Sons INC Thomas Telford Ltd, 12*(4), 67.
18. Cheng, Y. M., Li, L., & Chi, S. C. (2007). Performance studies on six heuristic global optimization methods in the location of critical slip surface. *Computers and Geotechnics, Elsevier, 34*(6), 462–484.
19. Greco, V. R. (1996). Efficient Monte Carlo technique for locating critical slip surface. *Journal of Geotechnical Engineering, American Society of Civil Engineers, 122*(7), 517–525.
20. Cheng, Y. M., Li, L., Chi, S., & Wei, W. B. (2007). Particle swarm optimization algorithm for the location of the critical non-circular failure surface in two-dimensional slope stability analysis. *Computers and Geotechnics, Elsevier, 34*(2), 92–103.
21. Singh, J., Banka, H., & Verma, A. K. (2019). Locating critical failure surface using meta-heuristic approaches: A comparative assessment. *Arabian Journal of Geosciences, Springer, 12*(9), 307.
22. Yamagami, T., & Ueta, Y. (1986). Noncircular slip surface analysis of the stability of slopes. *Landslides, The Japan Landslide Society, 22*(4), 8–16.
23. McCombie, P., & Wilkinson, P. (2002). The use of the simple genetic algorithm in finding the critical factor of safety in slope stability analysis. *Computers and Geotechnics, Elsevier, 29*(8), 699–714.
24. Das, S. K. (2005). Slope stability analysis using genetic algorithm. *Computers and Geotechnics, Elsevier, 10*, 429–439.
25. Zolfaghari, A. R., Heath, A. C., & McCombie, P. F. (2005). Simple genetic algorithm search for critical non-circular failure surface in slope stability analysis. *Computers and geotechnics, Thomas Telford Ltd, Elsevier, 32*(3), 139–152.

26. Sun, J., Li, J., & Liu, Q. (2008). Search for critical slip surface in slope stability analysis by spline-based GA method. *Journal of geotechnical and geoenvironmental engineering, American Society of Civil Engineers, 134*(2), 252–256.
27. Sengupta, A., & Upadhyay, A. (2009). Locating the critical failure surface in a slope stability analysis by genetic algorithm. *Applied Soft Computing, Elsevier, 9*(1), 387–392.
28. Kahatadeniya, K. S., Nanakorn, P., & Neaupane, K. M. (2009). Determination of the critical failure surface for slope stability analysis using ant colony optimization. *Engineering Geology, Elsevier, 108*(1), 133–141.
29. Khajehzadeh, M., Taha, M. R., & El-Shafie, A. (2012). A modified gravitational search algorithm for slope stability analysis. *Engineering Applications of Artificial Intelligence, Elsevier, 25*(8), 1589–1597.
30. Kashani, A. R., Gandomi, A. H., & Mousavi, M. (2016). Imperialistic competitive algorithm: a metaheuristic algorithm for locating the critical slip surface in 2-dimensional soil slopes. *Geoscience Frontiers, Elsevier, 7*(1), 83–89.
31. Yang, X.-S. (2012). *Flower pollination algorithm for global optimization* (pp. 240–249). Springer: International conference on unconventional computing and natural computation.
32. Balasubramani, K., & Marcus, K. (2014). A study on flower pollination algorithm and its applications. *International Journal of Application or Innovation in Engineering & Management (IJAIEM), 3*(11), 230–235.
33. Kayabekir, A. E., Bekdaş, G., Nigdeli, S. M., & Yang, X. S. (2018). *A comprehensive review of the flower pollination algorithm for solving engineering problems* (pp. 171–188). Springer: Nature-Inspired Algorithms and Applied Optimization.
34. Fister, I., Fister, I, Jr., Yang, X., & Brest, J. (2013). A comprehensive review of firefly algorithms. *Swarm and Evolutionary Computation, Elsevier, 13*, 34–46.
35. Singh, J., Verma, A. K., & Banka, H. (2018). Application of biogeography based optimization to locate critical slip surface in slope stability evaluation. *Fourth International Conference on Recent Advances in Information Technology (RAIT)* (pp. 1–5). IEEE.
36. Niu, W. J. (2014). *Determination of slope safety factor with analytical solution and searching critical slip surface with genetic-traversal random method*. Hindawi: The Scientific World Journal.

The manufacturer's authorised representative in the EU is Springer Nature Customer Service Centre GmbH, Europaplatz 3, 69115 Heidelberg, Germany. If you have any concerns regarding our products, please contact ProductSafety@springernature.com

Printed and bound by CPI Group (UK) Ltd, Croydon, CR0 4YY
30/09/2026
02225503-0009